Stevia

Medicinal and Aromatic Plants – Industrial Profiles

Individual volumes in this series provide both industry and academia with in-depth coverage of one major medicinal or aromatic plant of industrial importance.

Edited by Dr Roland Hardman

Stevia
The genus *Stevia*

Edited by

A. Douglas Kinghorn

Department of Medicinal Chemistry and Pharmacognosy
University of Illinois at Chicago
USA

Routledge
Taylor & Francis Group

LONDON AND NEW YORK

First published 2002 by Taylor & Francis

2 Park Square, Milton Park, Abingdon, Oxfordshire OX14 4RN
52 Vanderbilt Avenue, New York, NY 10017

Routledge is an imprint of the Taylor & Francis Group, an informa business

First issued in paperback 2019

Typeset in Garamond by
Integra Software Services Pvt. Ltd., Pondicherry, India

British Library Cataloguing in Publication Data
A catalogue record for this book is
available from the British Library

Library of Congress Cataloging in Publication Data
A catalog record for this book has been requested

ISBN 978-0-367-39660-2

Contents

Preface to the series

There is increasing interest in industry, academia and the health sciences in medicinal and aromatic plants. In passing from plant production to the eventual product used by the public, many sciences are involved. This series brings together information which is currently scattered through an ever increasing number of journals. Each volume gives an in-depth look at one plant genus, about which an area specialist has assembled information ranging from the production of the plant to market trends and quality control.

Many industries are involved such as forestry, agriculture, chemical, food, flavour, beverage, pharmaceutical, cosmetic and fragrance. The plant raw materials are roots, rhizomes, bulbs, leaves, stems, barks, wood, flowers, fruits and seeds. These yield gums, resins, essential (volatile) oils, fixed oils, waxes, juices, extracts and spices for medicinal and aromatic purposes. All these commodities are traded worldwide. A dealer's market report for an item may say 'Drought in the country of origin has forced up prices'.

Natural products do not mean safe products and account of this has to be taken by the above industries, which are subject to regulation. For example, a number of plants which are approved for use in medicine must not be used in cosmetic products.

The assessment of safe to use starts with the harvested plant material which has to comply with an official monograph. This may require absence of, or prescribed limits of, radioactive material, heavy metals, aflatoxin, pesticide residue, as well as the required level of active principle. This analytical control is costly and tends to exclude small batches of plant material. Large scale contracted mechanised cultivation with designated seed or plantlets is now preferable.

Today, plant selection is not only for the yield of active principle, but for the plant's ability to overcome disease, climatic stress and the hazards caused by mankind. Such methods as *in vitro* fertilization, meristem cultures and somatic embryogenesis are used. The transfer of sections of DNA is giving rise to controversy in the case of some end-uses of the plant material.

Some suppliers of plant raw material are now able to certify that they are supplying organically-farmed medicinal plants, herbs and spices. The Economic Union directive (CVO/EU No 2092/91) details the specifications for the *obligatory* quality controls to be carried out at all stages of production and processing of organic products.

Fascinating plant folklore and ethnopharmacology leads to medicinal potential. Examples are the muscle relaxants based on the arrow poison, curare, from species of *Chondrodendron*, and the anti-malarials derived from species of *Cinchona* and *Artemisia*. The methods of detection of pharmacological activity have become increasingly reliable and specific, frequently involving enzymes in bioassays and avoiding the use of laboratory animals. By using bioassay linked fractionation of crude plant juices or extracts, compounds can be specifically targeted which, for example, inhibit blood platelet aggregation, or have anti-tumour, or anti-viral, or any other required activity. With the assistance of robotic devices, all the members of a genus may be readily screened. However, the plant material must be *fully* authenticated by a specialist.

The medicinal traditions of ancient civilisations such as those of China and India have a large armamentaria of plants in their pharmacopoeias which are used throughout South-East Asia. A similar situation exists in Africa and South America. Thus, a very high percentage of the World's population relies on medicinal and aromatic plants for their medicine. Western medicine is also responding. Already in Germany all medical practitioners have to pass an examination in phytotherapy before being allowed to practise. It is noticeable that throughout Europe and the USA, medical, pharmacy and health related schools are increasingly offering training in phytotherapy.

Multinational pharmaceutical companies have become less enamoured of the single compound magic bullet cure. The high costs of such ventures and the endless competition from me too compounds from rival companies often discourage the attempt. Independent phytomedicine companies have been very strong in Germany. However, by the end of 1995, eleven (almost all) had been acquired by the multinational pharmaceutical firms, acknowledging the lay public's growing demand for phytomedicines in the Western World.

The business of dietary supplements in the Western World has expanded from the Health Store to the pharmacy. Alternative medicine includes plant-based products. Appropriate measures to ensure the quality, safety and efficacy of these either already exist or are being answered by greater legislative control by such bodies as the Food and Drug Administration of the USA and the recently created European Agency for the Evaluation of Medicinal Products, based in London.

In the USA, the Dietary Supplement and Health Education Act of 1994 recognised the class of phytotherapeutic agents derived from medicinal and aromatic plants. Furthermore, under public pressure, the US Congress set up an Office of Alternative Medicine and this office in 1994 assisted the filing of several Investigational New Drug (IND) applications, required for clinical trials of some Chinese herbal preparations. The significance of these applications was that each Chinese preparation involved several plants and yet was handled as a *single* IND. A demonstration of the contribution to efficacy, of *each* ingredient of *each* plant, was not required. This was a major step forward towards more sensible regulations in regard to phytomedicines.

My thanks are due to the staffs of Harwood Academic Publishers and Taylor & Francis who have made this series possible and especially to the volume editors and their chapter contributors for the authoritative information.

Roland Hardman

Preface

The South American species *Stevia rebaudiana* (Bertoni) Bertoni (Compositae) is an economically important plant which has attracted considerable controversy. In accumulating in its leaves very high concentration levels of the sweet-tasting secondary metabolites stevioside and rebaudioside A, *S. rebaudiana* is something of a chemotaxonomic curiosity. What makes this species controversial is that products made from the refined leaf extracts or the pure diterpene glycoside stevioside are used to substitute for sucrose in certain countries such as Japan and Korea, but have been subjected to past or present restrictive governmental legislation in other countries such as the United Kingdom and the United States. In the last century, well over 1,000 scientific articles and patents on *S. rebaudiana* and its sweet glycosidic constituents have been published, with some of these in languages other than English. While many technical aspects have been covered in this large body of literature, the three most frequently covered topics have been purification techniques for the *S. rebaudiana* sweet constituents, procedures for improving the quality of the sweet-taste response elicited by stevioside, and *in vitro* and *in vivo* safety studies on these natural sweeteners.

In the initial overview chapter of this volume, a number of aspects not dealt with later are presented, such as the history of the major scientific and regulatory developments which have led to the use of stevioside and its diterpene glycoside analogues as sweetening and flavoring agents. Stevioside is compared to other 'high intensity' natural sweetness in terms of its sweetness potency and other parameters. The physical properties of the sweet *S. rebaudiana* *ent*-kaurene glycosides are considered, as well as a summary of the various types of analytical methods developed for these compounds. Also discussed are the biosynthesis of steviol and stevioside, with information also provided on the cultivation, cell culture, and commercial production of *S. rebaudiana*. Finally, the potential of stevioside and its congeners in relation to dental caries is reviewed, and a summary of the worldwide regulatory status of stevioside as a sweetener is provided.

Chapters 2 and 3 deal in turn with the botany and ethnobotany of the genus *Stevia* as a whole and *S. rebaudiana* in particular. The first of these chapters describes the common botanical traits of the genus *Stevia*, which consists of 220–230 species and is entirely confined to North and South America. The ethnobotanical treatment traces back the relationship of the genus *Stevia* to man from the sixteenth century onwards, with particular emphasis on the early use of *S. rebaudiana* for sweetening by indigenous Paraguayan populations.

In the first of four chapters on phytochemical, chemical and biochemical aspects of the *S. rebaudiana* sweeteners, Chapter 4 provides a full listing of the presently known constituents of this species, inclusive of triterpenoids and steroids, flavonoids, and miscellaneous compounds, in addition to the diterpenoids. Chapter 5 summarizes the extensive phytochemical literature on *Stevia* species other than *S. rebaudiana*, with particular emphasis on the wide range of functionalized sesquiterpenoids that are elaborated by the members of this genus. In Chapter 6, procedures for the chemical synthesis of steviol and for the glycosylation of steviol to form stevioside are reviewed. In addition, synthetic applications using stevioside and steviol as starting materials are presented. In Chapter 7, chemical and biochemical methods developed in Japan to improve the taste of stevioside and the other sweet principles of *S. rebaudiana* are described.

Chapter 8 provides an in-depth evaluation of the safety of extracts of *S. rebaudiana* as well as the pure compounds stevioside and rebaudioside A, and their common *ent*-kaurene aglycone, steviol. Finally, in Chapters 9 and 10, the use of refined *S. rebaudiana* extracts and stevioside in Japan and Korea, respectively, is presented. The longest and most widespread use of products from *S. rebaudiana* have been in Japan, for the sweetening and flavoring of a many different types of food products. However, more recently, pure stevioside has been used increasingly as a sucrose substitute in Korea, particularly for the sweetening of a traditional distilled liquor called *soju*.

It is hoped that readers of this edited volume on *S. rebaudiana* and its sweet principles will be able to gain a better understanding of a fascinating topic from a wide perspective as a result of the various chapters provided, which have been written by acknowledged experts in each area. It seems likely that the current widespread interest in stevioside, rebaudioside A, and their various modified structural forms, will continue well into the twenty-first century.

A. Douglas Kinghorn

Acknowledgements

The Editor is grateful to all of the chapter authors, for lending their time and expertise to this project. In addition, he thanks the Book Series Editor, Dr Roland Hardman for his unwavering encouragement. Dr Aiko Ito is thanked for her helpful assistance in the editing of two of the chapters and the compilation of the index. Finally, the following are thanked for providing valuable information: Professors Jan M. C. Guens, Pier-Giorgio Pietta, Finn Sandberg and Vincente Oliverira Ferro, and Drs Gloria L. Silva and Luisella Verotta.

List of contributors

Carlos M. Cerda-García-Rojas
Departamento de Química
CINVESTAV
Instituto Politécnico Nacional
Apartado Postal 14-740
México, D.F. 07000, Mexico

Young Hae Choi
College of Pharmacy
Seoul National University
San 56-1, Sinlim-dong
Kwanak-ku
Seoul 151-742, Korea

Young-Hee Choi
College of Pharmacy
Ewha Women's University
11-1 Daehyun-dong, Seodaemun-gu
Seoul 120-750, Korea

Ryan J. Huxtable
Department of Pharmacology
College of Medicine
University of Arizona
P.O. Box 245050
Tucson, AZ 85724-5050

Edward J. Kennelly
Formerly Emeritus Professor
U.S. Food and Drug Administration
Center for Food Safety and Applied Nutrition
Office of Special Nutritionals
HFS-465, 200 C Street S.W.
Washington, DC 20204, USA
Present address:
Department of Biological Sciences
Lehman College
City University of New York
250 Bedford Park Boulevard West
Bronx, NY 10468, USA

Darrick S. H. L. Kim
Program for Collaborative Research in
the Pharmaceutical Sciences and
Department of Medicinal Chemistry
and Pharmacognosy
College of Pharmacy (M/C 877)
University of Illinois at Chicago
833 South Wood Street
Chicago, IL 60612, USA

Jinwoong Kim
College of Pharmacy
Seoul National University
San 56-1, Sinlim-dong
Kwanak-ku
Seoul 151-742, Korea

A. Douglas Kinghorn
Department of Medicinal Chemistry and
Pharmacognosy
College of Pharmacy (M/C 781)
University of Illinois at Chicago
833 South Wood Street
Chicago, IL 60612, USA

Rogelio Pereda-Miranda
Departamento de Farmacia
Facultad de Quimica
Universidad Nacional Autónoma de México
Coyaocan México D.F., 04510 Mexico

Kenji Mizutani
Research and Development Division
Maruzen Pharmaceuticals Co., Ltd.
14703-10 Mukaihigashi-cho
Onomichi, Hiroshima 722-0062, Japan

Kazuhiro Ohtani
Institute of Pharmaceutical Sciences
School of Medicine
Hiroshima University
1-2-3 Kasumi, Minami-ku
Hiroshima 734-8551, Japan

Djaja Djendoel Soejarto
Program for Collaborative Research in the
Pharmaceutical Sciences
College of Pharmacy (M/C 877)
University of Illinois at Chicago
833 South Wood Street
Chicago, IL 60612, USA
and
Botany Department, Field Museum
Roosevelt Road at Lake Shore Drive
Chicago, IL 60605, USA

Osamu Tanaka
Emeritus Professor
Institute of Pharmaceutical Sciences
School of Medicine
Hiroshima University
1-2-3 Kasumi, Minami-ku
Hiroshima 734-8551, Japan

Kazuo Yamasaki
Institute of Pharmaceutical Sciences
School of Medicine
Hiroshima University
1-2-3 Kasumi, Minami-ku
Hiroshima 734-8551, Japan

1 Overview

A. Douglas Kinghorn

HISTORICAL PERSPECTIVE AND INTRODUCTION

There can have been few botanical discoveries quite so dramatic as the realization that the leaves of *Stevia rebaudiana* (Bertoni) Bertoni (Compositae) are so highly sweet. It is not clear when this was made originally, but the observation was brought to the attention of the scientific community about a hundred years ago (Gosling 1901; Bertoni 1905; Lewis 1992). The British Consul at Asunción, Paraguay, Cecil Gosling, attributed the discovery of the sweetness of the plant *S. rebaudiana* to the Italian-Swiss botanist, Dr Moisés S. Bertoni (Gosling 1901). During the twentieth century, this native Paraguayan species and its sweet constituents have been the subject of well over 1,000 scientific papers and patent applications, and *S. rebaudiana* has become an important economic plant owing to its commercial use for sweetening and flavoring purposes. Stevioside, the most abundant sweet constituent of the leaves of species, and an *ent*-kaurene diterpene diglycoside, was first isolated in impure form in the first decade of the twentieth century (Bertoni 1905; 1918), but its final structure elucidation did not occur until nearly sixty years later (Mosettig *et al.* 1963). In the 1970s, the group of Professor Osamu Tanaka at the Hiroshima University in Japan isolated rebaudioside A, a second major sweet *ent*-kaurene diterpene glycoside from *S. rebaudiana* leaves (Kohda *et al.* 1976). Later on, six further less abundant sweet-tasting glycosides were isolated from this species, namely, dulcoside A, rebaudiosides B-E, and steviolbioside (Yamasaki *et al.* 1976; Kobayashi *et al.* 1977; Tanaka 1982). *Stevia rebaudiana* is certainly very unusual in accumulating secondary metabolites like stevioside and rebaudioside A at such high abundance in its leaves (Kinghorn and Soejarto 1985; Phillips 1987). Moreover, *S. rebaudiana* is an anomaly in the genus *Stevia*, since none of the other approximately 230 species in this North and South American genus has ever been found to produce these sweet compounds at high concentration levels (Soejarto *et al.* 1982; Kinghorn *et al.* 1984). Many secondary metabolites other than the sweet-tasting glycoside constituents have now been isolated and identified from *S. rebaudiana* (Kinghorn and Soejarto 1985; Phillips 1987).

Standardized extracts of *S. rebaudiana* and purified stevioside began to be utilized commercially for sweetening and flavoring foods and beverages in Japan in the mid-1970s, in order to substitute for several synthetic sweeteners which were banned from the market there at that time. By 1987, *S. rebaudiana* extracts containing stevioside occupied 41% of the 'high intensity' sweetener market in Japan, but that was before the development of aspartame (Anonymous 1988a). Currently, the largest and most diverse use of stevioside remains in Japan, although this compound is used increasingly in South Korea, where the primary use is in sweetening the alcoholic beverage, *soju* (Seon 1995). Stevioside is listed as an approved sweetener in Brazil and other South American countries (Bakal and O'Brien Nabors 1986; Kinghorn and Soejarto

1991). In contrast, in North America and in the 15 countries of the European Union, stevioside is not approved as a sucrose substitute. However, in practice, extracts of *S. rebaudiana* have been used fairly extensively in the United States and certain countries of western Europe as a dietary supplement (Bonvie and Bonvie 1996; Moroni 1999). Unfortunately, the use of *S. rebaudiana* products has not been without controversy. The import of *S. rebaudiana* leaves was banned by the United States Food and Drug Administration in 1991, but this import alert was lifted in 1995, in the aftermath of the Dietary Supplement Health and Education Act of 1994 (Blumenthal 1995). Currently, *S. rebaudiana* leaves, the extract of *S. rebaudiana* leaves, and stevioside are only permitted for import into the United States, if they are explicitly labeled as being of use as dietary supplements (Blumenthal 1995). Very recently, the Ministry of Agriculture, Fisheries, and Food in the United Kingdom has removed all products from health food stores containing *S. rebaudiana* leaves or their constituents, claiming that there is no evidence that these products are safe (Clark 2000).

As might be expected for such a fascinating topic, there have been many reviews and book chapters published dealing specifically with *S. rebaudiana* and its sweet constituents, including those by Bell (1954), Fletcher, Jr (1955), Jacobs (1955), Sumida (1973), Abe and Sonobe (1977), Akashi (1977), Felippe (1977), Fujita and Edahiro (1979), Toffler and Orio (1981), Kurahashi *et al.* (1982), Sakaguchi and Kan (1982), Tanaka (1982; 1997), Galperin de Levy (1984), Kinghorn and Soejarto (1985; 1991), Bakal and O'Brien Nabors (1986), Crammer and Ikan (1986; 1987), Pezzuto (1986), Phillips (1987), Yoshihira *et al.* (1987), Hanson and De Oliveira (1993), Brandle *et al.* (1998), and Kinghorn *et al.* (2000). In addition, the numerous articles and patents published on *S. rebaudiana* and stevioside have covered many botanical, biological, chemical, and pharmacological aspects. However, perhaps three particular areas have received the greatest degree of prominence in the scientific literature, namely, methods for the purification of stevioside and rebaudioside A, the development of procedures to enhance to sensory parameters of stevioside, and studies on the pharmacology and toxicology of stevioside and its enzymatically produced aglycone, steviol. These aspects will be described briefly in turn in the next few paragraphs.

Many procedures for the purification of extracts of *S. rebaudiana* containing stevioside in varying degrees of purity have appeared in the literature. Much of this type of information has appeared in the patent literature, especially from Japan. Most methods for the purification of stevioside involve initial extraction into an aqueous solvent, followed by refinement involving one or more of adsorption chromatography, coagulation, decolorization, electrolysis, ion exchange, precipitation, and/or solvent partition (Kinghorn and Soejarto 1985; 1991; Bakal and O'Brien Nabors 1986; Phillips 1987). For example, a new method involving the use of subcritical fluid extraction of the sweet glycosidic constituents of *S. rebaudiana* leaves was published recently, in which methanol was used as a modifier, and an overall extraction efficiency of 88% was obtained (Liu *et al.* 1997).

Stevioside has been rated as possessing about 300 times the relative sweetness intensity of 0.4% w/v sucrose, although its sweetness intensity decreases to only about 100 times that of sucrose at a 10% concentration. Unfortunately, the compound exhibits a menthol-like, bitter aftertaste (Bakal and O'Brien Nabors 1986). The sweetness intensities (i.e. sweetening power relative to sucrose, which is taken as = 1) of the other seven *S. rebaudiana* sweet principles have been determined as follows: dulcoside A, 50–120; rebaudioside A, 250–450; rebaudioside B, 300–350; rebaudioside C (previously known as dulcoside B), 50–120; rebaudioside D, 250–450; rebaudioside E, 150–300; and steviolbioside, 100–125 (Crammer and Ikan 1987). Rebaudioside A, the second most abundant *ent*-kaurene glycoside occurring in the leaves of *S. rebaudiana* is better suited than stevioside for use in foods and beverages, because it is not only more water soluble but it also exhibits a pleasanter taste (Kinghorn and Soejarto 1991; Tanaka 1997). Stevioside is often admixed with glycyrrhizin, and the resultant mixture is synergistic, with the taste profile of both sweeteners being improved. In addition, stevioside has been reported to be synergistic with aspartame, acesulfame-K, and cyclamate, but not with saccharin (Bakal and O'Brien Nabors 1986).

There have been numerous attempts to improve the sweetness hedonic (pleasantness) qualities of stevioside, which is regarded as a limitation that hinders the more widespread use of this sweetener. For instance, it is possible to formulate stevioside with a variety of flavor-masking and sweetness-enhancing agents, and many patents have appeared with improved sweetener compositions for *S. rebaudiana* extracts and stevioside (Bakal and O'Brien Nabors 1986; Phillips 1987; Kinghorn and Soejarto 1991). Efforts have been made to produce strains of *S. rebaudiana* which have a higher ratio of rebaudioside A to stevioside compared with wild Paraguayan populations, in order to harness the preferential properties of rebaudioside A in comparison to stevioside (Kinghorn and Soejarto 1991). Synthetic approaches have been taken to produce derivatives with enhanced sweetness parameters compared to natural stevioside (e.g. Esaki *et al.* 1984; DuBois *et al.* 1984; DuBois and Stephenson 1985). In an alternative strategy, enzymatic transglycosylation of stevioside has led to analogues with improved taste profiles over the parent substance (Mizutani *et al.* 1989; Tanaka 1997). For example, modification of the sugar moiety at C-19 of stevioside may be conducted with enzymic transglycosylation, with cyclomaltodextrin-glucanotransferase (CGTase) then used to catalyze *trans*-α-1,4-glucosylation (Tanaka 1997). A transglycosylated ('sugar-transferred') product of stevioside is sold commercially in Japan, produced by shortening the α-glucosyl chain of the mixture of compounds obtained by CGTase treatment using β-amylase (Tanaka 1997).

There is no doubt that the safety of *S. rebaudiana* extracts and stevioside is a controversial area. Several key papers have resulted in particular consternation over the years. In 1968, it was documented in *Science* magazine that the Matto Grosso Indians of Paraguay use a decoction prepared from the leaves and stems of dried *S. rebaudiana* leaves for contraceptive purposes. Furthermore, such a decoction was shown to produce a decrease in fertility in female rats (Planas and Kuć 1968). Subsequently, several groups of investigators have duplicated these experiments and found no evidence of contraceptive or antifertility effects of *S. rebaudiana* leaves in female rats (reviewed in Kinghorn and Soejarto 1985; 1991; Phillips 1987). Another contentious point regarding safety emanated from a report in 1985 from the University of Illinois at Chicago, indicating a mutagenic response of steviol, the aglycone of stevioside, which is produced from this sweet substance enzymatically rather than by treatment with mineral acids, whereupon isosteviol is produced (Pezzuto *et al.* 1985). In early work, it was reported that stevioside may be degraded by intestinal microflora in the rat to steviol and sugars, and these are then absorbed in the lower part of the intestine (Wingard *et al.* 1980; Nakayama *et al.* 1986). Steviol, when metabolically activated with a supernatant fraction derived from the livers of Aroclor-1254-pretreated rats, gave a reproducible mutagenic response in a forward mutation assay utilizing *Salmonella typhimurium* strain TM677, although untreated steviol and metabolically activated stevioside under the same assay conditions were inactive (Pezzuto *et al.* 1985; Pezzuto 1986). While an actual mutagenic metabolite of steviol has never been found, this study has stimulated additional genetic toxicity testing of stevioside and steviol, and it has been shown recently by scientists at the National Institute of Hygenic Sciences in Tokyo that steviol induces mutations of the guanine phosphoribosyltransferase (*gpt*) gene in *S. typhimurium* TM677 (Matsui *et al.* 1996). However, to offset the concerns on the mutagenicity of steviol somewhat, an extensive *in vivo* study, again carried out at the National Institute of Hygenic Sciences in Tokyo, concluded that stevioside was not carcinogenic for F344 rats of both sexes.

Stevioside was administered at high purity in this study, at doses up to 5% of the diet, and it was shown specifically that steviol was present in the contents of the large intestine (Toyoda *et al*. 1997).

Very recently, a negative opinion was expressed on the safety of stevioside by the Scientific Committee on Food of the European Commission (Anonymous 1999a). Several questions of concern were raised, including 'the specifications of the extracts (containing stevioside) that had been tested, questionable chronic toxicity and carcinogenicity studies, and possible effects on the male reproductive system that could affect fertility.' The report also pointed out that 'steviol, one metabolite of stevioside, that is produced by the human microflora, is genotoxic and induces developmental toxicity.' The Committee recommended additional studies in these areas. In addition, the leaves of *S. rebaudiana* were also evaluated as a food by the European Commission, Scientific Committee on Food. There was a concern expressed about the limited data submitted, and it was concluded that 'no appropriate data were presented to enable the safety of the commercial plant product to be evaluated' (Anonymous 1999b).

The present chapter will endeavor to cover *S. rebaudiana* and stevioside from perspectives not included in the rest of this edited volume, in an attempt to present to the reader as comprehensive and balanced a treatment of these topics as possible. Thus, in the next section of this chapter, a discussion will be provided as to how well stevioside compares with other 'intense' sweeteners of natural origin. This will be followed by short sections on the physical properties of stevioside, analytical methods for the *S. rebaudiana* sweet glycosides, the biosynthesis of stevioside and steviol, the production of *S. rebaudiana*, methods for the cultivation and cell culture of *S. rebaudiana*, concluded by brief considerations of the cariogenicity potential of stevioside, and the regulatory status of *S. rebaudiana* extracts containing stevioside and/or pure stevioside in various countries. Table 1.1 provides a chronological summary of the discovery and major events leading to the present utilization of the sweet *S. rebaudiana* principles as sweeteners and dietary supplements.

COMPARISON OF STEVIOSIDE TO OTHER NATURAL PRODUCT 'HIGH-INTENSITY' SWEETENING AGENTS

In 1998, the artificial sweetener market in the United States was worth about $610 million, and this is scheduled to increase 3.4% per year to $720 million by 2003 (Seewald 2000). There remains a strong demand for highly sweet, non-caloric, and non-cariogenic substances to substitute for sucrose in the diet. Such compounds, whether natural or synthetic, should exhibit a sucrose-like taste, and should also lack any offensive odor, exhibit satisfactory water solubility and hydrolytic and thermal stability, and should not be toxic or cariogenic, either in their unmodified or metabolized forms. Moreover, to be commercially viable, candidate sucrose substitutes should be relatively easy to manufacture or cultivate, and should fit in with the existing applications for other sweeteners (Hough *et al*. 1979). With the recent approval of sucralose in 1998, there are now four synthetic 'high-intensity' sweeteners on the market in the United States, with the others being aspartame, acesulfame potassium, and saccharin (Duffy and Anderson 1998). Thus far, there are no natural product 'high-intensity' sweetening agents on the US market (Duffy and Anderson 1998). Such compounds tend to be at least 50 times the sweetness intensity of sucrose, and are also known as 'non-nutritive sweeteners', and are usually placed in a separate category than the 'nutritive' or 'bulk' sweeteners, comprised by monosaccharides, disaccharides, and polyols, which have about the same sweetness intensity as sucrose (Kinghorn and Kennelly 1995; Kinghorn *et al*. 1998). In the United States, the monosaccharide

Table 1.1 Major events leading to the discovery and development of stevioside

Event	Reference
Report of the sweetness of *S. rebaudiana* leaves from Paraguay published in a major scientific journal	Gosling 1901
First chemical report on the sweet constituents of 'Kaá hê-é', noting that by 1899 the species was well-known in Paraguay and used by herbalists to sweeten teas	Bertoni 1905
Realization that stevioside ('eupatorin') is a glycoside	Dieterich 1908
Production of steviolbioside from stevioside	Wood, Jr *et al*. 1955
Evidence that stevioside is a sophoroside	Vis and Fletcher, Jr 1956
Final structures of steviol and isosteviol confirmed	Mosetigg *et al*. 1963
Contraceptive effect of *S. rebaudiana* leaves reported	Planas and Kuč 1968
Steviol chemically synthesized (by two groups independently)	Cook and Knox, 1970; Mori *et al*. 1970
S. rebaudiana from Brazil cultivated experimentally in Japan	Sumida 1973
Isolation and characterization of rebaudioside A	Kohda *et al*. 1976
Minor *S. rebaudiana* leaf sweet diterpene glycosides obtained	Yamasaki *et al*. 1976; Kobayashi *et al*. 1977
Advent of extensive use of *S. rebaudiana* extracts for sweetening and flavoring of foods and beverages in Japan	Abe and Sonobe 1977; Akashi 1977
First approval of *S. rebaudiana* products in Brazil	Anonymous 1980
From *ca.* 1982 onwards, large-scale cultivation of *S. rebaudiana* in the People's Republic of China	Kinghorn and Soejarto 1991
Demonstration of mutagenic activity of metabolically activated steviol in a forward mutation test	Pezzuto *et al*. 1985
During the 1980s *S. rebaudiana* leaves become a popular herbal tea in the United States	Blumenthal 1995
First approval of stevioside in South Korea	Anonymous 1988d
Import ban on *S. rebaudiana* leaves into the United States imposed by the Federal Drug Administration (FDA) (1991)	Blumenthal 1995
'Sugar-transferred' (enzymatically modified) and 'rebaudioside A-enriched' *S. rebaudiana* extracts become more widely used in Japan in the early 1990s	Anonymous 1993
FDA import ban on *S. rebaudiana* leaves rescinded in 1995	Blumenthal 1995
Long-term toxicity test, showing a lack of any carcinogenic effects, conducted in rats of both sexes on stevioside, by the National Institute of Hygenic Sciences, Tokyo, Japan	Toyoda *et al*. 1997

fructose, the disaccharide sucrose, and the polyols erythritol, hydrogenated starch hydrolysates, isomalt, lactitol, mannitol, sorbitol, and xylitol are all used, and have either GRAS (Generally Recognized As Safe) or other official regulatory status. The most widely used nutritive sweetener is high-fructose corn syrup (Duffy and Anderson 1998; Seewald 2000).

Prior to discussing the natural product 'high-intensity' sweetening agents used in certain countries abroad, it is pertinent to comment further about the synthetic compounds used in this manner in the United States. Many of the currently available sucrose substitutes are deficient in either their cost, thermal or hydrolytic stability, quality of taste, or perceived toxicity. For example, the general purpose dipeptide sweetener aspartame breaks down with heat to diketopiperazine, and loses its sweetness. This substance may not be used by persons with phenylketonuria, and therefore foods containing aspartame must contain a label indicating that they contain phenylalanine (Duffy and Anderson 1998). The oldest artificial sweetener, saccharin, has been used on an 'interim' basis for several years, and the container label must contain a cancer warning (Duffy and Anderson 1998). However, very recently, saccharin was removed from the list of carcinogens by the United States National Institute of Environmental Health Sciences (Seewald 2000). It is estimated that this change in status in saccharin will not boost its use in the coming years, however. In contrast, acesulfame potassium and sucralose are expected to increase their market share, as a result of their anticipated more extensive use in soft drinks (Seewald 2000).

Other synthetic sucrose substitutes approved for use outside the United States are cyclamate (used in about 50 countries) and alitame (approved in relatively few countries to date) (Duffy and Anderson 1998). Another synthetic compound of promise for market introduction as a high-intensity sweetener is neotame, an *N*-alkyl substituted aspartame derivative with a sweetness potency some 10,000 times that of aspartame (Walters *et al.* 2000). The search for improved synthetic sweeteners is continuing, with the *N*-cyclononylguanidine derivative, sucrononic acid being reported to exhibit a sweetness potency of about 200,000 times that of sucrose (Tinti and Nofre 1991).

The discovery that certain naturally occurring compounds are highly sweet has also become of commercial significance. There is considerable interest in natural food ingredients such as sweeteners among consumers and manufacturers alike. Plants are well-known as the sources of sweet-tasting monosaccharides, disaccharides, and polyols, but also about 75 highly sweet compounds ('intense sweeteners') have been reported to date. Such compounds are mainly in the terpenoid, flavonoid, and protein classes, although other types of plant constituent are known to be highly sweet, such as the amino acid, monatin, the steroidal saponins, osladin and polypodoside A, and various proanthocyandins ('condensed tannins') (Kinghorn *et al.* 1995). The highly sweet natural products reported to date have all been reported from green plants, and they appear to be randomly distributed among monocoyledons and dicotyledons (Kinghorn and Soejarto 1989). It is entirely possible that all of the more obvious candidate highly sweet plants have already been studied in the laboratory, and that in the future novel sweet-tasting natural product chemotypes will be discovered either by following up on ethnobotanical leads of plants used for sweetening by indigenous peoples in remote areas, or by random organoleptic evaluation of plants collected in the field for other purposes (Kinghorn *et al.* 1998).

Other than the *S. rebaudiana* sweet *ent*-kaurene glycosides, stevioside and rebaudioside A, the plant-derived sweeteners used commercially in Japan are the oleanane-type triterpene glycoside, glycyrrhizin (from *Glycyrrhiza glabra* L. and other species of *Glycyrrhiza*; Fabaceae), the cucurbitane-type glycoside, mogroside V [from *Siraitia grosvenorii* (Swingle) C. Jeffrey; Cucurbitaceae], the dihydroisocoumarin, phyllodulcin [from *Hydrangea macrophylla* Seringe var. *thunbergii* (Siebold) Makino; Saxifragaceae], and the protein thaumatin [from *Thaumatococcus daniellii* (Bennett) Benth.; Marantaceae] (Kinghorn *et al.* 1995; 1998). Thaumatin is also an available sweetener in other countries, including Australia and Switzerland, with the thaumatin product Talin® protein having GRAS status in the United States and employed as a flavor enhancer in chewing gum (Kinghorn and Compadre 1991). In addition, glycyrrhetic acid monoglucuronide (MGGR), which is prepared from glycyrrhizin by microbial enzymatic hydrolysis, has become a commercially available sweetener and flavor enhancer in Japan (Mizutani *et al.* 1998). Finally, two semi-synthetic plant-derived compounds also have some commercial use as sweeteners, namely, perillartine (based on the monoterpene perillaldehyde; used in Japan) and neohesperidin dihydrochalcone (based on the flavanone, neohesperidin;

Table 1.2 Species of origin of commercially used 'high-intensity' plant-derived sweetening agents

Compound type	Name	Plant name (family)	Sweetness potency[a]
Monoterpene	Perillartine[b]	*Perilla frutescens* Britton (Labiatae)	370
Diterpenes	Stevioside[c]	*Stevia rebaudiana* (Bertoni) Bertoni (Compositae)	210
	Rebaudioside A[c]	*S. rebaudiana*	242
Triterpenes	Glycyrrhizin[c]	*Glycyrrhiza glabra* L. (Leguminosae)	93
	Glycyrrhetic acid monoglucuronide[b]	*G. glabra*	941
	Mogroside V[c]	*Siraitia grosvenorii* (Swingle) C. Jeffrey (Cucurbitaceae)	392
Dihydroisocoumarin	Phyllodulcin[d]	*Hydrangea macrophylla* Seringe var. *thunbergii* (Siebold) Makino (Saxifragaceae)	400
Dihydrochalcone	Neohesperidin dihydrochalcone[b]	*Citrus aurantium* L. (Rutaceae)	1,000
Protein	Thaumatin[e]	*Thaumatococcus daniellii* (Bennett) Benth. (Marantaceae)	1,600

Sources: Data taken from Kinghorn *et al.* (1995) and Mizutani *et al.* (1998).

Notes

a Values of relative sweetness on a weight basis to sucrose (= 1). Such data vary with the concentration of the compound being evaluated.

b Semi-synthetic derivative of natural product.

c Major compound present in a refined extract of the plant of origin.

d Used in a ceremonial tea, with the plant of origin being crushed or fermented in order to generate the sweet compound.

e A complex of several related proteins, sold under name of Talin® protein.

used in Argentina and Belgium) (Kinghorn and Kennelly 1995; Kinghorn *et al.* 1995; 1998). Table 1.2 summarizes the species of origin and relative sweetness intensities of stevioside, rebaudioside A, and the other plant-derived compounds with commercial use as 'high-intensity' sweeteners.

The highly sweet plant constituents tend to be used as refined extracts rather than absolutely pure compounds, for reasons of cost competitiveness with synthetic sweeteners. Relatively few highly sweet plant constituents have been developed as sucrose substitutes, because such compounds tend to have low yields in the plant, making them unprofitable commercially. Another common problem is that the natural product sweet substances often have inferior taste quality properties such as a delay in the onset of the sweet-taste response and, in some cases, can cause an unacceptably prolonged sweetness effect in the mouth. However, of all the plant-derived sweeteners, there is little doubt that stevioside has attracted the greatest interest.

PHYSICAL PROPERTIES, SOLUBILITY, AND STABILITY OF STEVIOSIDE

In pure form, stevioside {(4α)-13-[2-*O*-β-D-glucopyranosyl-β-D-glucopyranosyloxy]kaur-16-en-18-oic acid β-D-glucopyranosyl ester} (*Chemical Abstracts* name) is a white crystalline material with a melting point of 198 °C, an optical rotation of −39.3° in water, an elemental composition of $C_{38}H_{60}O_{18}$, and a molecular weight of 804.88 (Anonymous 1996). Stevioside is only sparingly soluble in water (1 g dissolves in 800 ml water), but it is soluble in dioxane and

Table 1.3 Physical and solubility data for the eight sweet *ent*-kaurene glycoside from the leaves of *Stevia rebaudiana*

Compound	Mp (°C)	Specific rotation $[\alpha]_D^{25}$ (Degree)	Molecular weight	Solubility in water (%)
Stevioside	196–198	−39.3	804	0.13
Rebaudioside A	242–244	−20.8	966	0.80
Rebaudioside B	193–195	−45.4	804	0.10
Rebaudioside C	215–217	−29.9	958	0.21
Rebaudioside D	283–286	−22.7	1128	1.00
Rebaudioside E	205–207	−34.2	966	1.70
Steviolbioside	188–192	−34.5	642	0.03
Dulcoside A	193–195	−50.2	788	0.58

Sources: Data taken from Kinghorn *et al.* (1985), Crammer and Ikan (1987), and Phillips (1987).

slightly soluble in ethanol (Anonymous 1996). Rebaudioside A {mp 242–244 °C; $[\alpha]_D^{24}$ −20.8° (*c* 0.84, MeOH); $C_{44}H_{70}O_{23}$; mol. wt. 966}, the second most abundant sweet diterpene glycoside in *S. rebaudiana* leaves, is considerably more water soluble than stevioside, since it contains an additional glucose unit in its molecule (Kohda *et al.* 1976; Kinghorn and Soejarto 1991). Table 1.3 shows comparative melting point, specific rotation, molecular weight, and percentage solubility in water information for the eight sweet diterpene glycosides from *S. rebaudiana* leaves.

Stevioside is a stable molecule at 100 °C, when maintained in solution in the pH range 3–9, although it decomposes quite readily at alkaline pH levels of greater than ten under these conditions, (Kinghorn and Soejarto 1985). Detailed stability profiles have been determined for stevioside when treated with dilute mineral acids and enzymes, as has been reviewed previously (Kinghorn and Soejarto 1985). Both stevioside and its analogue, rebaudioside A, have been found to be stable when formulated in acidulated beverages at 37 °C for at least three months (Chang and Cook 1983). Solid stevioside is stable for one hour at 120 °C, but decomposition was noticed at temperatures exceeding 140 °C (Kroyer 1999). In coffee and tea sweetened with stevioside, the levels of caffeine and stevioside were both relatively unaffected (Kroyer 1999).

ANALYTICAL METHODS FOR THE SWEET *STEVIA REBAUDIANA* GLYCOSIDES

A large number of analytical methods are available for the determination of purity and stability of stevioside and rebaudioside A, and such procedures have been applied to *S. rebaudiana* samples cultivated in different countries and climates, as well as foods and beverages containing these sweeteners (Kinghorn and Soejarto 1985; 1991; Phillips 1987). Morever, sensitive analytical methodology is now available for the quantitation of stevioside and its metabolites such as isosteviol and steviol in blood, feces, and urine (Hutapea *et al.* 1999). The primary methods for the qualitative and quantitative analyses of the *S. rebaudiana* sweeteners were categorized previously as being based on colorimetric determination, an enzymatic procedure using crude hesperidinase, gas-liquid chromatography (GLC), high-performance liquid chromatography (HPLC), and thin-layer chromatography (TLC)/densitometry (Kinghorn and Soejarto 1985). More recently, a procedure has been published for stevioside, rebaudioside A, and steviolbioside in a commercial *S. rebaudiana* extract using capillary electrophoresis in the micellar mode (MEKC) (Mauri *et al.* 1996). Another new approach is the quantitative analysis of stevioside in

S. rebaudiana leaves using near infrared reflectance spectroscopy (Nishiyama *et al.* 1992). Gradually, analytical methods have evolved for stevioside, rebaudioside and their sweet analogues, decomposition products, and metabolites that do not require chemical derivatization. For example, this can be achieved by HPLC for the *S. rebaudiana* sweet glycosides using amino (NH_2) columns, which are usually used in the analysis of carbohydrates (e.g. Makapugay *et al.* 1984; Mauri *et al.* 1996).

BIOSYNTHESIS OF STEVIOSIDE AND STEVIOL

The leaves of *S. rebaudiana* are very unusual in biosynthesizing such high concentration levels of stevioside, rebaudioside A, and the other sweet-tasting *ent*-kaurene glycoside constituents, which may accumulate to the extent of 10% w/w or more (Kinghorn and Soejarto 1985; Phillips 1987). While the biosynthesis of stevioside takes place in the leaf chloroplasts of *S. rebaudiana*, there is still an incomplete picture of how this actually occurs. In an early study on *S. rebaudiana* leaves, it was found that $2\text{-}^{14}C$-acetate applied to the leaves of *S. rebaudiana* was incorporated into steviol [*ent*-13-hydroxykaur-16-en-19-oic acid], the enzymatically produced aglycone of stevioside, whereas $2\text{-}^{14}C$-mevalonic acid was not (Ruddat *et al.* 1965). It was then shown that steviol is biosynthesized from mevalonic acid through (−)-kaurene and (−)-kaur-16-en-19-oic acid (Hanson and White 1968). In more recent work, it was demonstrated that both the roots and leaves are required for cultured *S. rebaudiana* to biosynthesize stevioside from acetate (Swanson *et al.* 1992). A high activity of the enzyme 3-hydroxy-3-methylglutaryl CoA (HMG-CoA) reductase has been reported in the chloroplasts of *S. rebaudiana*, suggesting that the isopentenyl diphosphate (IPP) involved in the biosynthesis of steviol is produced via the acetate/mevalonate pathway (Kim *et al.* 1996). However, this conclusion has been refuted by more recent work by a group headed by Guens at Leuven in Belgium, in which it was shown that steviol is biosynthesized in *S. rebaudiana* leaves via a mevalonate-independent methylerythritol phosphate pathway. Their metholodology used $[1\text{-}^{13}C]$-glucose, and all the carbons C-1 through C-5 in the resultant IPP and dimethylallyl diphosphate (DMAPP) were appropriately labeled (Totté *et al.* 2000). Chloroplasts isolated from *S. rebaudiana* leaves contain the enzyme *ent*-kaurenoic acid 13-hydroxylase, which catalyzes the conversion of *ent*-kaurenoic acid to steviol (Kim and Shibata 1997). Thus far, the genes and enzymes involved in the assembly of geranylgeranyl diphosphate (GGPP, constituted by four IPP units) into the tetracyclic steviol unit have not been fully characterized, although it seems as though two cyclase enzymes are involved, namely, (−)-copalyl diphosphate synthase (CPS) and (−)-kaurene synthase (KS) (Richman *et al.* 1999). Enzyme fractions which catalyze the glycosylation of steviol to stevioside and other glycosylated products have been purified from *S. rebaudiana* leaves (Shibata *et al.* 1991).

CULTIVATION AND CELL CULTURE OF *STEVIA REBAUDIANA*

There is now a considerable body of literature on methods for the cultivation of *S. rebaudiana* leaves leading to the optimization of levels of their sweet *ent*-kaurene glycoside constituents. The native habitat of *S. rebaudiana* is at a latitude of 25° S in a subtropical region of northeastern Paraguay between 500–1,500 m above sea level, on soil having a low phosphate content, with an annual average temperature of 75 °F, and an average rainfall of about 55 in. per year (Sumida 1973; Shock 1982). This perennial plant has been severely restricted in its native habitat in northeastern Paraguay because of overgrazing, excessive harvesting, and transplantation to cultivated areas, so it has proven necessary to cultivate *S. rebaudiana* to meet the demand for sweetening and flavoring purposes of its glycoside constituents (Shock 1982). This species has proven to be adaptable to cultivation in many other parts of the world apart from its native Paraguay, and was first subjected to experimental cultivation in Japan, Korea, Taiwan, several countries in southeast Asia, southern Brazil, and southern California (Sumida 1973; Shock 1982). The plant requires frequent irrigation and is a poor competitor with weeds, and it requires a long growing season, warm temperatures with minimal frost, and high light intensities. *Stevia rebaudiana* occurs naturally on acid soils of pH 4–5, but will also grow on soils with pH levels of 6.5–7.5 (Shock 1982).

Stevia rebaudiana reaches a height of about 80 cm when fully grown, and contains between 7 and 10% w/w *ent*-kaurene glycosides (Bakal and O'Brien Nabors 1986). It was established in early agronomic investigations in Japan that when *S. rebaudiana* seedlings are transplanted in April or May, they can be harvested once in the summer and once in the autumn. The yield of dry leaves of this perennial plant increases from 30 kg/acre in the first year to 50 kg/acre in the second and subsequent years, with the standard density of plantation being 1,000–1,500 plants per acre (Akashi 1977). Investigators have examined the effects of the length of daylight, temperature variation, and the utility of fertilizers and growth regulators on the cultivation of *S. rebaudiana* and the resultant levels of stevioside (Valio and Rocha 1977; Metivier and Viana 1979; Kawatani *et al*. 1980; Mizukami *et al*. 1983).

There has been a considerable interest in the selection and breeding of new strains of *S. rebaudiana*, particularly those affording higher ratios of rebaudioside A relative to stevioside than found in the native plants of Paraguayan origin (Phillips 1987). It has been pointed out that plant breeding can be used to manipulate the proportions of the various sweet glycosides of *S. rebaudiana* leaves. Thus, leaf yield and stevioside concentration are both heritable (Brandle and Rosa 1992) and, the proportions of rebaudiosides A and C in *S. rebaudiana* are controlled by a single additive gene (Brandle 1999). A genetic linkage map has been devised for *S. rebaudiana*, to assist with future molecular breeding experiments (Yao *et al*. 1999).

A number of procedures have been developed for the production of stevioside and rebaudioside A by plant tissue culture of *S. rebaudiana*. Success in this type of endeavor could provide a continuous supply of these sweet compounds, without the need to rely on the cultivated plant. While some of the initial attempts at the plant tissue of *S. rebaudiana* did not result in the production of stevioside (Suzuki *et al*. 1976; Handro *et al*. 1977), early success in this regard first appeared in the patent literature (reviewed in Kinghorn and Soejarto 1985; Phillips 1987). Later success in producing stevioside in callus or stem-tip cultures was accomplished by a number of groups (Lee *et al*. 1982; Hsing *et al*. 1983; Tamura *et al*. 1984). A multiple shoot culture method using a bioreactor has permitted the successful propagation of *S. rebaudiana* seedlings (Nepovim and Vanek 1998).

COMMERCIAL PRODUCTION OF *STEVIA REBAUDIANA*

The commercialization of *S. rebaudiana* leaves for sweetening and flavoring purposes has been quite rapid since first being introduced to Japan. In a recent year, about 200 metric tons of purified stevioside and other sweetener products were prepared from about 2,000 metric tons of dried plant leaves for the Japanese market (Kinghorn *et al*. 2001). As has been the case for some time, the vast majority of the *S. rebaudiana* leaves for the Japanese market is cultivated in the People's Republic of China, especially in Fujian, Zhejiang, and Guangdong Provinces (Kinghorn and Soejarto 1991). The quantities of raw *S. rebaudiana* leaves grown in the People's Republic of China for the Japanese market rose from 200 metric tons in 1982 to 1,300 metric tons only five years later (Kinghorn and Soejarto 1991). Cultivation of *S. rebaudiana* for the Japanese market now also occurs to some degree in Taiwan, Thailand, and Malaysia (Kinghorn and Soejarto 1991). Formerly, however, this plant was produced for the Japanese market in Paraguay, its country of origin (Kinghorn and Soejarto 1985). About 115 metric tons of stevioside were consumed in Korea in 1995, with the majority of this being used in the sweetening of the alcoholic beverage, *soju*, and produced from leaves of *S. rebaudiana* grown in the People's Republic of China (Kinghorn *et al.* 2001). A refined extract of the leaves of *S. rebaudiana* (containing at least 60% stevioside) and pure stevioside (free from steviol and isosteviol) is approved in Brazil for the sweetening of chewing gum, dietetic foods and beverages, medicines, oral hygiene products, and soft drinks (Anonymous 1988b; 1998c). Production occurs in Parana Province in southern Brazil for the local market in that country (Hanson and De Oliveira 1993; Oliveira Ferro 1997). Publications have appeared recently describing the possibility of establishing the cultivation of *S. rebaudiana* in additional countries such as Canada (Brandle *et al.* 1998), the Czech Republic (Nepovim *et al.* 1998), India (Chalapathi *et al.* 1997), and Russia (Dzyuba and Vseross 1998).

CARIOGENICITY POTENTIAL OF STEVIOSIDE

In addition to their use in calorie control and for diabetics, alternative sweeteners offer another benefit in maintaining good dental health by reducing the intake of sucrose (Grenby 1997). Pure stevioside and rebaudioside A were tested for cariogenicity in an albino rat model at the College of Dentistry, University of Illinois at Chicago. In this study, 60 Sprague-Dawley rats were colonized with *Streptococcus sobrinus* and divided into groups fed basal diet 2,000 supplemented with either 0.5% stevioside, 0.5% rebaudioside A, 30% sucrose, or with no test compound at all. All four groups were sacrificed after five weeks of feeding, and viable *S. sobrinus* counts enumerated and caries evaluated according to Keyes' technique. It was concluded that stevioside and rebaudioside were not cariogenic under the conditions of the study (Das *et al.* 1992). In a more recent *in vitro* study, the eight sweet constituents of *S. rebaudiana* (stevioside, rebaudiosides A–E, dulcoside A, steviolbioside) and two hydrolytic products of stevioside (steviol and isosteviol) were tested against a panel of cariogenic and periodontopathic oral bacteria. Both their antibacterial activity and their ability to inhibit sucrose-induced adherence, glucan binding, and glucosyltransferase (GTF) activity were evaluated. None of these compounds suppressed the growth or acid production of the cariogenic organism, *Streptococcus mutans*, or affected sucrose-induced adherence, or GTF activity. However, rebaudiosides B, C, and E, steviol, and isosteviol inhibited the glucan-induced aggregation of mutans streptococci to some extent, and could provide oral health benefits by interference with cell surface functions of cariogenic bacteria (Wu *et al.* 1998).

REGULATORY STATUS

Stevia rebaudiana extracts containing stevioside and/or pure stevioside are approved or otherwise utilized as food additives in Japan, South Korea, Brazil, Argentina, and Paraguay, and are used in herbal preparations in other countries, inclusive of the People's Republic of China (where the great majority of this plant is produced for commerce), the United States and to a lesser extent, certain countries in western Europe.

CONCLUSIONS

Stevia rebaudiana powdered leaf has been used for over a century in Paraguay to sweeten beverages, although this practice seems only ever to have been conducted on a relatively small scale. Larger quantities of *S. rebaudiana* products have been used increasingly for about a quarter of a century in Japan, with there being many Japanese foods and beverages containing stevioside now available. Within the last few years stevioside has begun to be used extensively in South Korea in the formulation of the alcoholic beverage, *soju*, for domestic consumption. Stevioside is available for the sweetening of several different types of products in Brazil, including soft drinks. In addition, *S. rebaudiana* products are used as dietary supplements in the United States and countries such as Italy in western Europe. Despite this widespread use in several different parts of the world, no evidence of adverse reactions due to the ingestion of *S. rebaudiana* extracts of stevioside by humans has appeared in the biomedical literature. This situation may be contrasted with increasing evidence of toxic side effects appearing in the recent literature due to the excessive ingestion of licorice-flavored candy or chewing gum (of which the sweet-tasting triterpenoid glycoside, glycyrrhizin, is a major principle) (e.g. Chamberlain and Abolnik 1997; de Klerk *et al.* 1997). It is interesting to note that the Dutch Nutrition Information Board has advised against a daily glycyrrhizin intake of greater than 200 mg per day, equivalent to about 150 g of licorice candy (Fenwick *et al.* 1990). Along similar lines, the Ministry of Health in Japan has cautioned against a glycyrrhizin consumption in excess of 200 mg/day when used as a formulation excipient in medicines (Kinghorn and Compadre 1991).

It is remarkable that *S. rebaudiana* accumulates in its leaves such high levels of the most abundant sweet diterpene glycosides, stevioside and rebaudioside A, given that plant leaves are not normally regarded as storage organs for secondary metabolites. This phenomenon is all the more unusual because only one other species in the genus *Stevia* has ever been reported to biosynthesize sweet-tasting *ent*-kaurene constituents, namely, *S. phlebophylla* (Kinghorn *et al.* 1984). Accordingly, if *S. rebaudiana* were not such a chemotaxonomic curiosity, and unless stevioside and rebaudioside A did not possess such potent sweetness intensities relative to sucrose, there would be no widespread commercial use of *S. rebaudiana* products as sucrose substitutes, since they would not be cost effective. As has been pointed out in this chapter, and as will be elaborated on later in this book, a great deal of technical ingenuity has been applied in Japan towards improving the sweetness pleasantness of stevioside, which is another contributing factor in the increasing use of *S. rebaudiana*-based products.

There remains considerable conjecture about the safety of stevioside for human consumption, as will also be expanded on later in this volume. One significant omission from the current scientific literature is that a clinical study on the absorption and metabolism of stevioside in humans has not appeared in the literature thus far. This is important, because it needs to be established unequivocally if the *ent*-kaurene aglycone steviol is formed as a result of the consumption of stevioside. Increasingly, literature reports are showing that steviol is a more toxic compound than stevioside on a weight for weight basis (e.g. Wasuntarawat *et al.* 1998). However, taking all present knowledge into consideration, *S. rebaudiana* extracts and stevioside do not seem to present a potential toxicity risk for humans at the low consumption levels used in sweetening. An acceptable intake of stevioside for humans has been calculated as about 8 mg/kg/day (Xili *et al.* 1992). Recently, a non-profit health-advocacy group based in Washington, DC has come to the conclusion that if *S. rebaudiana* sweeteners are used sparingly, there seems to be no threat to public health, although caution should be exercised at higher daily intake levels (Schardt 2000).

REFERENCES

Abe, K. and Sonobe, M. (1977) Use of stevioside in the food industry. *New Food Industry* **19**, 67–72.

Akashi, H. (1977) Present status and prospect for stevioside for utilization. *Shokuhin Kogyo* **20**(24), 20–26.

Anonymous (1980) Concessão de registro e medicamento. *Diario Official*, Brazil, 19 September.

Anonymous (1988a) High intensity sweeteners – market size 7.2 billion yen. Stevia occupies 41%, but future gains will be made by aspartame. *Food Chemicals, Tokyo*, No. 6, 19–26.

Anonymous (1988b) Resolution No. 14. *Diário Oficial*, Brazil, 26 January.

Anonymous (1988c) Resolution No. 67. *Diário Oficial*, Brazil, 8 April.

Anonymous (1988d) Stevioside. In *The Korean Standards of Food Additives*, The Korea Foods Industry Association, Seoul, pp. 198–199.

Anonymous (1993) Stevia extract. In *Voluntary Specifications of Non-chemically Synthesized Food Additives* (2nd edn), Japan Food Additive Association, Tokyo, pp. 119–124.

Anonymous (1996) *The Merck Index. An Encyclopedia of Chemicals, Drugs, and Biologicals* (12th edn), Merck Research Laboratories, Whitehouse Station, NJ, pp. 1503–1504.

Anonymous (1999a) European Commission, Scientific Committee on Food, Opinion on stevioside as a sweetener (adopted on 17 June 1999), Brussels, Belgium pp. 1–7. (http://europa.eu.int/comm/dg24/health/sc/scf/index en/html).

Anonymous (1999b) European Commission, Scientific Committee on Food, Opinion on Stevia rebaudiana Bertoni plants and leaves (adopted on 17 June 1999), Brussels, Belgium, pp. 1–4. (http://europa.eu.int/comm/dg14 /health/sc/scf/index en/html).

Bakal, A. I. and O'Brien Nabors, L. (1986). Stevioside. In *Alternative Sweeteners*, O'Brien Nabors and R. C. Gelardi (Eds), Marcel Dekker, Inc., New York, pp. 295–307.

Bell, F. (1954) Stevioside: a unique sweetening agent. *Chemistry and Industry*, 17 July, pp. 897–898.

Bertoni, M. S. (1905) Le Kaá hê-é: sa nature ct scs propriélés. *Anales Científicos Paraguayos, Serie 1* **5**, 1–14.

Bertoni, M. S. (1918) La *Stevia Rebaudiana* Bertoni. La estevina y la rebaudina, neuvas substancias edulcorantes. *Anales Científicos Paraguayos, Serie II* **6**, 29–134.

Blumenthal, M. (1995) FDA lifts import ban on Stevia. Herb can be imported on as a dietary supplement; future use as a sweetener is still unclear. *HerbalGram* **35**, 17–18.

Bonvie, L. and Bonvie, B. (1996) Sinfully sweet? *New Age Journal*, January–February, pp. 60–64, 120, 122, 124, 126–128.

Brandle, J. E. (1999) Genetic control of rebaudioside A and C concentration in leaves of the sweet herb, *Stevia rebaudiana*. *Canadian Journal of Plant Sciences* **79**, 85–92.

Brandle, J. E. and Rosa, N. (1992) Heritability for yield, leaf:stem ratio and stevioside content estimated from a landrace cultivar of *Stevia rebaudiana*. *Canadian Journal of Plant Sciences* **72**, 1263–1266.

Brandle, J. E., Starratt, A. N. and Gizjen, M. (1998) *Stevia rebaudiana*: its agricultural, biological, and chemical properties. *Canadian Journal of Plant Sciences* **78**, 527–536.

Chalapathi, M. V., Thimmegowda, S., Sridhara, S., Parama, V. R. R. and Prasad, T. G. (1977) Natural non-calorie sweetener Stevia (*Stevia rebaudiana* Bertoni) – a future crop of India. *Crop Research* **14**, 347–350.

Chamberlain, J. J. and Abolnik, I. Z. (1997) Pulmonary edema following a licorice binge. *Western Journal of Medicine* **167**, 184.

Chang, S. S. and Cook, J. M. (1983) Stability studies of stevioside and rebaudioside A in beverages. *Journal of Agricultural and Food Chemistry* **31**, 409–412.

Clark, S. (2000) Sweet dreams. *London Sunday Times*, Style section, 1 April, p. 51.

Cook, I. F. and Knox, J. R. (1970) A synthesis of steviol, *Tetrahedron Letters*, pp. 4091–4093.

Crammer, B. and Ikan, R. (1986) Sweet glycosides from the *Stevia* plant. *Chemistry in Britain*, **22**, 915–916, 918.

Crammer, B. and Ikan, R. (1987) Progress in the chemistry and properties of the rebaudiosides. In *Developments in Sweeteners – 3*, T. H. Grenby (Ed.), Elsevier Applied Science, London, pp. 45–64.

Das, S., Das, A. K., Murphy, R. A., Punwani, I. C., Nasution, M. P. and Kinghorn, A. D. (1992) Evaluation of the cariogenic potential of the intense natural sweeteners stevioside and rebaudioside A. *Caries Research* **26**, 363–366.

de Klerk, G. J., Nieuwenhuis, M. G. and Beutler, J. J. (1997) Hypokalaemia and hypertension associated with liquorice flavoured chewing gum. *British Medical Journal* **314**, 731–732.

Dieterich, K. (1908) The constituents of *Eupatorium rebaudianum*, 'Kaa-he-e', and their pharmaceutical value. *Pharmazeutische Zentralhalle* **50**, 435–458. [*Chemical Abstracts* (1909) **3**, 2485–1].

DuBois, G. E. and Stephenson, R. A. (1985) Diterpenoid sweeteners. Synthesis and sensory evaluation of stevioside analogues with improved organoleptic properties. *Journal of Medicinal Chemistry* **28**, 93–98.

DuBois, G. E., Bunes, L. A., Dietrich, P. S. and Stephenson, R. A. (1984) Diterpenoid sweeteners. Synthesis and sensory evaluation of biologically stable analogues of stevioside. *Journal of Agricultural and Food Chemistry* **32**, 1321–1325.

Duffy, V. B. and Anderson, G. H. (1998) Position of the American Dietetic Association: use of nutritive and non-nutritive sweeteners. *Journal of the American Dietetic Association* **98**, 580–587.

Dzyuba, O. and Vseross, O. (1998) *Stevia rebaudiana* (Bertoni) Hemsley – a new source of natural sweetener for Russia. *Rastitel'nye Resursy (Plant Resources)* **34**, 86–95.

Esaki, S., Tanaka, R. and Kamiya, S. (1984) Synthesis and taste of certain steviol glycosides. *Agricultural and Biological Chemistry* **48**, 1831–1834.

Felippe, G. M. (1977) *Stevia rebaudiana* Bert.: uma revisão. *Ciência e Cultura (São Paulo)* **29**, 1240–1248.

Fenwick, G. R., Lutomski, J. and Nieman, C. (1990) Liquorice, *Glycyrrhiza glabra* L. – composition, uses, and analysis. *Food Chemistry* **38**, 119–143.

Fletcher, H. G., Jr (1955) The sweet herb of Paraguay. *Chemurgic Digest*, **14**, 7, 18–19.

Fujita, H. and Edahiro, T. (1979) Safety and utilization of Stevia sweetener. *Shokuhin Kogyo* **22** (20), 66–72.

Galperin de Levy, R. H. (1984) *Stevia rebaudiana* Bertoni: un singular edulcorante natural. *Acta Farmacéutica Bonarense* **3**, 47–50.

Gosling, C. (1901) *Caá-êhê* or *azuca-caá. Kew Bulletin*, pp. 183–194.

Grenby, T. H. (1997) Dental aspects of the use of sweeteners. *Pure and Applied Chemistry* **69**, 709–714.

Handro, W., Hell, K. G. and Kerbauly, G. B. (1977) Tissue culture of *Stevia rebaudiana*, a sweetening plant. *Planta Medica* **32**, 115–117.

Hanson, J. R. and White, A. F. (1968). Terpenoid biosynthesis. II. Biosynthesis of steviol. *Phytochemistry* **7**, 595–597.

Hanson, J. R. and De Oliveira, B. H. (1993) Stevioside and related sweet diterpenoid glycosides. *Natural Products Reports* **10**, 301–309.

Hough, C. A. M., Parker, K. J. and Vlitos, A. J. (Eds) (1979) Preface. In *Developments in Sweeteners – I*, Applied Science Publishers, London, pp. v–viii.

Hsing, Y. O., Su, W. F. and Chang, W. C. (1983) Accumulation of stevioside and rebaudioside A in callus culture of *Stevia rebaudiana* Bertoni. *Botanical Bulletin of the Academia Sinica* **24**, 115–119. [*Chemical Abstracts* (1983) **99**, 172877b].

Hutapea, A. M., Toskulkao, C., Wilairat, P. and Buddhasukh, D. (1999) High-performance liquid chromatographic separation and quantitation of stevioside and its metabolites. *Journal of Liquid Chromatography and Related Technologies* **22**, 1161–1170.

Jacobs, M. B. (1955) The sweetening power of stevioside. *The American Perfumer*, December, pp. 44–45.

Kawatani, T., Kaneki, Y., Tanabe, T. and Takahashi, T. (1980) On the cultivation of kaa he-e (*Stevia rebaudiana* Bertoni). VI. Response of kaa he-e to potassium fertilizer rates and to the three major elements of fertilizer. *Japanese Journal of Tropical Agriculture* **24**, 105–112.

Kim, K. K. and Shibata, H. (1997) Characterization of *ent*-kaurenoic acid 13-hydroxylase in steviol biosynthesis of *Stevia rebaudiana*. *Han'guk Nonghwa Hahoechi* **40**, 501–507. [*Chemical Abstracts* (1998) **128**, 214799].

Kim, K. K., Yamashita, H., Yoshihiro, S. and Shibata, H. (1996) A high activity of 3-hydroxy-3-methylglutaryl coenzyme A reductase in chloroplasts of *Stevia rebaudiana* Bertoni. *Bioscience, Biotechnology, and Biochemistry* **60**, 685–685.

Kinghorn, A. D. and Compadre, C. M. (1991) Less common high-potency sweeteners. In *Alternative Sweeteners* (2nd edn, Revised and Expanded), L. O'Brien Nabors and R. C. Gelardi (Eds), Marcel Dekker Inc., New York, pp. 197–218.

Kinghorn, A. D. and Kennelly, E. J. (1995) Discovery of highly sweet compounds from natural sources. *Journal of Chemical Education* **72**, 676–680.

Kinghorn, A. D. and Soejarto, D. D. (1985) Current status of stevioside as a sweetening agent for human use. In *Progress in Medicinal and Economic Plant Research*, H. Wagner, H. Hikino and N. R. Farnsworth, (Eds), Academic Press, London, **1**: 1–51.

Kinghorn, A. D. and Soejarto, D. D. (1989) Intensely sweet compounds of natural origin. *Medicinal Research Reviews* **9**, 91–115.

Kinghorn, A. D. and Soejarto, D. D. (1991) Stevioside. In *Alternative Sweeteners* (2nd edn, Revised and Expanded), L. O'Brien Nabors and R. C. Gelardi (Eds), Marcel Dekker, Inc., New York, pp. 157–171.

Kinghorn, A. D., Soejarto, D. D., Nanayakkara, N. P. D., Compadre, C. M., Makapugay, H. C., Hovanec-Brown, J. M. *et al.* (1984) A phytochemical screening procedure for sweet *ent*-kaurene glycosides in the genus *Stevia*. *Journal of Natural Products* **47**, 439–444.

Kinghorn, A. D., Soejarto, D. D., Katz, N. L. and Kamath, S. K. (1985) Studies to identify, isolate, develop, and test naturally occurring noncariogenic sweeteners that may be used as dietary sucrose substitutes. *Government Reports and Announcements Index (United States)* **85** (11), 47. [*Chemical Abstracts* (1985) **103**, 86674c].

Kinghorn, A. D., Fullas, F. and Hussain, R. A. (1995) Structure-activity relationship of highly sweet natural products. In *Studies in Natural Products Chemistry. Volume 15, Structure and Chemistry (Part C)*, Atta-ur-Rahman (Ed.), Elsevier, Amsterdam, pp. 3–41.

Kinghorn, A. D., Kaneda, N., Baek, N.-I., Kennelly, E. J. and Soejarto, D. D. (1998) Noncariogenic intense natural sweeteners. *Medicinal Research Reviews* **18**, 347–360.

Kinghorn, A. D., Wu, C. D. and Soejarto, D. D. (2001) Stevioside. In *Alternative Sweeteners* (3rd edn, Revised and Expanded), L. O'Brien Nabors (Ed.), Marcel Dekker, Inc., New York, pp. 167–183.

Kobayashi, M., Horikawa, S., Degrandi, I. H., Ueno, J. and Mitsuhashi, H. (1977) Dulcosides A and B, new diterpene glucosides from *Stevia rebaudiana*. *Phytochemistry* **16**, 1405–1407.

Kohda, H., Kasai, R., Yamasaki, K., Murakami, K. and Tanaka, O. (1976) New sweet diterpene glucosides from *Stevia rebaudiana*. *Phytochemistry* **15**, 981–983.

Kroyer, G. T. (1999) The low calorie sweetener stevioside: stability and interaction with food ingredients. *Lebensmittel-Wissenschaft und -Technologie* **32**, 509–512.

Kurahashi, H., Yamaguchi, Y., Tsuzuki, S. and Maehashi, H. (1982) Pharmacological studies of stevioside. *Matsumoto Shigaku* **8**, 56–62. [*Chemical Abstracts* (1982) **97**, 214437z].

Lee, K. R., Park, J. R., Choi, B. S., Han, J. S., Ph, S. L. and Yamada, Y. (1982) Studies on the callus culture of *Stevia* as a new sweetening source and the formation of stevioside. *Hanguk Sik'pum Kwahakhoe Chi* **14**, 179–183. [*Chemical Abstracts* (1982) **97**, 71003s].

Lewis, W. H. (1992) Early uses of *Stevia rebaudiana* (Asteraceae) leaves as a sweetener in Paraguay. *Economic Botany* **46**, 336–337.

Liu, J., Ong, C. P. and Li, S. F. Y. (1997) Subcritical fluid extraction of Stevia sweeteners from *Stevia rebaudiana*. *Journal of Chromatographic Science* **35**, 446–450.

Makapugay, H. C., Nanayakkara, N. P. D. and Kinghorn, A. D. (1984) Improved high-pressure liquid chromatographic separation of the *Stevia rebaudiana* sweet diterpene glycosides using linear gradient elution. *Journal of Chromatography* **283**, 390–395.

Matsui, M., Sofuni, T. and Nohmi, T. (1996) Regionally targeted mutagenesis by metabolically activated steviol: DNA sequence analysis of steviol-induced mutants of guanine phosphoribosyltransferase (*gpt*) gene of *Salmonella typhimurium* TM677. *Mutagenesis* **11**, 565–572.

Mauri, P., Catalano, G., Gardana, C. and Pietta, P. (1996) Analysis of *Stevia* glycosides by capillary electrophoresis. *Electrophoresis* **17**, 367–371.

Metivier, J. and Viana, A. M. (1979) The effect of long and short day length upon the growth of whole plants and the level of soluble proteins, sugars, and stevioside in leaves of *Stevia rebaudiana* Bert. *Journal of Experimental Biology* **30**, 1211–1222.

Mizukami, H., Shiba, K., Inoue, S. and Ohashi, H. (1983) Effect of temperature on growth and stevioside formation of *Stevia rebaudiana* Bertoni. *Shoyakugaku Zasshi* **37**, 175–179.

Mizutani, K., Miyata, T., Kasai, R., Tanaka, O., Ogawa, S. and Doi, S. (1989) Study on the improvement of sweetness of steviol bisglycosides: selective enzymatic transglycosylation of the 13-O-glycosyl moiety. *Agricultural and Biological Chemistry* **53**, 395–398.

Mizutani, K., Kambara, T., Masuda, H., Tamura, Y., Ikeda, T., Tanaka, O. *et al.* (1998) Glycyrrhetic acid monoglucuronide (MGGR): biological activities. In *Towards Natural Medicine in the 21st Century*, Excepta Medica, International Congress Series 1157, H. Ageta, N. Aimi, Y. Ebizuka, T. Fujita and G. Honda, (Eds), Elsevier, Amsterdam, pp. 225–235.

Mori, K., Nakahara, Y. and Matsui, M. (1970) Total synthesis of (±)-steviol. *Tetrahedron Letters* 2411–2414.

Moroni, L. (1999) Stevia: è naturale, sana, dietetica, e fino a 300 volte più doce dello zucchero. *Natura Scienza*, Bussolengo, Verona, Italy, **12**(1), 25–27.

Mosettig, E., Beglinger, U., Dolder, F., Lichti, H., Quitt, P. and Waters, J. A. (1963) The absolute configuration of steviol and isosteviol. *Journal of the American Chemical Society* **85**, 2305–2309.

Nakayama, K., Kasahara, D. and Yamamoto, F. (1986) Absorption, distribution, metabolism, and excretion of stevioside in rats. *Journal of the Food Hygiene Society of Japan* **27**, 1–8.

Nepovim, A. and Vanek, T. (1998) *In vitro* propagation of *Stevia rebaudiana* plants using multiple shoot culture. *Planta Medica* **64**, 775–776.

Nepovim, A., Drahosova, H., Valicek, P. and Vanek, T. (1998) The effect of cultivation conditions on the content of stevioside in *Stevia rebaudiana* plants cultivated in the Czech Republic. *Pharmacy and Pharmacology Letters* **8**, 19–21.

Nishiyama, P., Alvarez, M. and Vieira, L. G. E. (1992) Quantitative analysis of stevioside in the leaves of *Stevia rebaudiana* by near infrared reflectance spectroscopy. *Journal of the Science of Food and Agriculture* **59**, 277–281.

Oliveira Ferro, V. (1997) Faculty of Pharmaceutical Sciences, University of São Paulo, São Paulo, Brazil, private communication.

Pezzuto, J. M. (1986) Chemistry, metabolism and biological activity of steviol (*ent*-13–hydroxykaur-13-en-19-oic acid). In *New Trends in Natural Products Chemistry 1986. Studies in Organic Chemistry, vol. 26*, Atta-ur-Rahman and P. W. Le Quesne, (Eds), Elsevier Science Publishers BV, Amsterdam, pp. 371–386.

Pezzuto, J. M., Compadre, C. M., Swanson, S. M., Nanayakkara, N. P. D. and Kinghorn, A. D. (1985) metabolically activated steviol, the aglycone of stevioside, is mutagenic. *Proceedings of the National Academy of Sciences of the United States of America* **82**, 2478–2482.

Phillips, K. C. (1987) Stevia: steps in developing a new sweetener. In *Developments in Sweeteners – 3*, T. H. Grenby (Ed.), Elsevier Applied Science, London, pp. 1–43.

Planas, G. M. and Kuć, J. (1968) Contraceptive properties of *Stevia rebaudiana*. *Science* **162**, 1007.

Richman, A. S., Gijzen, M., Starratt, A. N., Yang, Z. and Brandle, J. E. (1999) Diterpene synthesis in *Stevia rebaudiana*: recruitment and up-regulation of key enzymes from the gibberellin biosynthetic pathway. *Canadian Plant Journal* **19**, 411–421.

Ruddat, M., Heftmann, E. and Lang, A. (1965) Biosynthesis of steviol. *Archives of Biochemistry and Biophysics* **110**, 496–499.

Sakaguchi, M. and Kan, T. (1982) Japanese research on *Stevia rebaudiana* (Bert.) Bertoni and stevioside. *Ciência e Cultura (São Paulo)*, **34**, 235–248.

Schardt, D. (2000) Stevia – a bittersweet tale. *Nutrition Action Healthletter*, April, p. 3.

Seewald, N. (2000) A steady diet of growth for sweeteners and fat substitutes. *Chemical Week*, 7 June, p. 4.

Seon, J. H. (1995) Stevioside as natural sweetener. *Report of the Pacific R & D Center, Seoul, Korea*, pp. 1–8.

Shibata, H., Sonoke, S., Ochiai, H., Nishihashi, H. and Yamada, M. (1991) Glycosylation of steviol and steviol glycosides in extracts from *Stevia rebaudiana* Bertoni. *Plant Physiology* **95**, 152–156.

Shock, C. C. (1982) Rebaudi's stevia: natural noncaloric sweeteners. *California Agriculture*, September–October, pp. 4–5.

Soejarto, D. D., Kinghorn, A. D. and Farnsworth, N. R. (1982) Potential sweetening agents of plant origin. III. Organoleptic evaluation of *Stevia* leaf herbarium samples for sweetness. *Journal of Natural Products* **45**, 590–599.

Sumida, T. (1973) Reports on *Stevia rebaudiana* Bertoni M. introduced from Brazil as a new sweetness resource in Japan. *Miscellaneous Publications of the Hokkaido National Agricultural Experiment Station*, No. 2, pp. 69–83.

Suzuki, H., Ikeda, T., Matsumoto, T. and Noguchi, M. (1976) Isolation and identification of rutin from cultured cells of *Stevia rebaudiana* Bertoni. *Agricultural and Biological Chemistry* **40**, 819–820.

Swanson, S. M., Mahady, G. B. and Beecher, C.W.W. (1992) Stevioside biosynthesis by callus, root, shoot, and rooted-shoot cultures *in vitro*. *Plant, Cell,Tissue and Organ Culture* **28**, 151–157.

Tamura,Y., Nakamura, S., Fukui, H. and Tabata, M. (1984) Comparison of *Stevia* plants grown from seeds, cuttings, and and stem-tip cultures for growth and sweet glycosides. *Plant Cell Reports* **3**, 180–182.

Tanaka, O. (1982) Steviol-glycosides: new natural sweeteners. *Trends in Analytical Chemistry* **1**, 246–248.

Tanaka, O. (1997) Improvement of taste of natural sweeteners. *Pure and Applied Chemistry* **69**, 675–683.

Tinti, J.-M. and Nofre, C. (1991) Design of sweeteners: a rational approach. In *Sweeteners: Discovery, Molecular Design, and Chemoreception*; Walters, D. E., Othoefer, F. T. and DuBois, G. E. (Eds), Symposium Series No. 450, American Chemical Society,Washington, DC, pp. 88–99.

Toffler, F. and Orio, O. A. (1981) Acceni sulla pianta tropicale kaá-hê-é o erba dolce. *La Revista della Società Italiana di Scienza dell'Alimentazione* **10**, 225–230.

Totté, N., Charon, L., Rohmer, M., Compernolle, F., Baboeuf, I. and Geuns, J. M. C. (2000) Biosynthesis of the diterpenoid steviol, and *ent*-kaurene derivative from *Stevia rebaudiana* Bertoni, via the methylerythritol phosphate pathway. *Tetrahedron Letters* **41**, 6407–6410.

Toyoda, K., Matsui, H., Shoda, T., Uneyama, C., Takada, K. and Takahashi, M. (1997) Assessment of the carcinogenicity of stevioside in F344 rats. *Food and Chemical Toxicology* **35**, 597–603.

Valio, I. F. M. and Rocha, R. F. (1977) Effect of photoperiod and growth regulator on growth and flowering of *Stevia rebaudiana* Bertoni. *Japanese Journal of Crop Science* **46**, 243–248.

Vis, E. and Fletcher, H. G., Jr (1956) Stevioside. IV. Evidence that stevioside is a sophoroside. *Journal of the American Chemical Society* **75**, 4709–4710.

Walters, D. E., Prakash, I. and Desai, N. (2000) Active conformations of neotame and other high-potency sweeteners. *Journal of Medicinal Chemistry* **43**, 1242–1245.

Wasuntarawat, C., Temcharoen, P., Toskulkao, C., Munkornkarn, P., Suttajit, M. and Glinsukon, T. (1998) Developmental toxicity of steviol, a metabolite of stevioside, in the hamster. *Drug and Chemical Toxicology* **21**, 207–222.

Wingard, R. E., Jr, Brown, J. P., Enderlin, F. E., Dale, J. A., Hale, R. L. and Sietz, C. T. (1980) Intestinal degradation and absorption of the glycosidic sweeteners stevioside and rebaudioside A. *Experientia* **36**, 519–520.

Wood, H. B., Jr, Allerton, R., Diehl, H.W. and Fletcher, H. G., Jr (1955) Stevioside. I. The structure of the glucose moieties. *Journal of Organic Chemistry* **20**, 875–883.

Wu, C. D., Johnson, S. A., Srikantha, R. and Kinghorn, A. D. (1998) Intense natural sweeteers and their effects on cariogenic bacteria. *Journal of Dental Research* **77**, 283.

Xili, L., Chengjiany, B., Eryi, X., Reiming, A., Yuengming, W., Haodong, S. and Zhiyian, H. (1992) Chronic oral toxicity and carcinogenicity study of stevioside in rats. *Food and Chemical Toxicology* **30**, 957–965.

Yamasaki, K., Kohda, H., Kobayashi, T., Kasai, R. and Tanaka, O. (1976) Structures of Stevia diterpene-glucosides: Applications of ^{13}C NMR. *Tetrahedron Letters* 1005–1008.

Yao,Y., Ban, M. and Brandle, J. (1999) A genetic linkage map for *Stevia rebaudiana*. *Genome* **42**, 657–661.

Yoshihira, K., Matsui, M. and Ishidate, M. (1987) Chemical characteristics and biological safety of a glycosidic sweetener, stevioside. *Tokishikaroji Forami* **10**, 281–289.

2 Botany of *Stevia* and *Stevia rebaudiana*

Djaja Djendoel Soejarto

INTRODUCTION

The genus *Stevia* belongs to the family Compositae[1] within the tribe Eupatorieae. One member of this genus, *S. rebaudiana* (Bertoni) Bertoni, is of worldwide importance today. Its economic importance lies in the fact that the plant contains sweet-tasting *ent*-kaurene diterpenoid glycosides, in particular, stevioside and rebaudioside A, currently used as non-nutritive high potency sweeteners, primarily in Japan. The common name of this plant in the Guaraní language of Paraguay is *Caá hê-é* or other renditions of this name, all of which mean *sweet herb*.

In view of the fact that only a portion (about 50%) of the members of the genus *Stevia* has been surveyed for the occurrence of the sweet-tasting compounds (Soejarto *et al*. 1982; 1983; Kinghorn *et al*. 1984), a review of the entire genus, in the context of its economic importance, will be useful in setting the stage for other discussions on chemistry, biology and economics. Thus, this chapter begins with an overview on the taxonomy of *Stevia* and species relationships within the genus, with the aim of providing a better perspective in the future search for sweet compounds in other members of the genus that have not been studied. Following a review of the genus, the discussion will shift into the subject plant, *S. rebaudiana*, for the remainder of the chapter.

Because of the author's personal acquaintance with the plant *S. rebaudiana* in its natural habitat (Soejarto *et al*. 1983) and in cultivation, much of the account about this plant in this chapter is based on his own first-hand knowledge and observations.

THE GENUS *STEVIA*

Stevia is a New World genus distributed from the southern United States to Argentina and the Brazilian highlands, through Mexico, the Central American States, and the South American Andes. It is one of five large genera of the Eupatorieae, of which none has been fully monographed in recent years. These five genera are *Ageratina*, *Chromolaena*, *Koanophyllon*, *Mikania*, and *Stevia* (Robinson and King 1977; King and Robinson 1987).

Members of *Stevia* comprise herbs and shrubs, found mostly at altitudes of 500–3500 m above sea level. Though they usually grow in semi-dry mountainous terrains, their habitats

---

1 According to the rules set down in the *International Code of Botanical Nomenclature* (Greuter *et al*. 1994), Compositae is the conserved name and Asteraceae is the alternative name. Use of either name is correct, but the author prefers the use of the conserved name.

range from grasslands, scrub forests, forested mountain slopes, conifer forests, to sub-alpine vegetation. Records indicate that the genus is not represented in the West Indies and Amazonia.

The taxonomy of *Stevia* is very complex. The *Index Kewensis* (Anonymous 1893–1993) lists 371 published Latin binomials (392, if homonyms are included) for this genus. Estimates on the number of species within the genus, however, range from 150 to 300 (King and Robinson 1967; Grashoff 1972; Robinson and King 1977). From the examination of a herbarium collection holding of *Stevia* in deposit at the John G. Searle Herbarium of the Field Museum, Chicago, and a review of the literature, as part of an organoleptic survey to detect sweet-tasting species (Soejarto *et al.* 1982), followed up by chemical screening for *ent*-kaurene diterpene glycosides (Kinghorn *et al.* 1984), the author was able to catalogue 220 species of *Stevia* considered to be in good taxonomic standing. In 1987, King and Robinson, specialists of the tribe Eupatorieae, recognized 230 species in good taxonomic standing as belonging to the genus *Stevia* (King and Robinson 1987). This means that at least 141 of the 371 published Latin binomials are mere synonyms. In fact, Grashoff (1972) in his revision of the North American species of *Stevia* relegated 126 Latin binomials into synonymy. All of these indicate the lack of unifying criteria for species delimitation in the genus.

Of the 220–230 species considered to be in good standing, the greater portion is found in South America. Unfortunately, there is currently no comprehensive taxonomic monograph that covers the entire South American range of the genus. The area with the greatest concentration of species appears to be located within the triangle formed by Peru–Bolivia, southern Brazil–Paraguay, and northern Argentina, where at least 120 species occur. For North America, our knowledge on the species of *Stevia* is better. According to Grashoff, who monographed the genus as his doctoral dissertation (Grashoff 1972), 79 species of *Stevia* are known to occur in North America, north of the Colombia–Panamanian border, of which at least 70 species are found in Mexico alone. Two more new species were added by Grashoff in 1974 (Grashoff 1974).

We owe our understanding of the taxonomy of *Stevia* to many botanists who have studied this genus in the past, but particularly to the work of Bonpland *et al.* (1820), De Candolle (1836), Schultz-Bipontinus (1852), Grisebach (1874; 1879), Baker (1876), Hemsley (1881), Hoffmann (1894), Hieronymus (1897), Hassler (1912), Blake (1926), Robinson (1930a–d; 1931a–c; 1932a–b), Grashoff (1972; 1974), King and Robinson (1967; 1968; 1987), and Robinson and King (1977). As a genus, the taxon was consolidated in 1816 by Lagasca, though the generic name was coined by Cavanilles in 1797 (see Robinson 1930a). Since no type species was originally established, King and Robinson (1969) selected and designated *Stevia serrata* Cav. as a lectotype species of the genus. The generic name *Stevia* was given by Cavanilles in memory of 'Petri Iacobi Stevii', a noted medical practitioner and botany professor of Valencia, Spain (King and Robinson 1987).

CLASSIFICATION OF *STEVIA*

Members of *Stevia* are characterized by the following. *Plant habit*: Annual, perennial or rhizomatous perennial herbs or shrubs. *Leaves*: Simple, opposite, rarely alternate, mostly petiolate, but exstipulate, penninerved or reticulately nerved. *Inflorescences*: Capitula (flower heads) are arranged on a corymbose, paniculate or thyrsoid clusters at tips of branches or stem. *Capitulum*: The capitulum consists of a cylindrical to funnelform involucre of five linear to oblong phyllaries (bracts) arranged in one series or whorl and in a valvate to subimbricate aestivation, and five disk florets (flowers) on flat, glabrous (hairless) receptacle; ray flowers are absent. *Floret*: The five florets are normally white, but may be purplish to reddish purple, tubular, the color of the upper and lower parts of the florets may be different; they are arranged into a 5-cornered whorl, each opposite a phyllary. The length of individual florets at anthesis exceeds that of the involucre, while the corolla lobes are unequal and much shorter than the tube; a ring of trichomes is present at the corolla throat, at the junction of the lobes and the tube; the stamens are inserted, with appendaged anther, whereas the style branches, which are filiform and usually papillate, exceed or protrude far beyond the corolla tube. *Fruit (Achene)*: The achenes are cylindrical to fusiform or prismatic, 5-angled, 5-ribbed or nerved, with concave faces; the pappus (modified, persistent calyx) is represented by awns (aristae) or scales, or by a ring of scale-like structure or by a combination of these.

On the basis of pappus type, inflorescence, and growth habit, members of *Stevia* may be classified into subgeneric groups, a common practice by taxonomic monographers to indicate relationships and to facilitate species identification. In this respect, the subgeneric classification of *Stevia* is rather confusing.

For example, Grashoff (1972) grouped the North American species into three Series (a category of subgeneric grouping): Series *Corymbosae* (38 species; mostly perennials with corymbose inflorescences and with flower pedicels shorter than or equal to the length of the involucre); Series *Podocephalae* (16 species; annuals, inflorescences lax, pedicels longer than the involucre); and Series *Fruticosae* (25 species; shrubs, inflorescences corymbose). He tentatively assigned the South American species into two Series, Ser. *Breviaristatae* and Ser. *Multiaristatae*.

Much earlier, B. L. Robinson (1930a–d; 1931a–c; 1932a–b) grouped the North American species into two subgeneric sections, *Corymbosae* and *Podocephalae*, with the South American species (specifically, the Peruvian species) grouped into three sections, namely, *Eustevia*, *Breviaristatae*, and *Multiaristatae*. Obviously, a comprehensive taxonomic treatment that would embrace all species of South American *Stevia* is greatly needed.

In either of the two schemes of subgeneric classification mentioned above, *S. rebaudiana* falls within the subgeneric group *Multiaristatae*, a group characterized by the presence of a pappus with numerous awns of about the same length on the achene (Figure 2.1). At least 45 other species also possess a multiaristate pappus, although in some species, two types of pappus awns may be present. Table 2.1 lists species of *Stevia* possessing a multiaristate pappus, all of which occur only in South America, with concentrations in the Peruvian and Bolivian Andes and in Paraguay and southern Brazil.

SPECIES RELATIONSHIPS IN *STEVIA*

Although sound phylogenetic relationships can be established only following a thorough systematic study of the genus, a discussion on relationships of species and subgeneric groups based on currently available data, and with emphasis on chemotaxonomic relationships, will provide guidance in continuing and future efforts to search for sweet-tasting molecules within the genus.

On the basis of information available to him, Grashoff (1972) postulated two hypotheses for the origin and evolution of *Stevia*. *Hypothesis 1*: The genus initially developed and evolved in central South America into Series *Multiaristatae* and Series *Breviaristatae* and, following evolutionary radiation, further developed in northwestern South America, and later Mexico. *Hypothesis 2*: The genus initially developed and evolved in northwestern South America

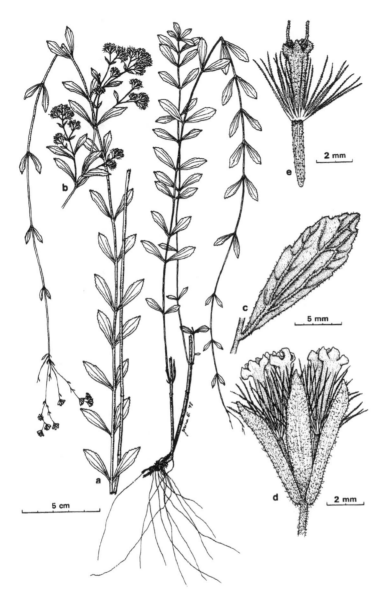

Figure 2.1 Analytical drawing of *Stevia rebaudiana*. (a) Based on collection *Soejarto 5174* (a specimen collected from wild population in Paraguay). (b) Based on *Soejarto 5417* (a cultivated specimen). (c) A leaf of from *Soejarto 5174*, enlarged. (d) A capitulum with open florets. (e) A floret at anthesis. (Illustration by Zorica Dabich, Field Museum, Chicago.)

and, following evolutionary radiation, further evolved into Series *Podocephalae* and Series *Fruticosae* in North America, and into Series *Multiaristatae* and Series *Breviaristatae* in South America. The following evidence was provided by Grashoff as the basis for the formulation of his hypothesis.

Table 2.1 Species of *Stevia* with a multiaristate pappus[a], including *S. rebaudiana*

Species	Ecua[a]	Peru	Boli[a]	Chile	Arge[a]	Para[a]	Braz[a]	Urug[a]
S. amambayensis B. L. Robinson						•		
S. ammotropha B. L. Robinson						•		
S. amplexicaulis Hassler						•		
S. andina B. L. Robinson	•	•						
S. apensis B. L. Robinson						•		
S. aristata D. Don					•			•
S. balansae Hieron.						•		
S. bangii Rusby			•					
S. cajabambensis Hieron.		•						
S. cinerascens Sch.-Bip.							•	
S. collina Gardn.							•	
S. crenata Benth.	•							
S. crenulata Baker							•	
S. cuneata Hassler						•		
S. cuzcoensis Hieron.		•						
S. discolor B. L. Robinson			•					
S. entreriensis Hieron.					•	•	•	
S. estrellensis Hassler					•			
S. gardneriana Baker							•	
S. herrerae B. L. Robinson		•						
S. hoppii B. L. Robinson		•						
S. involucrata Sch.-Bip. ex Baker							•	
S. kuntzei Hieron.			•					
S. leptophylla Sch.-Bip. ex Baker							•	
S. lundiana DC.							•	
S. maimarensis (Hieron.) Cabrera					•			
S. mandonii Sch.-Bip.		•	•					
S. melancholica B. L. Robinson			•					
S. mercedensis Hieron.					•			
S. minor Griseb.					•			
S. multiaristata Spreng.					•		•	•
S. parvifolia Hassler						•		
S. pearcei B. L. Robinson		•						
S. petiolata (Cass.) Sch.-Bip.		•						
S. philippiana Hieron.				•				
S. pohliana Baker							•	
S. puberula Hook.		•						
S. punensis B. L. Robinson		•						
S. rebaudiana (Bertoni) Bertoni						•		
S. rojasii Hassler						•		
S. samaipatensis B. L. Robinson			•					
S. sarensis B. L. Robinson			•					
S. satureifolia Sch.-Bip.					•		•	•
S. selloi (Spreng.) B. L. Robinson						•	•	•
S. setifera Rusby			•					
S. tarijensis Hieron.			•		•			
S. weberbaueri B. L. Robinson		•						
Total	2	11	9	1	9	12	12	4

Note

a Country abbreviations: Ecua = Ecuador; Boli = Bolivia; Arge = Argentina; Para = Paraguay; Braz = Brazil; Urug = Uruguay.

1   Multiaristate and breviaristate Stevias are both well represented in the central portion of South America.

2   Multiaristate, breviaristate and exaristate (absence of awns on pappus) Stevias are all represented in the northwestern part of South America.

3   Series *Podocephalae* is almost equally represented in North and South America, but is scarce in central South America.

4   Series *Corymbosae* is best represented in Mexico, but members occur and are scattered in the northwestern and central portions of South America.

5   Series *Fruticosae* is restricted to North America, except for one species, which extends to northwestern South America.

6   Apomictic reproduction occurs in members of the primarily North American Series *Podocephalae* and Series *Corymbosae*, but is absent in the South American Stevias.

Since our knowledge on the taxonomy of the South American species of the genus and on the subgeneric grouping based on a thorough study of the South American species is still incomplete, the above comparison of disparate knowledge between North and South American *Stevia* does not truly reflect the most probable evolutionary trends among members in the two geographic regions. A complete overall picture may eventually be obtained only when members of both elements are equally well studied. Therefore, in order to postulate a more sound hypothesis, further study of the South American members of the genus throughout their entire range is imperative.

Data from studies other than taxonomic are also still far from adequate to support any postulation of a hypothesis.

1   *Reproduction.* Based on his examination of the pollen characteristics of more than 50% of all North American species of the genus, supplemented with detailed studies of the sporogenesis and gametogenesis of selected species, Grashoff (1972) found that a large number of North American species have an agamospermous (apomictic) mode of reproduction, compared to none for South America. This fact leads into the assumption that North American species are evolutionarily derived, and that the more primitive members of the genus should be sought in South America. Since only 37 (26%) of the South American species were examined by Grashoff, further studies may change Grashoff's conclusions and may help point out new evolutionary trends, based on reproductive mode, among the South American species.

2   *Chromosome number.* Grashoff *et al.* (1972), who performed extensive chromosome counts on many species of *Stevia* (*Corymbosae* 16 species counted out of 38, *Fruticosae* 9/25; *Podocephalae* 5/16) summarized their findings as follows. For North America, the shrubby species all have a gametic chromosome number of $n = 12$, herbaceous species with flower heads in a lax paniculate cluster have $n = 11$ (with no aneuploidy), while those in compact corymbose cluster have $2n = 34$ univalents with aneuploidy. Species with $2n - 34$ univalents with aneuploidy may be interpreted as a triploid derivative of $n = 11$ (King and Robinson, 1987), due to the widespread occurrence of apomixis among the North American species. Grashoff *et al.* (1972), who found the occurrence of $2n = 34$ univalents and $2n = 17$ bivalents in the same species, *S. jorullensis*, interpreted $2n = 34$ as $n = 17$ (Grashoff *et al.* 1972). The occurrence of $2n = 43 + 1$ univalents and $2n = 44$ univalents in *S. origanoides* and *S. plummerae* var. *durangensis* (Keil and Stuessy 1975), respectively, strengthens the interpretation that $2n = 34$ is a triploid derivative of $n = 11$. This

allows for the interpretation of basic chromosome numbers as $x = 11$ and $x = 12$ for the North American species.

For South America, Grashoff *et al.* (1972) established the following basic chromosome numbers: *Breviaristatae* $x = 12$; *Eustevia* $x = ?$; *Multiaristatae* $x = 11$. However, the number of species for which chromosome numbers were actually counted is small and far from adequate: *Breviaristatae* 1/35?; *Eustevia* ?/55?; *Multiaristatae* 4/50?. Future chromosome counts, especially on the South American species, may provide additional information that could shed a new light on evolutionary relationships among species and subgeneric groups.

3   *Pollen study.* Pollen morphology of *Stevia* was examined by King and Robinson (1967; 1968) using light microscopy and was found to show two patterns. The South American species have uniform grain size and morphology, whereas the North American species have more variable size, shape and symmetry of grains, not only between, but also within, species. Since exine morphology was not studied, it was concluded that pollen size, shape and symmetry of grains do not seem to have any or are only of limited significance in the taxonomy of the genus. According to Grashoff (1972), the variable nature of the pollen of North American species may be explained by the apomictic nature of the species. Thus, based on the currently available data, there is little that can be said on the significance of pollen grain in contributing to the postulation of relationships between species and subgeneric groups.

4   *Chemical data.* The most extensive available chemical data on *Stevia* are those for the Paraguayan species *S. rebaudiana*, on the occurrence of its sweet- and non-sweet-tasting diterpene glycoside constituents (see Chapter 4). Chemical data on other species are more limited and a review and discussion on the importance of chemical data in establishing trends of evolutionary relationships within the genus, except for the sweet-tasting glycosides, is outside the scope of this chapter. It can be said, however, that phytochemical screening studies to determine the infrageneric distribution of a particular class of compounds have been performed only on the sweet *ent*-kaurene diterpene glycosides (Kinghorn *et al.* 1984).

Organoleptic tests of the leaves of 110 species of *Stevia* (including *S. rebaudiana*) taken from herbarium specimens, followed by field studies, and phytochemical screening for the sweet *ent*-kaurene diterpene glycosides (Soejarto *et al.* 1982; 1983; Kinghorn *et al.* 1984) led to the discovery of stevioside in the leaves of a specimen of *S. phlebophylla* A. Gray (*Pringle 2291*) (Figure 2.2), collected in Guadalajara, Mexico, in 1889; stevioside was also detected in the dried leaves of a specimen (*Gosling s.n.*) collected in Paraguay in 1919. The discovery of stevioside in dried leaves of *S. phlebophylla* was somewhat unexpected, because this species is distantly separated from *S. rebaudiana*, both taxonomically and geographically. Taxonomically, *S. phlebophylla* belongs to the Ser. *Fruticosae*, while *S. rebaudiana* belongs to the Ser. *Multiaristatae* (*sensu* Grashoff, 1972). Morphologically, the two species are also very distinct: *S. phlebophylla* is a large shrub with a corymbosely arranged capitula and non-aristate achenes, while *S. rebaudiana* is a rhizomatous perennial herb, with paniculately arranged capitula and multiaristate achenes. The evolution of the enzyme systems in *Stevia* to effect the accumulation of stevioside and related *ent*-kaurene glycosides appears to have achieved maximal development in *S. rebaudiana*. The finding of stevioside in a species unrelated taxonomically and geographically is evidence that similar enzyme systems may have been developed in other species as well. Thus, in view of the fact that only about 50% of all species of *Stevia* have been screened for the occurrence of sweet *ent*-kaurene glycosides, it is reasonable to expect that stevioside or other sweet *ent*-kaurene glycosides may occur somewhere in the other 50% of the species that have not been examined.

Figure 2.2 A photograph of a specimen of *Stevia phlebophylla* from Guadalajara, Mexico, collected in 1899. Evidence indicates that this species, which contains stevioside, may now be extinct (see text for details). (Photo by D. D. Soejarto.)

STEVIA REBAUDIANA

History of classification

The Latin binomial *S. rebaudiana* first appeared in the literature in 1905, coined by Moisés Santiago de Bertoni (Bertoni 1905). The account of the events that led to the publication of this name is given by Bertoni in his 1905 and 1918 papers (Bertoni 1905; 1918).

Towards the end of 1887, at the early stage of his botanical explorations in northeastern Paraguay, Bertoni (1857–1929), a Italian-Swiss botanist-naturalist who immigrated to Paraguay in 1882 and became a citizen of Paraguay (Gorham 1973a; Anonymous 1998), was told by *yerbateros* (herbalists) from the Northeast and by native Indians of Mondaíh (Monday River, west of Ciudad de Este, formerly Puerto Stroesner/Franco/Bertoni, at the border with Brazil), on the existence of a sweet plant in those areas. The Mondaíh Indians knew the plant from the grasslands of *Mbaeverá* and *Kaá Guasú* (today known as Caaguazú) (see Figure 2.3 for localities; map has been adapted from Kleinpenning 1987). Bertoni (1905; 1918) mentioned that the plant was rare, and that he was not able to obtain a specimen of the plant at that time. Years later, in Asunción, the capital city of Paraguay, a Customs Officer, Daniel Candía, gave him a sample of this sweet plant, consisting of pieces of stems, branches, leaves and inflorescences, in a form ready to be used to sweeten a *maté* drink, by adding a small portion of the crushed leaves to the bitter drink. The Customs Officer who gave the sample had apparently received it from a herbalist from the northern part of the country. Bertoni was able to identify the plant in the crushed sample and place it in its correct taxonomic position, in the Tribe Eupatorieae of the Compositae. However, due to the fragmentary nature of the specimen, he could not pinpoint the generic affiliation, namely, whether the plant belonged to *Stevia* or to *Eupatorium*. Nevertheless, he decided that the plant was a species new to science and in 1899

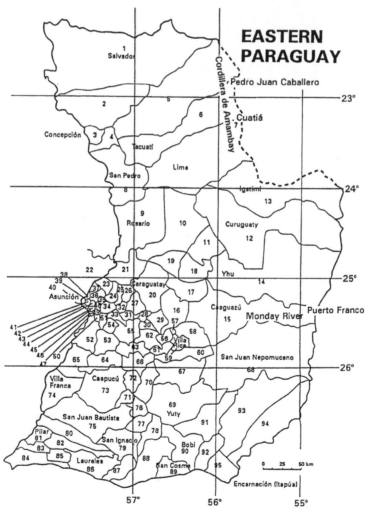

Figure 2.3 Districts in Eastern Paraguay in the 1860s. [Adapted from Kleinpenning (1987)].

communicated it to the scientific community under the name of *Eupatorium rebaudianum* Bertoni in the *Boletin de la Escuela Agrícola de Asunción*, vol. 2: 35 (not seen by the author). The specific epithet *rebaudianum* was dedicated to a Paraguayan chemist, Ovidio Rebaudi, whom he admired and who later performed the first chemical study of the plant, from samples provided by Bertoni.

Following the publication of the new species, the British Consul in Asunción, H. B. M. Cecil Gosling, informed the Royal Botanic Gardens at Kew, England, through a letter, of the existence of the sweet plant *E. rebaudianum*, and sent with his letter, some fragments of the plant together with the Latin description of the new species. Gosling's letter and the Latin description were published in the 1901 *Kew Bulletin* (pp. 173–174) under an anonymous authorship (Anonymous 1901), which later authors have cited as 'Gosling 1901'.

In 1904, Bertoni was able to obtain the first dried flowering specimen of the plant, but because of civil war in the country, he had to travel to Upper Paraná. On his return to Asunción, he received a live specimen of the plant from a resident of San Pedro (Bertoni referred to this person simply as 'M. R.'; Bertoni 1918). It was this specimen that enabled him to assign the correct taxonomic placement of the plant within the genus *Stevia* and published it as *S. rebaudiana* Bertoni (Bertoni 1905). [From the nomenclatural point of view, the complete authority citation would be: *S. rebaudiana* (Bertoni) Bertoni.] Unaware of the publication of Bertoni's 1905 paper, in 1906, D. Prain (Prain 1906), editor of the *Hooker's Icones Plantarum*, published a line drawing of *Stevia rebaudiana* accompanied by a Latin description of this species, and attributed the Latin binomial authorship to Hemsley (Hemsley 1906). Since the name *S. rebaudiana* (Bertoni) Hemsley appeared one year after that of *S. rebaudiana* (Bertoni) Bertoni, for reasons of priority, the former becomes a later homonym, an illegitimate name, which must be rejected (based on rules established in the *International Code of Botanical Nomenclature*; Greuter *et al.* 1994) and relegated to synonymy.

The plant

From 28 April to 2 May 1981, the author visited the native land of *S. rebaudiana* in the northeastern part of Paraguay (Departament of Amambay) to collect specimens of this species for botanical and chemical studies. Permission was obtained from Paraguayan authorities and the field work was cleared by the Dean of the Faculty of Sciences of the University of Asunción. The field team consisted of the author and a Paraguayan botanist (Eugenia Bordas) from the University of Asunción, and a businessman (Ceferino Aranda) from Pedro Juán Caballero (PJC), the capital city of the Department of Amambay.

After obtaining clearance from the Amambay Government authority at PJC, two police sergeants (one, driving the pick-up truck that was rented, the other, providing the escort) accompanied the team to Cerro Cuatiá, the locality of *S. rebaudiana*, approximately 80 km south of PJC on the western slopes of the Cordillera de Amambay (500–700 m altitude above sea level). Based on the collections made during this field trip (*Soejarto and Bordas 5170, 5172, 5174* and *5182*), and on other collections (including *Gosling s.n.*, 1919) in deposit at the John G. Searle Herbarium of the Field Museum in Chicago, as well as on plants cultivated at the Pharmacognosy Field Station of the University of Illinois at Chicago, and on the original characterization of the species given by Bertoni (1905), the following taxonomic description has been prepared.

Stevia rebaudiana (Figures 2.1, 2.4, 2.5, and 2.6) is a perennial herb with filiform roots, whose 30–50 cm tall, erect and slender **stem**, which easily produces secondary shoots (suckers) from its base, dies off and is renewed annually. In its wild habitat the plant is

Figure 2.4 Stevia rebaudiana plants among grasses in their wild habitat in Cerro Cuatiá, Cordillera of Amambay, Paraguay. (Photo by D. D. Soejarto.)

slender and little-branched, the stem, branches and leaves are green and covered with very fine, short and whitish hairs. In cultivation, the stem usually produces numerous lateral branches, thus, forming a more-or-less roundish and dense crown. The crushed leaves exude a strong odor, and all the green parts of the plant taste sweet.

The **leaves** are simple, opposite, subsessile, internodes 2–4 cm long, blades subcoriaceous, very variable in shape and size, narrowly elliptic to oblanceolate or spatulate– oblanceolate, to linear–oblong or ovate, 2–3 cm long, 0.6–1 cm wide, apex obtuse to subacute, base cuneate, margins entire, often toothed (crenate to serrate) on the upper half, entire on the lower half, three primary veins arise from the leaf base, raised and prominent on the blade's lower side, somewhat immersed on the upper side, secondary venation reticulate, somewhat immersed, the blades in dry state olive-green to brownish green, usually darker on the upper side, both surfaces subscabrous, with black glandular dots on the lower side, the leaves subsessile or the petiole to 3–4 mm long.

The capitula are arranged into loose, paniculate–corymbose *inflorescences* at the terminal ends of the branches, peduncle 1–4 cm long, very slender, pedicels of each capitulum slender, 1–4 mm long, bracts linear–lanceolate, 1–2 mm long; each **capitulum** (*ca*. 8 cm long) is enveloped by an involucre that is almost as long as the pedicel, light green on the lower half, yellowish on the upper half in fresh state; phyllaries 5, finely hairy, green when fresh, linear to subulate, 4–5 mm long, acute to rounded at apex, each capitulum is made up of 5 disk florets.

Figure 2.5 A flowering specimen of *Stevia rebaudiana*, aerial portion. (Photo by D. D. Soejarto.)

Figure 2.6 *Stevia rebaudiana* plant in a cultivated plot at the Pharmacognosy Field Station, University of Illinois at Chicago, Downer's Grove, Illinois (1982). (Photo by D. D. Soejarto.)

Each *floret* is exserted above the involucre, corolla actinomorphic, white, corolla tube slender, equal in length to or longer than the pappus awns (*ca.* 4 mm long), greenish below, dirty white to purplish above, covered with very fine hairs on the inside, almost glabrous outside, the lobes ovate to ovate–lanceolate, unequal, white with purplish throat, obtuse to subacute, 0.7–1 mm long, ciliolate, style branches twice the length of the corolla lobes or longer, divergent and usually recurved, densely covered by clear-brownish glands and very fine short hairs, achenes subglabrous, 2.5–3 mm long, but finely barbed along the ridges, pappus awns straw-colored, of 9–17 subequal, *ca.* 3 mm long, somewhat rigid and finely barbed awns ('aristae'), base of awns often broader and flattish, barbs rigid, linear, arising from the awns at an acute angle and pointing upwards.

In cultivation, the plant can grow much taller, normally, to 70–80 cm, though a height of 140 cm under the most optimal conditions has been recorded (Mitsuhashi as cited in Sakaguchi and Kan 1982). The cauline leaves normally are also larger, 3–5 cm long, 1–2.5 mm wide, and the texture more papery.

Geographic range

Stevia rebaudiana is native to Paraguay and is probably endemic to this country. Natural populations, though scarce, are still found in the northeastern part of the country, in the highlands of the Department of Amambay, in particular, in the slopes and valleys of the Cordillera of Amambay, such as in Cerro Cuatiá, near Capitán Bado, a town about 90 km south of PJC (Pedro Juan Caballero). In 1918, Bertoni (1905) gave a distribution range from Amambay south to the Monday River, especially in the *maté* plantation zones of *San Pedro, Alto Jejuy, Vaca'reta and Ihu* or *Yhu* (see Figure 2.3 for localities), approximately from 22° 30' to 25° 30' south latitudes, and from 55° to 57° west longitudes, within 200–700 m altitudinal zones. Thus, Bertoni placed the geographic range of the species entirely within the Paraguayan territory.

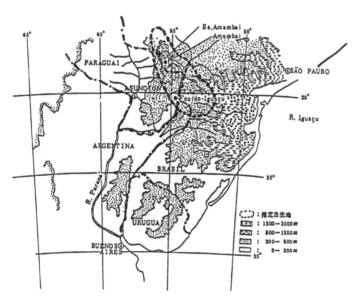

Figure 2.7 Distribution range of *Stevia rebaudiana* according to Sumida (1973).

In his 1973 paper, Sumida, who took up residence in Brazil at the Northern Agricultural Institute from 1969 to 1971, gave a distribution range from 20° to 26° south latitudes and 52° to 57° west longitudes, an area delimited by a more or less elliptical shape that runs from north-northwest to south-southeast (Figure 2.7). Further, he stated that '*S. rebaudiana* grows naturally or is cultivated in the Amambay and Iguaçu Districts on the border of Brazil, Paraguay and Argentina . . .' (Sumida 1973: 81). Clearly, Sumida included Brazilian and Argentinian territories bordering with the Paraguayan Amambay Cordillera as part of the distribution range of *S. rebaudiana*. However, in his 1975 paper (as cited in Sakaguchi and Kan 1982), Sumida gave a range from 22° to 24° south latitudes and from 55° to 56° west longitudes, a narrower range, all within the Paraguayan territory.

Unfortunately, a thorough botanical field study that covers the area delimited by both Bertoni and Sumida has not been performed. Thus, the question remains as to whether the natural distribution of *S. rebaudiana* is limited to the Paraguayan side (thus, endemic to Paraguay) or is shared by Brazil and Argentina.

Botanical literature and herbarium records to which the author has had access, including some material examined by B. L. Robinson (1930a–c), indicate that specimens of *S. rebaudiana* have never been collected in the Brazilian or Argentinian territories. King and Robinson in their book *The Genera of the Eupatorieae (Asteraceae)* (1987) give the distribution of *S. rebaudiana* simply as Paraguay.

Pio Corrêa (1926) in his *Diccionario das Plantas Uteis do Brasil*, without documentation, mentioned that a sweet plant referred to by him as *S. collina* Gardn., under the name of 'Caá-Ehé', and a variety of this species, var. *rebaudiana*, are found in the Brazilian Mato Grosso and Minas Gerais. This report is erroneous. First, *S. collina*, a definitely Brazilian species, and *S. rebaudiana* are two different species; second, organoleptic tests by the author of leaves of *S. collina* did not present any sweet taste (Soejarto *et al.* 1982), while TLC and HPLC, followed by GC/MS analysis (Kinghorn *et al.* 1984) did not show the presence of steviol glycosides. In a 1967 paper, von Schmeling described a search of *S. rebaudiana* in Mato Grosso, the Brazilian territory bordering the northeastern Paraguay's Cordillera of Amambay, where the species occurs. She reported that the search resulted in the collection of the plant in Ponta-Pora, the Brazilian town adjacent to PJC, the capital city of the Department of Amambay of Paraguay. She stated: *Encontramo-la, de fato, em Ponta-Pora, divisa do Brasil com o Paraguai* (von Schmeling 1967: 68) ('Indeed, we found the plant in Ponta-Pora, in the border between Brazil and Paraguay'). The search crew returned to São Paulo with a 'beautiful plant in a pot, a proof that the *caá-heê* plant is also found in Brazil.' According to her account, this potted plant eventually gave off beautiful flowers in São Paulo and, based on the examination performed on 23 November 1944, at the Institute of Botany of the State of São Paulo, the identity of the potted plant as 'the same *Stevia* plant' ('. . . *se tratava mesmo da Stevia*') (sic!) was confirmed. Unfortunately, no mention at all is made anywhere in the paper that this discovery of the *Stevia* plant in Ponta-Pora was documented, namely, whether herbarium specimen(s) were prepared and whether it(they) have been deposited in a herbarium institution, giving the name of the institution, for future reference, as the botanical evidence of the finding. During the author's 1981 trip, he crossed the border to Ponta-Pora, and made inquiries on the possible existence of natural populations of *Caá-hê-hê* or *Caá-enhem* in Ponta-Pora. He also went to an excursion to the fields some distance away from this town. To his disappointment, no sign of the presence of *S. rebaudiana* was found. To complicate matters, Sumida in his 1973 paper states that *S. rebaudiana*, with local names of '*Caá-hê-hê* or *Caá-enhem*', has been used '. . . in some localities of Brazil from old times as a sweetening material'. Like von Schmeling, no botanical documentation to support the statement appears to exist. Interestingly, Felippe (1977), a biologist at the Instituto de Botánica in São Paulo, stated that '*A planta* [*S. rebaudiana*] *foi introduzida no Brasil na dêcada de 60. A instituciâo responsavêl pel a introducâo foi o Instituto do Botânica de São Paulô.* ('The plant was introduced to Brazil in the decade of the 1960s. The institution responsible for its introduction was the Institute of Botany of São Paulo.') This statement contradicts the finding described by von Schmeling and the statement made by Sumida.

Based on the geographic range of *S. rebaudiana* as indicated by Bertoni (1918), the climatic requirements for the natural occurrence of *S. rebaudiana* may be given as a region with an annual precipitation range of between 1300 and 1700 mm, and a mean annual temperature of 22 °C (see Fariña Sanchez 1973; Kleinpenning 1987; 1992, for these climatic parameters in Paraguay) (Figure 2.8). Such a region corresponds to a subtropical *mesothermal and humid climatic zone, with a rainfall-deficient winter* (Fariña Sanchez 1973). The data (climatogram) for Amambay presented by Sumida (1973: 72) indicates an annual precipitation of about 1600 mm and a mean annual temperature of about 22.5 °C, with a minimum and a maximum of *ca.* 17 °C and 26 °C, respectively, based on 1946–1967 meteorological data.

According to Sumida (1980; and as cited by Sakaguchi and Kan 1982), who carried out extensive physiological experimentations with *S. rebaudiana* under different greenhouse and phytotron conditions in Japan (1973–1980), the optimal temperature range for the growth of *S. rebaudiana* is 15–30 °C, though the plant can tolerate critical temperatures of 0–2 °C. Miyasaki (as cited by Sakaguchi and Kan 1982) showed the absolute limit as −3 °C.

As to soil moisture, Sumida (1980) demonstrated that under culture conditions, the plant attains high productivity even under excess soil moisture. Oxygen consumption by the roots is also less compared to other crops, such as soybean. His conclusion was that *S. rebaudiana* is

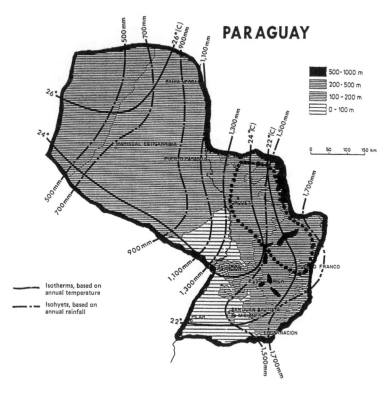

Figure 2.8 Distribution range of *Stevia rebaudiana* based on information provided by Bertoni (1905; 1918). (Map has been adapted from Gorham (1973b) and Kleinpenning (1992).)

highly tolerant to the wet soils. However, this plant is not drought resistent. In the vigorously growing stage and the late growing stage, the leaves wilt easily when exposed to a drought.

Natural habitat

The following account, transcribed and edited from the author's diary during his visit to Cerro Cuatiá in 1981 (pp. 27, 28) should provide a good picture on the specific conditions and habitat of this species and on the current extent of the natural populations of *S. rebaudiana*.

On April 30, we were off to Cerro Cuatiá by 9:35 a.m. The road was semi-paved but in broken-down condition in parts. We stopped several times along the way to examine the vegetation, to search for *S. rebaudiana*, and to take photographs. The vegetation cover in this region is woodland savanna ('cerrado' vegetation), but highly modified through cultivation and cattle ranching, with secondary forests. No *Stevia* of any species were encountered along the way. Further towards Cuatiá along the ridge of the Sierra, we saw cultivations of soy bean, coffee, and *Cajanus cajan* (pigeon pea). *Grevillea robusta* was frequent along the roadside. The weather is cooler here, since the altitude (500–700 m) is somewhat higher than PJC. Soil is black and sandy and appears to be more fertile.

At km 72, before the settlement of Cuatiá, we took a right turn, the road winding down somewhat and becoming much worse, rocky and sandy in places. We were traveling along a valley between two ridges covered by forests, some of them had been cleared, especially along the slopes. Several kilometers from the split from the PJC-Capitán Bado road, the soil became even more sandy and the terrain continued to wind down, through occasional stretches of plains. Estancia 'La Sirena' (approximately, 55° 42′ west longitude, 23° 10′ south latitude), the ranch of Mr Ceferino Aranda, our destination, is about 18 km from the road split. Vegetation cover in this area consists of tall subtropical forests with stands of 'peroba' (*Aspidosperma* sp.) and 'lapacho' (*Tabebuia* sp.) of 25–30 m canopy. These plants have straight boles and a diameter of 1–2 m at breast height. Forest clearing consisted of cutting understory trees, leaving the tree trunks to dry, burning the forest, and then, cutting the huge trees into logs of 10 m long, loading them into big trucks and bringing them to PJC, where they are cut and processed, for sale to Ponta Pora on the Brazilian side of PJC. We stopped at the entrance to Aranda's ranch, about 2 km from his ranch house. There was a road track to his ranch house, but not to the *campos* (grasslands) where *Kaá Hè-é* is found wild.

First, we walked through a field with tall (1–3 m tall) grasses, then, remnants of forests, corn and cassava plantations, second growth forests and into the *campo*. Two Pay Tavytera Indians joined us as guides. All this walk was downhill. The *campo* slopes down on gently rolling hills towards a stream. When one stands at the lowest point somewhere near the stream, one can see around him the *campos* gently sloping up into distant forests along the higher ridges. Low shrubs are occasionally seen in this grassland, but not a tree. Higher up the hill, the grasses are taller (1–3 m), but lower down the slopes, they are shorter (50–80 cm tall). *S. rebaudiana* plants are found here at the lower end of the slope, among the grasses (Figure 2.4), where the soil is sandy with an almost blackish color, but humid and well drained. At the time of our visit, the larger plants (30–50 cm tall) had dried fruiting inflorescences, which were mostly empty, remains of the previous year's fruiting (November–December). Many small *S. rebaudiana* plants growing in clumps were found around and nearby the bigger ones. Presumably, they have originated from seed germination. We were told by Mr Aranda and the two Pay Tavytera natives that the *Kaá Hè-é* plants flower and fruit only once a year in the wild, but that they can yield three crop harvests (cutting of the aerial parts) per year under cultivation. Usually, the *Kaá Hè-é* plants grow well after fire in August, which cleans the grasses and other competing plants. There were countless *S. rebaudiana* plants on this site.

Further examination of this site and other similar grassland areas showed that *Kaá Hè-é* is common in that one spot only. It is absent in the grassland uphill, with taller grasses, as well as in areas where there are shrubs and trees. Its natural habitat appears to be rather specific. Examination of grasslands where cattle grazed indicates that the plants disappear from such places. During the time of the visit, the natural habitat of *S. rebaudiana* probably had shrunk from its original range and is presently limited only in remote and difficult-to-access places. In fact, no other species of *Stevia* was found.

The next day we arranged for a trip to La Estrella settlement (approximately 55° 48′ west longitude, 22° 25′ south latitude), approximately 40 km north of PJC. Herbarium and literature records indicate that at least four different species of *Stevia* had been collected in this locality. The habitat where these four species appear to have been collected is described as sandy plains, with grasses, near Estrella. These were exactly the areas we visited. Indeed, the areas consisted of sandy plains and gently rolling hills, with open grasslands, broken by depauperate *cerrado* forest cover and islands of slightly denser forests. Much of the lands had been sold to private owners, and cattle roams the areas, all the way north to the border with Brazil. Some areas have been converted to cultivated lands. A whole-day search in various locations and habitat types did not yield a single species of *Stevia*; we suspected that they had been wiped out by cattle grazing and by the changed habitat that resulted.

The sensitivity of *S. rebaudiana* to habitat modification, as described above, is consistent with the behavior of other species of *Stevia* seen by the author in other localities in Paraguay, such as the *campos* of Luque, Caaguazú, Villarica, and Foz de Iguaçú, and in other places in Argentina, Peru, Colombia, and Mexico (Soejarto *et al.* 1983). In areas within the range of distribution of a particular species, populations of that species had been reduced or wiped out, where habitat modifications, either through human or animal activities, had taken place. Sensitivity of species of *Stevia* to habitat modifications results in rarity of species and, eventually, disappearance from its natural habitat. The case of the disappearance of *S. phlebophylla* A. Gray (see Figure 2.2) from its natural habitat in Guadalajara, Mexico (Grashoff 1972: 501; Soejarto *et al.* 1982), provides a disquieting example.

In the case of *S. rebaudiana*, the rarity of this species in its natural distribution range was already stressed by Bertoni (Bertoni 1905; 1918). Cattle grazing throughout the 20th century (see some data on cattle production in northeastern Paraguay, in Kleinpenning 1987; 1992), encroachment of the natural habitat through agricultural activities (also, see some data on crop production, in Kleinpenning 1987; 1992), together with intense activities of removal of plants from its natural habitat for transplant to newly established plantations (Figure 2.9) in various locations in Paraguay during the early 1970s, have helped reduce the extent of the plant's natural habitat and accelerate the rarity of this species within its original range of distribution.

Large scale removal of the wild *S. rebaudiana* plants was triggered by the widely publicized information in Paraguay, in the late 1960s and early 1970s, on the commercial potential of the plant as a source of sugar substitute. For example, the Bulletin *El Agricultor* published in Asunción, carried detailed descriptions on the botany, chemistry, properties, habitat, and methods of propagation and culture of the plant, in articles titled '*La Yerba Dulce (Caá – Jjheé)*' published in 1967 and 1969 (Aranda 1967; 1969). A *Manual de Agricultor Paraguayo*, in its chapter on '*Importancia del Cultivo del Caá-jheé*' promoted the commercialization of the plant (Mengüal 1970). While such information is important for the economy of the country, its effect has been negative on the well

Figure 2.9 A cultivated field of *Stevia rebaudiana* in Paraguay (1981). (Photograph provided through the courtesy of Mr Luis Enrique de Gasperi of Asunción, Paraguay.)

being of *S. rebaudiana* in its natural habitat, due to the emphasis given on the ease of propagation through the splitting of the plant, rather than seed germination. Indeed, the short time span of seed (achene) viability was already widely known at that time (also, see Felippe 1971).

A document written by Akira Sugii, a Japanese immigrant who worked on commission with Toyomenka Kaisha, Ltd, Japan, and filed with the Ministry of Agriculture and Animal Husbandry of Paraguay in Asunción in 1977, a copy of which was made available to the author by Mr Luis Enrique de Gasperi, a *S. rebaudiana* cultivator and businessman in Asunción who co-hosted the author's 1981 and subsequent visits, indicates that shipment of samples of wild *S. rebaudiana* to Japan started in 1969, while transplant activities began in earnest in 1973. According to this document (Sugii 1977), the first transplant operation was performed at a site located 150 km south of PJC, and 50 km west of Capitán Bado. Between August and December 1973, 20,000 wild plants were removed from their natural location and transplanted into the first lot of *S. rebaudiana* plantation in Paraguay. In 1974, the first shipment, from harvest of the first hectare of plantation, was made to Japan. From seeds produced from this plantation, as well as from stem cuttings, more plantations in other parts of Paraguay (PJC, Colonia Iguazú, Colonia Fram in Itapuá, and Luque) were established. He reported satisfactory results, despite differences of soil types. It was Akira Sugii's goal to promote widespread cultivation of *S. rebaudiana* in Paraguay by other farmers, so that within five years, the extent of plantation in Paraguay would reach 10,000 hectares, producing an export of 20,000 tons of *S. rebaudiana* leaves and a revenue of US$50 million annually. As we know it today, this dream never materialized, because of the shift of leaf production from Paraguay to Asia in the mid-1970s (Anonymous 1988; Kinghorn and Soejarto 1991).

SUMMARY AND CONCLUSIONS

The sweet-tasting plant *S. rebaudiana* (Bertoni) Bertoni belongs to the genus *Stevia* (Compositae – Eupatorieae), which comprises 220–230 species of herbs and shrubs found entirely in the

Americas. The distribution range of *Stevia* extends from southern United States and Mexico, in the north, to Argentina, Chile and the Brazilian highlands, in the south, through the Central American States and the South American Andes. The North American species of the genus has recently been monographed (Grashoff 1972), in which 79 species were accepted as species in good standing. The South American species, which are more numerous, have not been monographed throughout their entire range, except by country by country treatments of B. L. Robinson in the early 1930s. Based on review of the literature and on examinations of herbarium specimens of *Stevia* in repository at the John G. Searle Herbarium of the Field Museum in Chicago, 51 species of *Stevia* occurring in Ecuador, Peru, Bolivia, Chile, Argentina, Paraguay, Brazil, and Uruguay, including *S. rebaudiana*, were found to belong to the subgeneric grouping *Multiaristatae*, a group characterized by numerous pappus awns on the achene. This information, and knowledge on the taxonomy and distribution of other species, should be useful in guiding the effort in the future search for other sweet-tasting species, and the steviol glycoside non-nutritive sweet compounds, and in establishing taxonomic and evolutionary relationships among species and groups of species. It is evident that comprehensive taxonomic studies are badly needed.

Stevia rebaudiana is a 30–60 cm tall, pubescent herbaceous plant with perennial rhizomes, simple, opposite and narrowly elliptic to oblanceolate or spatulate–oblanceolate or linear–oblong leaves, trinerved venation, paniculate–corymbose inflorescences with white flowers, and achenes bearing numerous, equally long pappus awns. Although botanical records indicate a natural distribution range in the northeastern part of Paraguay, in the Cordillera of Amambay, various literature reports indicate a natural occurrence of the species in the Brazilian and Argentinian territories, bordering the Amambay region of Paraguay. Further field studies are needed, in order to establish more accurately the natural distribution range of *S. rebaudiana*.

The natural habitat of *S. rebaudiana* is subtropical grasslands (mesothermal-humid climatic zone) at altitudes of about 200–600 m above sea level, in the Amambay Cordillera (mountain range) of northeastern Paraguay, where the soil is blackish, humid, sandy or loamy, in well drained terrains. Various types of habitat disturbances, including cattle grazing, encroachment of the natural habitat through agricultural activities, together with intense activities of removal of plants from its natural habitat for transplant to newly established plantations, have accelerated the reduction of *S. rebaudiana*'s natural habitat and led to the rarity of this species within its original distribution range. The long term result of such a process is predictable, namely, a threat to the continuing existence of the species. It is time to evaluate the situation and to take measures to protect certain areas of Amambay, where the natural habitat of this species still exists. Because of its importance as a source of non-nutritive high potency sweeteners, and because it appears to be endemic to one locality (Cordillera of Amambbay) of Paraguay, *S. rebaudiana* represents not only a national patrimony for Paraguay, but a heritage for humanity's future generation, that must be protected.

ACKNOWLEDGEMENTS

Field work cited in the text was supported, in part, by contract NIH-N01-DE-02425 with the National Institute of Dental Research, National Institutes of Health, Bethesda, Maryland, USA.

The author wishes to thank Mr Ceferino Aranda, resident of Pedro Juán Caballero, who hosted the author's visit in 1981, Prof Eugenia Bordas of the University of Asunción, who accompanied the author in his field work in Amambay, and who facilitated field communications through her fluency in the Guaraní language, and to Mr Luis Enrique de Gasperi of Asunción, for his hospitality and stimulating discussions during the author's several visits to Paraguay, and for making available to the author a numbers of documents cited in this chapter, as well as photographs of his *S. rebaudiana* plantations.

To Ms Jodi Slapcinsky, M.Sc., Research Assistant at the Field Museum of Natural History, Chicago, who kindly assisted the author in the search of certain data used for the writing of this chapter, and to Ms Zorica Dabich, who prepared the line drawing of *S. rebaudiana* appearing as Figure 2.1, the author wishes to express his thanks and appreciation.

REFERENCES

Anonymous (1893–1993) *Index Kewensis*, The Clarendon Press, Oxford, U.K., vols. 1 and 2, and Supplements 1–20 (CD-ROM, 1996).

Anonymous (1901) Ca-á êhê or azucá ca-á. *Bulletin of Miscellaneous Information*, Royal Botanic Garden, Kew, England, pp. 173–174.

Anonymous (1988) High intensity sweeteners – Market size 7.2. billion yen: Stevia occupies 41%, but future gains will be made by aspartame. *Food Chemicals, Tokyo*, No. 6, 19–26.

Anonymous (1998) Paraguay: 'Bertoni Project'. *Helvetas – World Wide Web* (http://www.helvetas.ch/helvetas_projects/eparaguay_pa3l.html).

Aranda, J. B. (1967) La Yerba Dulce (Caá-jheê) (*Stevia rebaudiana* Bert.). *El Agricultor* (Asunción), Year VII, No. 68, 28 February 1967, p. 6; Year VII, No. 69, 31 March 1967, p. 4.

Aranda, J. B. (1969) La Yerba Dulce (Caá-jheê). *El Agricultor* (Asunción), Year IX, No. 94, 31 August 1969, p. 18; Year IX, No. 95, 30 September 1969, pp. 8–10; Year IX, No. 96, 31 October 1969, p. 11.

Baker, J. G. (1876) Compositae II. Eupatoriaceae. In *Flora Brasiliensis*, C. F. P. von Martius, A. W. Eichler, and I. Urban (Eds), Fleischer, Munich, Lehre, Germany, **6**, part 2(1), pp. 199–211, plates 55–58.

Bertoni, M. S. (1905) Le Kaá Hê-é – Sa nature et ses propriétés. *Anales Científicos Paraguayos, Serie I*, No. 5, 1–14.

Bertoni, M. S. (1918) La *Stevia rebaudiana* Bertoni. *Anales Científicos Paraguayos, Serie II*, No. 2, 129–134.

Blake, S. F. (1926) *Stevia*. In *Trees and Shrubs of Mexico*, P. C. Standley (Ed.), Contributions of the U.S. National Herbarium **23**, 1424–1429.

Bonpland, A., Humboldt, A. von and Kunth, C. S. (1820) *Nova Genera et Species Plantarum*, Lutetiae Parisiorum, Apud N. Maze, Bibliopolam, Tomus Quartos, pp. 139–149, plates 351–353.

De Candolle, A. P. (1836) *Prodromus Systematis Naturalis Regni Vegetabilis*, Treuttel and Würtz, Paris, **5**, 115–124.

Fariña Sanchez, F. (1973) The climate of Paraguay. In *Paraguay: Ecological Essays*, J. R. Gorham (Ed.), Academy of the Arts and Sciences of the Americas, Miami, Florida, pp. 33–38.

Felippe, G. M. (1971) Observações a respeito da germinação de *Stevia rebaudiana* Bert. *Hoehnea* **1**, 81–93 (Figures 1–8).

Felippe, G. M. (1977) Erva-do-Paraguai. *Suplemento Agrícola o Estado de São Paulo* **22**, 14.

Gorham, J. R. (1973a) The history of natural history in Paraguay. In *Paraguay: Ecological Essays*, J. R. Gorham (Ed.), Academy of the Arts ands Sciences of the Americas, Miami, Florida, pp. 1–8.

Gorham, J. R. (1973b) The Paraguayan Chaco and its rainfall. In *Paraguay: Ecological Essays*, J. R. Gorham (Ed.), Academy of the Arts ands Sciences of the Americas, Miami, Florida, pp. 58–60.

Grashoff, J. E. (1972) *A Systematic Study of the North and Central American Species of Stevia*. Ph.D. Dissertation, University of Texas, University of Michigan Microfilm 73–7556, pp. 1–608.

Grashoff, J. E., Bierner, M. W. and Northington, D. K. (1972) Chromosome numbers in North Central American Compositae. *Brittonia* **24**, 379–394.

Grashoff, J. E. (1974) Novelties in *Stevia* (Compositae. Eupatorieae). *Brittonia* **26**, 347–384.

Greuter, W., Barrie, F. R., Burdet, H. M., Chaloner, W. G., Demoulin, V., Hawksworth, D. L., Jorgensen, P. M., Nicolson, D. H., Silva, P. C., Trehane, P. and McNeill, J. (1994) *International Code of Botanical Nomenclature (Tokyo Code)*, p. 103.

Grisebach, A. (1874) Plantae Lorentzianae. *Abhandlungen der Königliche Gesellschaft der Wissenschaften, Göttingen* **19**, 164–166.

Grisebach, A. (1879) Symbolae ad Floram Argentinam. *Abhandlungen der Königliche Gesellschaft der Wissenschaften, Göttingen* **24**, 166–168.

Hassler, E. (1912) XXXVI. Ex Herbario Hassleriana: Novitates Paraguarienses. XV. *Fedde's Repertorium Specierum Novarum Regni Vegetabilis* **11**, 165–167.

Hemsley, W. B. (1881) *Biologia Centrali-Americana, Botany*, R. H. Porter and Dulau Co., London, pp. 84–90 (*Stevia*).

Hemsley, W. B. (1906) Tabula 2816: *Stevia rebaudiana* Hemsley. In *Hooker's Icones Plantarum*, D. Prain (Ed.), Dulau and Co., London (4th Ser.), **9**, 2816.

Hieronymus, G. (1897) Erster Beitrag zur Kenntnis der Siphonogamenflora der Argentina und der angrenzenden Länder, besonders von Uruguay, Paraguay, Brasilien und Bolivien. *Botanischer Jahrbücher für Systematik* **22**, 708–741.

Hoffmann, O. (1894) Compositae. In *Die Natürlichen Pflanzenfamilien*, Berlin, **4**(5), 137 (*Stevia*).

Keil, D. J. and Stuessy, T. F. (1975) Chromosome counts of Compositae from the United States, Mexico and Guatemala. *Rhodora* **77**, 171–195.

King, R. M. and Robinson, H. (1967) Multiple pollen forms in two species of the genus *Stevia* (Compositae). *Sida* **3**, 165–169.

King, R. M. and Robinson, H. (1968) Studies in the Compositae – Eupatorieae VIII. Observations on the microstructure of *Stevia*. *Sida* **3**, 257–269.

King, R. M. and Robinson, H. (1969) Studies in the Compositae – Eupatorieae. Typification of the genera. *Sida* **3**, 329–342.

King, R. M. and Robinson, H. (1987) *The Genera of the Eupatorieae (Asteraceae). Monographs in Systematic Botany*, Missouri Botanical Garden, St. Louis, October, **22**.

Kinghorn, A. D., Soejarto, D. D., Nanayakkara, N. P. D., Compadre, C. M., Makapugay, H. C., Hovanec-Brown, J. M. *et al.* (1984) A phytochemical screening procedure for sweet *ent*-kaurene glycosides in the genus *Stevia* (Compositae). *Journal of Natural Products* **46**, 439–444.

Kinghorn, A. D. and Soejarto, D. D. (1991) Stevioside. In *Alternative Sweeteners* (2nd edn, Revised and Expanded), L. O'Brien Nabors, and R. C. Gelardi (Eds), Marcel Dekker, Inc., New York, pp. 157–171.

Kleinpenning, J. M. G. (1987) *Man and Land in Paraguay*, CEDLA, Amsterdam.

Kleinpenning, J. M. G. (1992) *Rural Paraguay, 1870–1932*, CEDLA, Amsterdam.

Mengüal, L. (Ed.) (1970) *Manual del Agricultor Paraguayo*, Sociedad Nacional de Agricultura, Asunción, pp. 227–229 ('Importancia del Cultivo del Caá-jheé').

Pio Corrêa, M. (1926) *Diccionario das Plantas Uteis do Brasil e das Exoticas Cultivadas*, Imprensa Nacional, Rio de Janeiro, **1**, 348.

Prain, D. (Ed.) (1906) *Icones Plantarum*, London, **9**, pl. 2816.

Robinson, B. L. (1930a) Observations on the genus *Stevia*. *Contributions of the Gray Herbarium of Harvard University* **90**, 36–58.

Robinson, B. L. (1930b) The Stevias of the Argentine Republic. *Contributions of the Gray Herbarium of Harvard University* **90**, 58–79.

Robinson, B. L. (1930c) The Stevias of Paraguay. *Contributions of the Gray Herbarium of Harvard University* **90**, 79–90.

Robinson, B. L. (1930d) The Stevias of North America. *Contributions of the Gray Herbarium of Harvard University* **90**, 90–159.

Robinson, B. L. (1931a) The Stevias of Colombia. *Contributions of the Gray Herbarium of Harvard University* **96**, 28–36.

Robinson, B. L. (1931b) The Stevias of Venezuela. *Contributions of the Gray Herbarium of Harvard University* **96**, 37–43.

Robinson, B. L. (1931c) The Stevias of Ecuador. *Contributions of the Gray Herbarium of Harvard University* **96**, 43–49.

Robinson, B. L. (1932a) The Stevias of Peru. *Contributions of the Gray Herbarium of Harvard University* **100**, 20–36.

Robinson, B. L. (1932b) The Stevias of Bolivia. *Contributions of the Gray Herbarium of Harvard University* **100**, 36–39.

Robinson, H. and King, R. M. (1977) Eupatorieae – systematic review. In *The Biology and Chemistry of the Compositae*, V. H. Heywood, J. B. Harborne and B. L. Turner (Eds), Academic Press, New York, **1**, 437–485.

Sakaguchi, M. and Kan, T. (1982) As pesquisas Japonesas com *Stevia rebaudiana* (Bert.) Bertoni e o estevio-sideo. *Ciência e Cultura (São Paulo)* **34**, 235–248.

Schultz-Bipontinus, C. H. (1852) *Stevia. Linnaea* **25**, 268–292.

Soejarto, D. D., Kinghorn, A. D. and Farnsworth, N. (1982) Potential sweetening agents of plant origin. III. Organoleptic evaluation of *Stevia* leaf herbarium samples for sweetness. *Journal of Natural Products* **45**, 590–599.

Soejarto, D. D., Compadre, C. M., Medon, P. J., Kamath, S. K. and Kinghorn, A. D. (1983) Potential sweetening agents of plant origin II. Field search for sweet-tasting *Stevia* species. *Economic Botany* **37**, 71–78.

Sugii, A. (1977) *Plan de Cultivo de Caá-jheé*. A Type-written document filed by A. Sugii with the Ministerio de Agricultura y Ganaderia, Asunción, Paraguay, pp. 1–2.

Sumida, T. (1973) Reports on *Stevia rebaudiana* Bertoni introduced from Brazil as a new sweetness resource in Japan. *Miscellaneous Publications of the Hokkaido National Agricultural Experimental Station*, No. 2, 69–83 (in Japanese, with a shorter English text).

Sumida, T. (1980) Studies on *Stevia rebaudiana* Bertoni as a possible new crop for sweetening resource in Japan. *Journal of the Central Agricultural Experimental Station* **31**, 1–71 (in Japanese with a shorter English text).

von Schmeling, G. A. (1967) Caá-heé: edulcorante natural nâo calórico (*Stevia rebaudiana* Bert). *Boletim do Sanatório Sâo Lucas* **29**, 67–78.

3 Ethnobotany of *Stevia* and *Stevia rebaudiana*

Djaja Djendoel Soejarto

INTRODUCTION

A review of the botany of the genus *Stevia* has been presented in Chapter 2. Of the 220–230 species of this genus, only one has an outstanding ethnobotanical record, namely, the use of *S. rebaudiana* (Bertoni) Bertoni to impart sweetness to foods and drinks (Gosling 1901; Bertoni 1905; 1918; Ulbricht as quoted by von Schmeling 1967; Soejarto *et al*. 1983a; 1983b; Lewis 1992). It is this sweetness property which led to the discovery of this species and, eventually, to the tremendous attention given to it. This is especially true from the late 1960s to the present, when it was established that the non-sugar sweet principles in this plant, the *ent*-kaurene diterpene glycosides, particularly stevioside and rebaudioside A, could be used as an industrial substitute for sucrose. This tremendous interest is demonstrated by the fact that in 1970s, at least 96 papers on *S. rebaudiana* or its sweet principles were published, mostly on the chemistry and biology (including patents), while in 1980s, at least 104 papers of similar nature were published. These statistics have been based on data from the NAPRALERT database, as of June 1998 (see Farnsworth 1996, for details about NAPRALERT).

Because of such an outstanding record, the ethnobotanical importance of other species of *Stevia* has been eclipsed by that of *S. rebaudiana*. Yet, there are interesting aspects on the relationships between *Stevia* and man, which may also lead to new discoveries, if natural product scientists would spend their time and efforts to follow-up on such leads. As is the case with the rise on the importance of *S. rebaudiana*, interest towards other species of the genus was awakened only in late 1970s and early 1980s, as part of the efforts in searching for sweet *ent*-kaurene diterpene glycosides in other species within the genus, as well as in searching compounds with potential biological activity. Among such efforts has been the screening of 110 species of the genus for the possible presence of the sweet principles by our research group at the University of Illinois at Chicago (Soejarto *et al*. 1982; Kinghorn *et al*. 1984), followed by a field search (Soejarto *et al*. 1983a) and a review of the ethnobotanical records of the genus (Soejarto *et al*. 1983b).

Evidence for the rise of interest towards other species of *Stevia* is given by the number of papers published in the 1970s and 1980s. In the 1970s, only ten papers on the chemistry and biology of species other than *S. rebaudiana* were published, as compared to 47 papers of a similar nature published in 1980s. These statistics have also been based on data from the NAPRALERT database, as of June 1998.

RELATIONSHIP BETWEEN *STEVIA* AND MAN

Through the ages, a relationship between man and *Stevia* plants has developed. Such a relationship took the form of recognition of the plants as expressed in their common names and the discovery of their utility in various ways. The knowledge discovered became part of the folklore of a culture or a population and, occasionally, came to the attention of botanists, explorers, or even casual travelers, who recorded it and communicated it in a written form. Some of the older records are not easy to interpret, primarily because of the large diversity of species and the deficient botanical knowledge at that time.

Members of *Stevia* comprise mostly herbaceous perennial plants, or rarely large shrubs to small trees (those belonging to the subgeneric group *Fruticosae*) (see Chapter 2). They occur primarily in open vegetation formations, such as grasslands or scrub forests. Within the distribution range of a given species, the plants are found on plains, river valleys, slopes, hills, and interfaces between open fields and wooded or forested areas. They usually grow in patches, with white to pinkish or purplish flower heads, depending on the species. Some species are found close to areas of human habitation.

There are many more species in some countries than in others, as a result of different types of *Stevia* habitats found in a particular country, suitable for the existence and evolution of the species. Abundance of species is reflected in the number of Latin binomials that have been published for the country; for example, there are at least 130 Latin binomials that have been published for Mexico, 41 for Bolivia, 35 for Brazil, 31 for Argentina, and 20 for Peru, while there are only two for Honduras, five for Guatemala, three for Venezuela, six for Ecuador, and four for Chile (Anonymous 1893–1993). Because of their commonness in some localities, and because of certain unique characteristics, *Stevia* plants are bound to be noticed by people who live in the area where the plants grow.

One of the characteristic features of *Stevia* plants is a strong odor, sometimes referred to as a 'goat's smell' (see also, Grashoff 1972), when the leaves are crushed. This strong odor varies among species and may not be entirely unpleasant. The presence of such odors has induced plant collectors to record them in various expressions on herbarium labels, such as 'strong fragrant odor', 'pleasantly aromatic', 'plant fragrant', 'leaves emitting strong-scented gum when crushed', 'fragrant shrub', 'herbage gives aromatic odor', 'herbage with pungent odor', 'foliage fragrant', and so on (von Reis Altschul 1973).

Another characteristic feature is stickiness, when the plant or the leaves are crushed, due to the presence of glands or glandular hairs on the epidermal surface, containing volatile oils. Still another characteristic is bitterness or sweetness of the leaves and other plant parts, when one chews these parts. Such characteristics appear to have induced different cultures or populations to describe *Stevia* plants under various descriptive common names. Examples are 'amargo' (bitter), 'anís de ratón' (rat's anise), 'caá-hê-ê' (sweet herb), 'flor de María' (Maria's plant, a beautiful plant), 'hierba de la pulga' (flea's herb), 'matapulgas' (flea killer), 'mejorana' (name of a condiment plant, *Origanum* sp.), 'tomilho silvestre' (wild thyme), etc. (Table 3.1).

EARLY RECORDS

Early ethnobotanical records of *Stevia* are represented by the account in the book *Natural History of Plants of the New Spain* written by Francisco Hernandez, a physician, between 1570 and 1576. This book has been published several times in Latin and Spanish, between 1605 and 1959 (Hernández 1959).

As already described previously (Soejarto *et al.* 1983b), in this book the plants are listed under their Nahuatl names and are given brief descriptions concerning botanical characteristics and medicinal virtues, and sometimes provided with illustrations. The book contains such a rich information that Schultes (1962) called it an 'incredible treasury'. Del Pozo (1979)

Table 3.1 Species of *Stevia* with an ethnobotanical history

Species (Common name)	Country (Locality)	Part used	Use or disease treated	Route of administration	Reference
S. balansae Hieron. (**Charúa kaá**)	Paraguay (Nueva Colombia)	Root	Populace drinks decoction from the root in cases of diarrhea	Oral (drink)	Soejarto *et al.* 1983a; Soejarto *et al.* 1983b
S. bogotensis Tr. ex Cortés (**Clavito, Eupaloria, Eupatoria, Jarilla**)	Colombia (Bogotá)	Whole plant	Used as a febrifuge and diaphoretic	Not stated	Cortés 1919; Pérez-Arbeláez 1937; Soejarto *et al.* 1983b
S. caracasana DC. (**Molinillo**)	Venezuela		Herbarium label of *Jahn 1098* at Harvard University Herbarium states that this plant is known locally as **Molinillo**; no uses are given		von Reis Altschul 1973; Soejarto *et al.* 1983b
S. cardiatica Perkins	Bolivia	Whole plant	Used in the treatment of heart diseases	Not stated	Robinson 1932; Cardenas 1943; Soejarto *et al.* 1983b
S. collina Gardn. (**Caá-ehé**)	Brazil	Not stated	Recommended for use as a sweetener and as a stomachic	Oral	Pio Corrêa 1926
S. connata Lag. (**Pericón de monte**)	Guatemala (Huehuetenango)	Not stated	Reputed to be used to treat stomachache (note on the herbarium label of *Steyermark 50493* collected in 1942, in deposit at Field Museum)	Not stated	Soejarto *et al.* 1983b
S. cuzcoensis Hieron. (**Chipi-cuca**)	Peru (Cuzco)		Uses are not mentioned; common name only is given		Herrera 1930;1939; Soukup 1970; Soejarto *et al.* 1983b
S. dictiophylla B. L. Robins. (**San Marcos**)	Mexico		Known as **San Marcos** in the State of Jalisco, but no uses are mentioned		Martínez 1979; Soejarto *et al.* 1983b
S. elatior HBK. (**A-cí**)	Mexico (Oaxaca)	Leaf	Used to soothe burns and scratches	External	Lipp 1971; Soejarto *et al.* 1983b

Species	Locality	Preparation	Uses/Notes	References
S. eupatoria (Spreng.) Willd. (**Hierba del borrego, Yerba del borrego, Cola del borrego**)	Mexico	Not stated	Reported to be used as a diuretic and antimalarial; reported to be toxic to animals; has been introduced to and cultivated in Cuba under the name of *S. purpurea* Pers. (**Estevia**)	Flores 1907; Roig y Mesa 1953; Martínez 1979; Soejarto *et al.* 1983b
S. glandulosa Hook. et Arn. (**Hierba de la pulga** /Oaxaca; **Merba** /Nayarit)	Mexico	Not stated	Reported to be 'used medicinally in fevers' (Jalisco) on the herbarium label of *Mexía 147a* in deposit at Harvard University Herbarium	von Reis Altschul 1973; Martínez 1979; Soejarto *et al.* 1983b
S. hirsuta DC. (**Oreja de ratón**)	Guatemala (Los Encuentros)		A note on the herbarium label of *Molina and Molina 26641* at Field Museum states that this species is known as **Oreja de ratón**; no uses are reported	Soejarto *et al.* 1983b
S. jorullensis HBK. (**Mejorana**)	Guatemala (Chimaltenango)		Note from label of herbarium specimen *Molina and Molina 26700* at the Field Museum, Chicago states that this plant is known as **Mejorana**; no uses are given	Soejarto *et al.* 1983b
S. jorullensis HBK. (**Roselina**)	Mexico		Known as **Roselina**; no use is given	Martínez 1979; Soejarto *et al.* 1983b
S. lehmannii Hieron. (**Oreja de ratón**)	Guatemala (Huehuetenango)		The common name is given on the herbarium label of *Molina and Molina 26296* at the Field Museum, Chicago; no uses are given	Soejarto *et al.* 1983b
S. lucida Lag. (**Yerba de aire, Hierba de la araña**)	Mexico	Not stated	Used to treat conditions caused by *bad air* (e.g. chilly cramps)	von Reis Altschul 1973; Díaz 1976b; Latorre and Latorre 1977; Martínez 1979; Soejarto *et al.* 1983b

Table 3.1 (Continued)

Species (Common name)	Country (Locality)	Part used	Use or disease treated	Route of administration	Reference
S. lucida Lag. (**Ma-li-too**)	Mexico	Leaf	Poultice used to cure wounds	External	Zamora-Martínez and Pascual-Pola 1992
S. lucida Lag. (**Kebuj, Mariposa**)	Guatemala (Totonicapán)	Not stated	Used to treat rheumatism	Not stated	Anonymous 1929; Soejarto et al. 1983b
S. lucida Lag. (**Chilca, Javillo, Jarilla**)	Colombia	Not stated	Used to soothe pains (under S. glutinosa HBK., a synonym)	External	Pérez-Arbeláez 1936; Soejarto et al. 1983b
S. lucida Lag. (**Golondrina de la sabanera**)	Colombia (Bogotá)	Aerial part	Decoction is used to treat inflammation	Not stated	von Reis Altschul 1973; Soejarto et al. 1983a, 1983b
S. lucida Lag. (**Chilca, Chirca**)	Venezuela	Leafy stem	Boiled, the resinous decoction is employed to alleviate rheumatism	Not stated	von Reis Altschul 1973; Morton 1975; Soejarto et al. 1983b
S. macbridei B. L. Robins. var. anomala B. L. Robins. (**Gualamoco**)	Peru (Jauja-Huancayo)	Whole plant	Decoction is used as a bath by women	External	Soejarto et al. 1983a, 1983b
S. monardifolia HBK. (**Mara Antonia**)	Mexico (Michoacán, Zitacuaro)		No uses are mentioned; common name only is given		Soejarto et al. 1983a, 1983b
S. nelsonii B. L. Robins. (**Hierba de la paloma**)	Mexico (Michoacán)		No uses are mentioned; common name only is given		Soejarto et al. 1983a, 1983b
S. nepetifolia HBK. (**Zazal**)	Mexico (Sinaloa)		Only common name is mentioned; no uses are reported		Martínez 1979; Soejarto et al. 1983b
S. nepetifolia HBK. (**Peracón, Anís de ratón**)	Guatemala	Not stated	Tea is prepared to treat dysmenorrhea (under the name of S. rhombifolia HBK., a synonym of S. nepetifolia)	Oral (drink)	von Reis Altschul 1973; Soejarto et al. 1983b
S. origanoides HBK. (**Amargo**)	Mexico (Temascaltepec)		Uses are not mentioned; only common name is given		Soejarto et al. 1983b

Species	Location	Plant part	Notes/Uses	Internal/External	References
S. ovata Willd. (**Cuali-aquinina**)	Mexico (Huejutla)		Only common name is known (under the name of *S. rhombifolia* HBK. var. *typica*)		Martínez 1979; Soejarto *et al.* 1983b
S. ovata Willd. (**Flor de Plata, Tuán**)	Guatemala		Common name only known, as stated on label of herbarium specimen *Steyermark 51753* at Field Museum, Chicago, identified as *S. rhombifolia* HBK.		von Reis Altschul 1973; Soejarto *et al.* 1983b
S. palmeri Gray (**Raníweri, Raníwori**)	Mexico (Chihuahua/Tarahumara)		'Odoriferous plant'; a note on herbarium label of *R. A. Bye, Jr 3115* (Oakes Ames Economic Herbarium, Harvard University); no uses are given		Regalado 1998
S. petiolata (Cass.) Sch.-Bip. (**Guarmi-guarmi**)	Peru (Rio Blanco, near Lima)	Whole plant	In the process of baking meat in a covered pit (a practice known as *pachamanca*) the plant is added to give flavor to the meat		Soejarto *et al.* 1983a; 1983b
S. pilosa Lag. (**Flor de Maria**)	Mexico		Common name only is given; no uses are reported		von Reis Altschul 1973; Soejarto *et al.* 1983b
	Mexico	Not stated	Used as an antimalarial, antipyretic, cathartic and diuretic	Not stated	Díaz 1976b; Soejarto *et al.* 1983b
S. plummerae Gray (**Romino**)	Mexico (Chihuahua/Tarahumara)	Whole plant	'Avoided by grazing animals'; a note on herbarium label of *R. A. Bye, Jr 2817* (Oakes Ames Economic Herbarium, Harvard University)	Internal	Regalado 1998
S. plummerae Gray (**Rorino**)	Mexico (Chihuahua/Tarahumara)	Whole plant	'... roots used to make a wash and poultice of open wounds of animals and man'; a note on herbarium label of *R. A. Bye, Jr 2817* (Oakes Ames Economic Herbarium, Harvard University)	External	Regalado 1998

Table 3.1 (Continued)

Species (Common name/Country)	Country (Locality)	Part used	Use or disease treated	Route of administration	Reference
S. polycephala Bertol. var. *polycephala* (**Mejorana**)	Guatemala (Huehuetenango)		Only common name is given; no uses are reported		Soejarto *et al.* 1983b
S. puberula Hook. (**Lima-lima**)	Peru	Not stated	Used as a 'tea substitute and stomach medicine'	Oral	von Reis Altschul 1973; Soejarto *et al.* 1983b
S. punensis Bl. Robins. (**Enduchuina**)	Peru		Common name only is given; the name **Enduchuina** may have been derived from the Spanish word 'endulzar' (= to sweeten); no uses are given		Herrera 1939; Soukup 1970; Soejarto *et al.* 1983b
S. rebaudiana Bertoni (**Caá Hé-é or Kaá Hé-é, Caá-éhé, Caá heé or Kaá Héé, Caá-ehe, Caá-hé-hé, Caá-enhem, Azucá-caá**)	Paraguay	Leaf	The various renditions of the common name **Caá-hé-é** in the Guaraní language of Paraguay means 'sweet herb'; leaves are sometimes used to sweeten the Paraguayan traditional drink *maté*	Oral	Gosling 1901; Bertoni 1905; Bell 1954; Soejarto *et al.* 1983a; 1983b; Lewis 1992
	Paraguay	Leaf	Infusion used to prevent pregnancy among native populations throughout the range of the plant (post-1960 claim)	Oral	Planas and Kuc 1968; Brondegaard 1973; Felippe 1977
	Paraguay	Leaf	Decoction is used for diabetes by natives (post-1960 claim)	Oral	Oviédo *et al.* 1970
S. rebaudiana Bertoni (**Caá-ehé, Caá-hé-hé, Caá-enhem**)	Brazil (Mato Grosso)	Not stated	Stated to be a variety of *S. collina* used as a sweetener; misidentification is involved here	Oral	Pio Corrêa 1926

Species	Country	Part	Use	Administration	Reference
	Brazil	Leaf	Long used by the local people of Paraguay and bordering Brazilian State of Mato Grosso as a sweetening agent; no documentation is provided	Oral	Mors and Rizzini 1966
	Brazil	Not stated	Used some localities in Brazil since old times a sweetening material; no documentation is provided	Oral	Sumida 1973
S. rhombifolia HBK. var. *stephanocoma* Sch.-Bip.	Peru	Root	Decoction is used for stomachache	Oral	Velasco-Negueruela *et al.* 1995
S. rhombifolia HBK. var. *stephanocoma* Sch.-Bip. (**Manka pak'i**; manka = a kind of pot; pak'i = break)	Peru (Chincheros)	Root	'Roots used for stomachache. Leaves steeped for vomiting'. A note on herbarium label of *King et al. 251* (Oakes Ames Economic Herbarium, Harvard University)	Oral	Regalado 1998
S. rhombifolia HBK. var. *stephanocoma* Sch.-Bip. (**Pirq'a**)	Peru (Chincheros)	Root	'Used for maté'. A note on herbarium label of *King et al. 251* (Oakes Ames Economic Herbarium, Harvard University)	Oral	Regalado 1998
S. rhombifolia HBK. var. *stephanocoma* Sch.-Bip.	Peru	Root	Decoction is used as an emetic	Oral	Velasco-Negueruela *et al.* 1995
S. salicifolia Cav. (**Hierba del aire**/Hidalgo, **Hierba de la mula**/Guanajuato, **La envidia**/Michoacán and Mexico, **Yerba de la mula, Zazal, Zazal**/Sinaloa, **Zazale de olor**)	Mexico	Not stated	**Zazal** is offered for sale in Juárez market of Toluca, for the preparation of a water decoction or an alcoholic infusion to be used as a rub for rheumatism. In Morelia (Michoacán), the plant is used for the same purpose, but under the name of **Yerba de la mula**	External	von Reis Altschul 1973; Martínez 1979; Soejarto *et al.* 1983b

Table 3.1 (Continued)

Species (Common name)	Country (Locality)	Part used	Use or disease treated	Route of administration	Reference
S. salicifolia Cav. (**Hierba de la Santa Rita**)	Mexico (San Luis Potosí)		Reported as **Hierba de la Santa Rita** in the State of San Luis Posotí, under the name of S. stemphylla A. Gray		Martínez 1979
S. salicifolia Cav.	Mexico	Root	Decoction of dried roots is used as a cathartic	Oral	Mata et al. 1991
		Root	Infusion of dried roots taken for intestinal upset due to parasites	Oral	Mata et al. 1991
S. salicifolia Cav.	Mexico (Chihuahua / Tarahumara)	Root	'. . . roots . . . are used as a purgative: roots mashed and placed in lukewarm water, drunk.' Note on a herbarium label of R. A. Bye, Jr 4916 (Oakes Ames Economic Herbarium, Harvard University)	Oral	Regalado 1998
S. salicifolia Cav. (**Ritunáwa;** rituku = snow or frost)	Mexico (Chihuahua / Tarahumara)	Leaf	'Leaves used in making tea for fevers and colds; given this name because more snow and frost piles up on this plant than others.' Note on herbarium label of R. A. Bye, Jr 4913 (Oakes Ames Economic Herbarium, Harvard University)	Oral	Regalado 1998
S. salicifolia Cav. (**Wagúsari**)	Mexico (Chihuahua / Tarahumara)	Root	'. . . root used as a fish poison'. Note on herbarium label of R. A. Bye, Jr 4889 (Oakes Ames Economic Herbarium, Harvard University)	Not stated	Regalado 1998

Species (common name)	Country/Region	Plant part	Uses	Administration	Reference
S. satureifolia Sch.-Bip. (**Guaco**)	Brazil (Rio Grande do Sul)	Not stated	A number of species of *Eupatorium, Mikania* and *Stevia* are referred to by Pío-Corréa as 'melíferas' or honey-bearing, including *S. satureifolia* from Rio Grande do Sul	Not stated	Pío-Corréa and de Azeredo Pena 1952; Soejarto *et al.* 1983b
S. satureifolia Sch.-Bip. (**Charrúa, Tomilho silvestre, Yerba del Charrúa**)	Uruguay		Application of the common names listed by Pío-Corréa for *S. satureifolia* needs reexamination, due to the wide species concept adopted; no uses are mentioned		Pío-Corréa and de Azeredo Pena 1952; Soejarto *et al.* 1983b
S. serrata Cav. (**Roninc**)	Mexico (Chihuahua/Tarahumara)	Root	'... roots used to make a wash and poultice of open wounds of animals and men.' Note on herbarium label of *R. A. Bye, Jr 2802* (Oakes Ames Economic Herbarium, Harvard University)	External	Regalado 1998
S. serrata Cav. (**Uriki**)	Mexico (Chihuahua/Tarahumara)	Root	'Root mashed, applied to cuts on feet'. Note on herbarium label of *R. A. Bye, Jr 2590, 2717, 2728* (Oakes Ames Economic, Herbarium, Harvard University)	External	Regalado 1998
S. serrata Cav. (**Otoninawa**)	Mexico (Chihuahua/Tarahumara)	Root	'Occasionally added to corn *teguino* in order to make it stronger'. Note on herbarium label of *R. A. Bye, Jr 4904* (Oakes Ames Economic Herbarium, Harvard University)	Oral	Regalado 1998

Table 3.1 (Continued)

Species (Common name)	Country (Locality)	Part used	Use or disease treated	Route of administration	Reference
S. serrata Cav. (**Chapo**)	Mexico (Chihuahua/Tarahumara)	Whole plant	'Young plants eaten as *quelites*'; a note on herbarium label of *R. A. Bye, Jr 4064* (Oakes Ames Economic Herbarium, Harvard University)	Oral	Regalado 1998
S. serrata Cav. (**Yerba picete**)	Mexico (Chihuahua/Tarahumara)	Whole plant	'Plant crushed and rubbed on snake bite as a cure.' Note on herbarium label of *R. A. Bye, Jr 2328* (Oakes Ames Economic Herbarium, Harvard University)	External	Regalado 1998
S. serrata Cav. (**Hipericón, Tlalacxoyatl, Tlalchichinole**)	Guatemala		The names **Tlalacxoyatl** and **Tlalchichinole** refer to *S. linoides*, a synonym of *S. serrata*		Díaz 1976a; Martínez 1979; Soejarto et al. 1983b
S. serrata Cav. (**Q'ang'aj**)	Guatemala	Not stated	Used as a cough remedy (a note on herbarium specimen *Cominsky 74* at the Field the Museum, Chicago)	Oral	Soejarto et al. 1983b
S. serrata Cav. (**Anís silvestre**)	Guatemala (Huehuetenango)		Common name is from a note on field label of herbarium specimen *Molina and Molina 26288* at the Field Museum, Chicago		Soejarto et al. 1983b

Species	Distribution	Plant part	Notes / Uses	Administration	References
S. subpubescens Lag. (**Zezal, Hierba de la mula**)	Mexico (Michoacán)	Aerial part	In Zitaquaro, decoction of the aerial parts is recommended as a bath by women after parturition, while the leaves are used for stomachache; in Patzcuaro, the plant is used to treat pains in the joints by rubbing fried plants on the affected parts	External	Soejarto *et al.* 1983a, 1983b
S. trifida Lag. (**Manzanilla de agua**)	Mexico	Root, Flower	Infusion is used to treat dysentery. Note on label of herbarium specimen *E. Langlassé 33* collected in 1898	Oral	von Reis Altschul 1973; Soejarto *et al.* 1983b
S. viscida HBK. (**Matapulgas and Pipizhuatl** in Jalisco, **Hierba de la pulga** in Mexico)	Mexico (Jalisco and Mexico)		The name **Pipizhuatl** is derived from **Pipitzáhuatl**, a name used by Sessé and Mociño to baptize *Eupatorium purpureum* in 1893, now a synonym of *S. viscida* (Grashoff, 1972; no uses are cited)		Hernandez 1946; Martínez 1979; Sessé and Mociño 1893; Grashoff 1972; Soejarto *et al.* 1983b

considered it a very reliable source of information. There are three editions of the book (Hernández 1942: vol. 1; Hernández 1943: vol. 2; and Hernández 1946: vol. 3), in which the editors have provided a tentative identification of or, more appropriately, interpretation to a great number of the plants listed therein. Thus, the equivalent Latin binomials are provided. Yet, despite these Latin binomials, and the remarkable accuracy of interpretation, the final and correct botanical identity of many of the plants, including species of *Stevia*, remains the subject of conjecture.

Stevia or possible *Stevia* species mentioned in Hernandez's book are referred to under the names of **Anonima Mechoacanense, Camopaltic, Cihuapatli Yacapichtlense, Cihuapatli Pitzahoac, Tlacochichic de ocopetlayuca**, and **Tonalxihuitl** (Soejarto *et al*. 1983b).

Anonima Mechoacanense

A plant under this name has been interpreted as a species of *Stevia* in Hernández' book (1943: 481). Since the illustration provided depicts a plant with verticillate (in 5's) leaves, this interpretation is incorrect. No species of *Stevia* has verticillate leaves (see Chapter 2 of this volume).

Camopaltic

In Mexico, 'Camopaltic' is a Nahuatl word which means 'herb with purple color'. Several plants in Hernández' book are called the first, second and third **Camopaltic**. The identity of the third **Camopaltic** is given as *S. clinopodia* DC. (Hernández 1946: 844), a synonym of *S. jorullensis* HBK. (Grashoff 1972). Hernández noted that the plant is a 'cold herb', and that 'half an ounce of the roots taken with water evacuates the urine'.

In an early work by Urbina (1906), the third **Camopaltic** was given a different interpretation as either *S. linoides, S. laxiflora, S. purpurea* or *S. paniculata*, with *S. linoides* to be the most likely species, due to the similar form of its leaves to flax. All these Latin binomials, however, are illegitimate names (Grashoff 1972); their legitimate or correct names are *S. serrata* Cav. (for *S. linoides* Sch.-Bip.), *S. viscida* HBK. (for *S. laxiflora* DC.), and *S. ovata* Willd. (for *S. paniculata* Lag.). According to Grashoff, the Latin binomial *S. purpurea* is an ambiguous name, since it includes elements belonging to *S. viscida* HBK., *S. pilosa* Lag., *S. eupatoria* (Spreng.) Wild., and *S. porphyrea* McVaugh.

Since criteria for species delimitation in *Stevia* include not only vegetative and inflorescence characters, but also minute floral and achene characters, which cannot be verified from an illustration of the plant in Hernández' book (see also the illustration reproduced in Soejarto *et al*. 1983b), the correct species identity of the third **Camopaltic** may never be determined with finality.

Cihuapatli Yacapichtlense

A plant under this name is listed in Hernández' book (1946: 888) and has been interpreted as either one of the following: *Baccharis glutinosa* Pers., *S. viscida* HBK. or *S. salicifolia* Cav. The illustration accompanying the description (Hernández 1946: 889, Figure 215), however, depicts a plant with alternate leaves, characteristic of either *Baccharis* or *S. viscida*, but certainly not of *S. salicifolia*, which has opposite leaves. Further, the description mentions a plant with purple flowers, a common flower color among species of *Stevia*, but never in species of *Baccharis*, which is yellowish.

The word 'Cihuapatli' in the Nahuatl language means 'medicine for women', whereas the epithet 'yacapichtlense' refers to the locality of the plant in the State of Morelos. Since there are 22 other 'Cihuapatli's' mentioned in Hernández's text, the use of the word 'yacapichtlense' is a logical way distinguish it from the other 21. This plant is also referred to as **Pitzahoaccihuapatli** in the same text.

Cihuapatli Yacapichtlense is stated to have the following medicinal virtues: 'Its decoction drunk or applied resolves admirably the humours which have penetrated the joints and nerves, and calms any pains by eliminating their cause' (Hernández 1946: 888).

Cihuapatli Pitzahoac

In Hernández' book (1946: 894), a plant with this name has been interpreted as *S. viscida* HBK. The plant is described as having '. . . a slightly bitter taste. The leaves applied prevent the fall of hair'.

It appears that the interpretation of this plant as *S. viscida* has been based on the fact that in some parts of Mexico, the common name for *S. viscida* is **Pipitzáhuatl** (see Table 3.1). Since this common name also applies to species of *Perezia* and *Eupatorium* (Martínez 1969), the true identity of **Cihuapatli Pitzahoac** remains in question.

Tlacochichic de ocopetlayuca

On page 583 of Hernández' 1943 book, a plant under this name is interpreted as belonging to a species of *Stevia*. 'Tlacochichic' means 'bitter stem' in the Nahuatl language. This plant has the following medicinal virtues: '. . . the juice of an ounce of the root drunk with water evacuates the phlegmatic humours through the inferior conduit. It is hot and tastes acrid and bitter. The people from the Panuco region (Veracruz, Mexico) say that it cures scabies and abdominal pains, and reduces the spleen.'

Tonalxihuitl

One of 12 plants listed in Hernández's 1942 book under this name (Hernández 1942: 56), has been interpreted as *S. salicifolia* Cav. The description provided fits well the characteristics of this species, but no illustration is given for purposes of comparison. The text states that the roots of this plant are '. . . excellent in reducing fevers, when crushed and rubbed or drunk, in a dose of an ounce.' Since another plant, referred to as **Tonalxihuitl Yacapichtlense**, has been interpreted as *Veronica americana* Schwein. ex Benth., a species belonging to a completely different family, Scrophulariaceae, the identity of 'Tonalxihuitl' may never be elucidated.

In this context, the book *Florentine Codex* should be mentioned, written in sixteenth century in the Nahuatl language under the direction of Bernardino Sahagún, based on information provided by Aztec physicians. In this book, three mentions are made of **Tonalxihuitl**. Following Hernández's book, in the 1961 and 1963 English translation of the book (Sahagún 1961; 1963), they are interpreted as either *S. salicifolia* or *V. americana*. The illustration provided on Figure 574 in this *Codex* (Sahagún 1963), however, depicts a plant with alternate leaves, a characteristic that does not fit *S. salicifolia*. Thus, doubt on the true identity of **Tonalxihuitl** continues.

Interesting uses are attributed to 'Tonalxihuitl' (Soejarto *et al.* 1983b), however, because of the confusion on the possible identity of the plant(s) referred to under this name, these uses may not be directly relevant to *Stevia*.

RECENT ETHNOBOTANICAL RECORDS

Records with definite species identity

A list of species of *Stevia* for which ethnobotanical records have been found to date is presented in Table 3.1. This list has been based, in the main part, on a previous review (Soejarto *et al.* 1983b), and on a number of additional references that had been missed previously.

The author believes that this list represents what is presently known about the ethnobotany of this genus. Further field search and inquiries, however, are still expected to yield new information, and additional new species, presently not appearing in the list. Such new data may come from areas with a large number of species, such as Mexico, Peru, Bolivia, Brazil, Paraguay, and Argentina, and from areas with a low number of species, but where no ethnobotanical records are presently registered, such as the southern United States, Honduras, Costa Rica, Panama, Ecuador, and Chile.

Records with doubtful species identity

Uses of several species of *Stevia* are reported in the literature, in which the specific identities of the plants are presently either unknown or questionable. Among such reports are:

Albahaca del campo

A species of *Stevia* collected in Argentina in 1936 and referred to as 'Albahaca del campo' is listed in von Reis Altschul's book (1973), but no other details were provided.

Kaadyupé

This name was used by Bertoni (1914) to refer to a species of *Stevia* from Paraguay, but without further details.

Molinillo

Pittier (1926: 300) used this name to refer to *S. urticaefolia* Thunb. from Venezuela. A case of species misidentification appears to have been involved here, since the herbarium specimen referred to by Pittier appears to be the same specimen (*Jahn 1098*) cited by von Reis Altschul (1973), under the Latin binomial *S. caracasana* (Table 3.1). *S. urticaefolia* is a Brazilian species. Readers are referred to a previous paper (Soejarto *et al.* 1983b) for further discussion on uses of 'Molinillo'.

Pericón Blanco

Martínez (1969) considered that a plant known under this name in the Tasco region of Mexico is a species of *Stevia*, and that '. . . dried leaves and flowers, boiled in water with salt, are used against colic'.

Salvia

A catalogue of medicinal plants of Guatemala published in 1929 (Anonymous 1929) refers 'Salvia' to *S. rhombifolia*. However, the Latin binomial *S. rhombifolia* has been used by different authors to refer to three different species, *S. ovata* Willd., *S. triflora* DC., and *S. jorullensis* HBK. (Grashoff 1972). Based on this concept, the true species identity of the Guatemalan 'Salvia' may never be known with certainty.

Tlalchichinole

Martínez (1969) uses the name 'Tlalchichinole', for a plant from the Guerrero State of Mexico, to refer to a species of *Stevia*. A decoction of this plant is used to wash infected pimples.

According to Díaz (1976a), the name 'Tlalchichinole' refers to *S. linoides*, a synonym of *S. serrata* Cav. (Grashoff 1972).

ETHNOBOTANY OF *STEVIA REBAUDIANA*

Although data on the ethnobotany of *S. rebaudiana* are given in Table 3.1, because of the importance of this species, and because of the author's personal acquaintance with the plant in its native land, Paraguay (Soejarto *et al.* 1983a), further comments concerning this species are justified.

The oldest record that points out the sweet-tasting properties of *S. rebaudiana* is a paper by Moisés S. Bertoni published in *Revista de Agronomía de Asunción*, vol. 1: 35, 1899 (not seen by the author), when he described this species under the name of *Eupatorium rebaudianaum* Bertoni. At the beginning of 1901, the description of the new species, in Latin, was sent to the Royal Botanic Gardens at Kew, England, with a letter, by the British Consul in Asunción, H.B.M. Cecil Gosling. Gosling's letter together with the Latin description were published in the *Kew Bulletin* (1901: 174), as an anonymous note. This anonymous note is the second oldest record; it has been cited in many reviews on *S. rebaudiana* as 'Gosling 1901' (Felippe 1977; Soejarto *et al.* 1983b; Kinghorn and Soejarto 1985; Lewis 1992; among others). In his 1905 paper, Bertoni renamed the species as *S. rebaudiana* (Bertoni) Bertoni, and cited his 1899 paper (Bertoni 1905), but not the *Kew Bulletin* note. In 1906, unaware of Bertoni's 1905 paper, Hemsley published a new species of *Stevia* under the name of *S. rebaudiana* Hemsley (Hemsley 1906). For reason of priority, this later homonym *S. rebaudiana* Hemsley is illegitimate; it becomes a synonym of *S. rebaudiana* (Bertoni) Bertoni (see Chapter 2).

Two intriguing questions that invite answers are, 'When were the sweet-tasting properties of *S. rebaudiana* discovered by the Guaraní Indians of Amambay?', and, 'For how long had the plant been known and used by the native populations of northeastern Paraguay, in the region of Amambay, when the news of its existence reached Bertoni in 1887?' As pointed out by Kinghorn and Soejarto (1986: 80), a substance which tastes sweet gives a pleasant oral gratification that is not easily forgotten, thus attracting special attention, in particular, if having a sweetness so intense as *S. rebaudiana* leaves. Through word of mouth, such an experience is transmitted and eventually comes to the attention of other persons. Conceivably, the knowledge eventually becomes widely known within the community or population. Such a diffusion of information could have been rapid or slow, depending on the ready availability or rarity of the plant. Using this scenario as the possible event that may have taken place following the discovery of the sweetness property of *S. rebaudiana*, the following points of conjecture are presented.

First, *the plant's sweet-tasting property may have been known since time immemorial by the peoples of Amambay, perhaps, even before the arrival of the Spaniards in the sixteenth century, but these peoples did not want to share such a knowledge with others they did not know, hence, keeping the information secret*. Although this seems to be a logical argument (see also, Gosling 1901; Lewis 1992), it is difficult to visualize this situation, since a sweet taste is nothing of immense value to the person who discovered it or to the population in which the knowledge was known. Thus, the discovery of this sweet plant did not present an extraordinary and immediate value that compelled the discoverer and others who knew about it to 'protect' such a knowledge. Even today, its use to sweeten *maté* drink is only sporadic (Soejarto *et al.* 1983a) among the populations where the plant is found naturally. In other words, the discovery of the sweet taste of *S. rebaudiana* was more of a curiosity, than anything else, to the discoverer. There is an even stronger reason to believe that this was the case, if the discovery took place after 1537, the date of the arrival of the Spaniards in Paraguay and the establishment of the city of Asunción, the capital city of Paraguay (Washburn 1871). By then, sugar, which was introduced to Paraguay in 1541 (Rivarola Paoli 1986), was already widely known. Relevant literature records from 1905, the date when *S. rebaudiana* became known to the scientific community, through 1940 did not mention other uses of *S. rebaudiana* among the peoples of Amambay and Paraguay in general, other than for its sweet taste to sweeten *maté* drink, a tea prepared from crushed/milled leaves of *Ilex paraguariensis* St.-Hill., Aquifoliaceae (Mabberley 1996) and other foods (Gosling 1901; Bertoni 1905; 1918; Robinson 1930a: 50; Robinson 1930b: 87; Lewis 1992).

Use of the plant or its sweet principles relating to its potential antidiabetic action appeared for the first time in an unpublished memorandum by Melville in 1941, addressed to the Director of Kew Gardens in the United Kingdom, as a response to the latter's request for information on possible sugar and saccharin substitutes, due to the increasing scarcity of sweeteners during the Second World War (Lewis 1992). In this memorandum, Melville states that '. . . the possibility of employing the leaves or the extracted sweet principles was considered for sweetening diabetic foods, but in normal times this natural product is unable to compete with saccharin' (Melville, as quoted by Lewis 1992). No reference was made on the use of *S. rebaudiana* to treat diabetes by the native Guaraní populations of Paraguay. According to Planas and Kuc (1968), J. B. Aranda of Asunción reported the use of *S. rebaudiana* for the treatment of diabetes in 1945. In 1966, Miquel declared the existence of 'Un nuevo hipoglicemiante oral' ('a new oral hypoglycemic agent') (Miquel 1966). In a 1967 paper, von Schmeling (1967: 67) cited Ulbricht, who wrote a travel book in 1930 (not seen by the author of this chapter), who reported that '. . . enquanto os homens se entretinham com bebidas, durante as festas da tribo, as mulheres e crianças preferiam preparados das dôces fôlhas de Caá-heê . . .' ('. . . while the men entertain themselves with drinks, during festivities of the tribe, the women and children prefer drinks prepared using the sweet leaves of *Caá-heê* . . .'). Von Schmeling, in this paper, further stated (her own wording) that 'Os bebedores do maté observaram, em breve, que a ingestao dêsse chá, quando adocado com fôlhas de *caá-heê*, era seguida de persistente sensaçao de bem estar. Assim, elas passaram a integrar o ról das plantas medicinais no Paraguai, sendo indispensáveis ao arsenal dos *'mineros'* e recomendadas pelos *médicos de yúyos*, por sus propiedades tônicas, disgestivas e úteis aos diabéticos' ('In short, the maté drinkers observed that, when the leaves of *caá-heê* were added, the ingestion of this tea was followed by a persistent sensation of well-being. Thus, it passed to become part of medicinal plants of Paraguay, being indispensable to the arsenal of the herbalists and was recommended by the herbal doctors for their tonic, digestive and useful properties to diabetics'). Please note that this last quoted passage is an interpretation by von Schmeling, not the words of Ulbricht. It was her perception that the plant had tonic, digestive and useful properties to diabetics, since in the paragraph that follows the above annotation, she stated (p. 67) 'Com a impressao de que a *Stevia* podería vir a ser um recurso a mais nas dietas dos diabéticos, devido às interessantes propriedades de edulcorantes nao calórico 'in natura', procuramos conhecê-la' ('With the impression that *Stevia* could serve as one more resource in the diets of diabetics, due to the interesting properties of being a natural, non-caloric sweetener, we attempted to know it'). Von Schmeling is a physician and knew the potential of the sweet *S. rebaudiana* plant, which contains no sugar, as a drug for diabetics. Further studies on the hypoglycemic action of *S. rebaudiana* then followed (Oviédo *et al*. 1970; von Schmeling *et al*. 1977; see also review by Kinghorn and Soejarto 1985). Indeed, the author could not find confirmation, either in the Amambay region or in Paraguay in general on the use of *S. rebaudiana* in the treatment of diabetes. The sale of *S. rebaudiana* leaves as a drug to treat diabetes, started, perhaps, in the 1940s.

As to the use of *S. rebaudiana* as a contraceptive, this is not found in the 1901–1960 literature, until Planas and Kuc, and, later, Brondegaard, mentioned it in 1968 and 1973, respectively (Planas and Kuc 1968; Brondegaard 1973). As is the case with the antidiabetic claim, the author could not find confirmation of this use among contemporary and native Indian populations of Paraguay (Soejarto *et al*. 1983a).

Second, *early Spaniard arrivals in Paraguay may have ignored such information* (see also, Lewis 1992), *thus, this knowledge remained unnoticed*. In support of this supposition, the author sought historical (Washburn 1871; Gonzalez and Ynsfran 1929; Schuster 1929; Williams 1979; Herken Krauer 1984; Rivarola Paoli 1986; Hanratty and Meditz 1990; Kleinpenning 1987; 1992; Reed 1995) and folklore-related (Bejarano 1960; Carvalho Neto 1961; Villanueva 1967; Cadogan 1973; Bejarano 1980; Barreto 1989; Cardozo Ocampo 1989) literature of Paraguay, but none of the references examined indicates any mention or any sign of possible knowledge on the existence of the sweet plant *S. rebaudiana*, either by the Spanish-descended or the native Indian populations of Paraguay, before Moisés S. Bertoni's arrival in Paraguay. One piece of direct evidence of the existence of the relationship between man and a plant, especially, a species with such a spectacular taste properties as *S. rebaudiana*, would be the presence of a mention of such a plant in the folklore of a people or culture. Such evidence is strikingly absent in the case of *S. rebaudiana*.

As to an indirect evidence, this may be traced to the folklore of the *maté* drink. A natural tea of *maté* is normally steeped in cold or hot water and is drunk without sugar or any other additive; this tastes bitterish. It is called *maté* when it is steeped in hot or warm water, and *tereré* when steeped in cold water. *Maté* and *tereré* are popular recreational beverages in Paraguay, Argentina and Uruguay. They are part of the culture in these countries. Barreto says that 'Tomar maté no es la misma cosa que tomar un café o un vaso de vino. Tomar maté requiere un clima, un tiempo, un ambiente espiritual definido. No se toma maté en cualquier parte o con cualquiera' ('Drinking *maté* is not the same as drinking coffee or a glass of wine. Drinking *maté* requires a mood, a time, a definite spiritual state. *Maté* is not drunk in just any place or with anybody') (Barreto 1989: 15). Some people, especially women, like to drink *maté* with additives (sugar, milk, fragrant herbs, etc.) (Barreto 1989; Cardozo Ocampo 1989). In case a sweetening additive is added, sugar is the sweetener used. None of these practices indicates the addition of any sweet substance, leaves or herbs, other than sugar. Among folklore expressions regarding *maté* are: bitter maté = indifferent; sweet maté = friendship; very sweet maté = talk with my parents; maté with 'toronjil' (aromatic herb) = displeasure; maté with cinnamon = you are in my thoughts; maté with orange peel = come find me; maté with milk = esteem; very hot maté = I, too, am burning with love; maté with citron = I accept; maté with honey = marriage proposal, foaming maté = I love you very much (Barreto 1989: 77). With such a rich folklore, it is conceivable to expect that, if *S. rebaudiana* was commonly used as a *maté* additive 'since time immemorial', both in urban and rural settings, its folklore should have become embedded in the culture of *maté*. This has not been the case. There are only two pre-1940 authors who mention the use of *Caá-heé* to sweeten *maté* drink, based on first-hand knowledge; Bertoni (1905; 1918) and Ulbricht (as quoted by von Schmeling 1967, already cited above). Gosling in his letter to the Royal Botanic Gardens at Kew, mentioned the use of *Caá-heé* 'to sweeten a strong cup of tea or coffee' (Gosling 1901). It took a scientist, in 1887, who was inquisitive enough to pursue the search for a plant to be rumored to have a sweet taste, in the name of Moisés S. Bertoni, who, then, revealed this information to the world. Bertoni, an Italian-Swiss, immigrated to Paraguay in 1882, shortly after graduating with a doctorate in natural science from the University of Zürich (Gorham 1973; Anonymous 1998).

Once the plant was rediscovered and studied, and it was shown that the sweet taste did not come from sucrose, but from a substance 'resembling glycyrrhizin' (Bertoni 1905), immediate attention was given to the plant as a potential source of a new sweetener. A flurry of activities ensued, namely, further studies on the chemical properties, and on the potential use as a sweetener, and as a substitute for sugar. If there was indeed 'ignorance' in the early phases of the discovery of the sweet taste of *S. rebaudiana*, this could have been the result of the rarity of the plant (see Chapter 2), thus, difficulty was experienced in getting access to a specimen for actual sensory experiments. The sweetness of the leaves is so intense that anyone who has actually tasted them would never forget the experience and would relate this with great excitement.

Third, *the discovery of the sweet taste may have taken place within a relatively short time, before the spread of the news about the existence of the sweet plant* S. rebaudiana *in 1887.* Most likely, this happened after and, perhaps, long after, the arrival of the Spaniards in the sixteenth century. This possibility is reflected in the letter of Gosling published in the *Kew Bulletin* (Gosling 1901), in which he stated that 'This plant, which has probably been known to the Indians since a hundred years or more . . .', implying a post-sixteenth century discovery. Thus, all evidence points to a relatively recent discovery (post-sixteenth century) of the sweet-tasting property of *Caá-heé*. This, together with the fact that the plant is rare within its natural distribution range (see Chapter 2), appears to have been responsible for the relatively late information as received by Bertoni in 1887, on the existence of this sweet plant.

ETHNOMEDICAL IMPORTANCE OF *STEVIA* AND ITS SCIENTIFIC BASIS

Ethnomedical importance

Perusal of Table 3.1 shows that *Stevia* species play an important role in the healthcare practices of different cultures and populations within the range of the distribution of the genus.

Because of the complex nature of a disease or health complaint, it is difficult or impossible to pinpoint the etiology of the diseases or health complaints listed in Table 3.1, without patient examination and medical diagnosis. Nevertheless, one can infer on the causative agent(s) of certain complaints, and group the diseases into broad disease categories. Table 3.2 presents the correlation of diseases and health complaints that have been treated using *Stevia* preparations listed in Table 3.1, without regard to the route of administration, based on possible causative agents of each disease or complaint. Obviously, this is a loosely structured grouping, but this is intended solely to show the wide spectrum of diseases or health complaints, in which at least 17 species of *Stevia* have played a role in their therapy.

From Table 3.2, one can get an idea as to how important plants belonging to this genus have been in therapy against infectious diseases. Thirteen of the 16 species with a history of medicinal use are employed for this purpose (GI problems, skin problems, malaria, fever/colds). For other health complaints, ten of the 16 species are used in therapy against metabolic, CNS-related, cardiac-related and women's health-related problems.

Table 3.2 Ethnomedical importance of *Stevia* (Extracted from Table 3.1)

Disease treated or therapeutic use	Complaint/Use	Species	Country
Arthritis/ rheumatism/ inflammation	• To treat inflammation	*S. lucida*	Guatemala
	• To treat rheumatism	*S. lucida*	Colombia
	• To treat rheumatism	*S. lucida*	Venezuela
	• To treat rheumatism (rub)	*S. salicifolia*	Mexico
	• Pains in joints	*S. subpubescens*	Mexico
Cardiac problems	• Heart diseases	*S. cardiatica*	Bolivia
CNS-related problems	• To soothe pains	*S. lucida*	Colombia
	• As a cathartic	*S. pilosa*	Mexico
	• As an emetic	*S. rhombifolia*	Peru
	• For vomiting	*S. rhombifolia*	Peru
	• As a cathartic/purgative	*S. salicifolia*	Mexico
Diabetes	• For diabetes	*S. rebaudiana*	Paraguay
Fever/Colds	• For fevers and colds	*S. glandulosa, S. salicifolia*	Mexico
	• Conditions caused by 'bad air'	*S. lucida*	Mexico
	• Cough remedy	*S. serrata*	Guatemala
GI problems	• Diarrhea	*S. balansae*	Paraguay
	• Stomachache	*S. connata, S. rhombifolia*	Guatemala
	• Stomach medicine	*S. puberula*	Peru
	• Intestinal upsets/parasites	*S. salicifolia*	Mexico
	• Stomachache	*S. subpubescens*	Mexico
	• Dysentery	*S. trifida*	Mexico
Malaria	• Malaria	*S. eupatoria*	Mexico
	• Antipyretic/anti-fever	*S. pilosa*	Mexico
Poison antidote	• Rubbed on snake bite	*S. serrata*	Mexico
Skin problems	• Burns/scratches	*S. elatior*	Mexico
	• Wounds	*S. lucida*	Mexico
	• Open wounds	*S. plummerae*	
	• Cuts on feet/open wounds	*S. serrata*	Mexico
Urinary problems	• Diuretic	*S. eupatoria, S. pilosa*	Mexico
	• Diuretic	*S. pilosa*	Mexico
Women's complaints	• Bath by women	*S. macbridei*	Peru
	• Dysmenorrhea	*S. nepetifolia*	Guatemala
	• Contraceptive	*S. rebaudiana*	Paraguay
	• Bath after parturition	*S. subpubescens*	Mexico

Geographically, members of *Stevia* are also important throughout the distribution range of the genus, from southern Mexico south to Paraguay. Because of the large number of species in Mexico, it is expected that many more species of *Stevia* are employed medicinally in this country than in others. It is interesting to note that one widely distributed species, *S. lucida*, a large shrub, which grows from Mexico through the Central American States to Colombia and Venezuela, has similar or related uses for the treatment of inflammation-related diseases (pains, rheumatism, inflammation, and 'bad air') in Mexico, Guatemala, Colombia and Venezuela. Such observations should present a challenge for the scientific investigation of this species for potentially interesting biologically active compounds.

Biological/pharmacological basis

Of the approximately 32 species appearing in Table 3.1, only four appear to have been subjected to biological evaluation, other than for their (potential) sweetening properties. As stated in this Chapter, 110 species of *Stevia* have been screened for the presence of sweet compounds (Soejarto *et al.* 1982). These four species are *S. eupatoria, S. rebaudiana, S. satureifolia*, and *S. serrata*. Biological evaluations on *S. eupatoria* and *S. rebaudiana* have been performed in an attempt to validate the folk medicinal claims (Flores 1907; Planas and Kuc 1968). Biological evaluations that have been performed on *S. serrata* (see Table 3.3) do not relate to their claimed ethnomedical uses, however. In the case of *S. eupatoria*, results of an early pharmacological study (Flores 1907) has been summarized and discussed in a previous paper (Soejarto *et al.* 1983b). This means that, aside from *S. eupatoria* and *S. rebaudiana*, no attempt at ethnomedical validation has been performed with other species.

In order to appreciate the scope of pharmacological studies that have been performed on these four species, a summary collated from the literature is presented in Table 3.3. This table serves to point out the scope of research that still needs to be done for the remainder of the genus.

Clearly, *S. rebaudiana* is the only species in the genus that has been investigated thoroughly in terms of botany, phytochemistry, and pharmacology (Felippe 1977; Angelucci 1981; Sakaguchi and Kan 1982; Kinghorn and Soejarto 1985). The reason for this special attention given to this species is obvious, namely, the sweet-tasting properties of the plant and its use as a commercial sweetener.

Chemical basis

Many chemical studies have been performed on many species of *Stevia*. As a result, numerous different secondary metabolites are now known to occur in plants of this genus (see Chapters 4 and 5).

This chemical diversity appears to explain the diverse folk medicinal claims in many species of *Stevia*. For example, one noticeable characteristic of *Stevia* plants is a bitter taste (Soejarto *et al.* 1983a), a property often associated with medicinal virtues. The bitter taste is associated with the presence of sesquiterpene lactones (Rodriguez *et al.* 1976; Rodriguez 1977) or certain diterpenes (Cocker 1966; Sticher 1977). The presence of sesquiterpenes appears to be characteristic in members of the Compositae (Hegnauer 1977; Herz 1977) and diverse types have been isolated from species of *Stevia* (Ríos *et al.* 1967; Salmón *et al.* 1973; Rodriguez *et al.* 1976; Salmón *et al.* 1977; Bohlmann *et al.* 1979; 1982). Recently (Mata *et al.* 1992), an extremely bitter-tasting *ent*-atisene diterpene glycoside was also reported from the roots of *S. salicifolia*. The extent to which these principles are responsible for alleged therapeutic activities in members of *Stevia* remains to be investigated.

The diversity of compounds also explains the characteristic pungent to aromatic odor present in members of *Stevia*, sometimes characterized as 'goat's smell' in certain species. Strong and aromatic odors produced by plants are due to the presence of essential oils, which possess rubefacient, counter-irritant, and anti-microbial properties (Harvey 1975; Bruneton 1995). Perhaps this is one reason for the use of many species of *Stevia* for treating infectious diseases (Table 3.2). They may also be responsible for certain other therapeutic uses, such as anti-pyretics, anti-rheumatics, and diaphoretics.

For other therapeutic activities, for which species of *Stevia* have been allegedly used, such as uterine relaxant (anti-dysmenorrheics), treatment of cardiac problems, as analgesics, to prevent hair from falling, etc., the phytochemical basis is not known at present.

Table 3.3 Studies on biological activities of species of *Stevia*

Species	Biological evaluation/Investigation[a] (Reference)
S. eupatoria Willd.	• Anti-malarial effect in human of various root preparations (oral) (Flores 1907) • Diuretic effect in human of root decoction (Flores 1907) • Diuretic effect in rabbits (Flores 1907) • Toxic effects in human (nausea, vomiting, diarrhea) of decoction (Flores 1907) • Toxicity tests in rabbits and sheep of dried plants (Flores 1907)
S. rebaudiana Bertoni[b]	• Anti-bacterial effect (Anonymous 1980; Dao and Le 1995) • Anti-caries effect (Pinheiro *et al.* 1987) • Anti-edema effect (Yasukawa *et al.* 1993) • Anti-fertility/contraceptive and/or interceptive effects (Planas and Kuc 1968) • Anti-fungal/antiyeast effect (Takaki *et al.* 1985) • Anti-hyperglycemic effect in rabbits (Suzuki *et al.* 1977; Harmaini 1986) • Bradycardia, positive inotropic and hypotensive effects (Boeckh 1981) • Cytotoxic effect (Arisawa 1994) • General toxicity effect (Suzuki *et al.* 1977; Boeckh 1981) • Hypoglycemic effect in human (Oviédo *et al.* 1970; Alvares *et al.* 1981) • Natriuretic effect (Melis 1995) • Oxidative phosphorylation inhibition effect (Bracht *et al.* 1981) • Peroxidase inhibition effect (Dao and Le 1995) • Plant root growth stimulant effect (Randi and Felippe 1981) • Protease stimulation effect (Dao and Le 1995)
S. satureifolia Sch.–Bip.	• Anti-malarial effect (Spencer *et al.* 1947)
S. serrata Cav.	• Anti-tumor activity effect (Abbott *et al.* 1966)

Notes

a Due to its voluminous nature, studies on the sweetening effect of the plant or its isolates are excluded.

b Refers to *S. rebaudiana* extracts, sweet glycoside constituents, and their derivatives. For comprehensive reviews, see Kinghorn and Soejarto 1985, and Chapter 8 of the present volume.

SUMMARY AND CONCLUSIONS

Of the approximately 220–230 species of *Stevia*, only about 34 (15%) have some type of ethno-botanical record that relates to common names and uses of the species. Of these 34 species, only one has an outstanding record of use, namely, *S. rebaudiana* (Bertoni) Bertoni, whose sweet leaves are used to impart sweetness to foods and beverages. This information, first learned by a western scientist in 1887, and recorded and communicated to the scientific community for the first time in 1899, led to research to this date on the potential use of the sweet compounds derived from this plant, stevioside and rebaudioside A, as non-nutritive high-potency sweet-eners. Today *S. rebaudiana* extracts, *ent*-kaurene glycoside constituents, and derivatives of these compounds have found commercial applications as sweeteners in several countries, and hundreds of scientific papers have been published in this area in the twentieth century.

The oldest record on the ethnobotanical relationship between *Stevia* and man is registered in the book *Natural History of Plants of the New Spain* written by Francisco Hernández, a Spanish physician, between 1570 and 1576. Unfortunately, the specific identity of *Stevia* plants in the Nahuatl language listed in this book cannot be determined with certainty. Similarly, the specific identity of a number of plants having more recent ethnobotanical records cannot be established unequivocally. Ethnobotanical data for the 34 species of *Stevia* whose specific identity is established are listed in Table 3.1. Except for *S. rebaudiana* and *S. eupatoria*, none of the ethnobotanical records referring to other species has generated any studies towards the validation of these alleged ethnomedical uses, or towards further pharmacological research in the discovery of biologically active compounds as potential therapeutic agents. Obviously, this rather dismal situation represents a challenge for action.

Based on the available evidence, some conjecture has been made as to the timeframe on the discovery of the sweetness property of *S. rebaudiana*. It is suggested that the time of discovery by the native Paraguayan populations in the Amambay region on the sweetness property of the plant was made between the sixteenth century and the year 1887, perhaps, some time after the arrival of the Spanish conquistadores in Paraguay. This knowledge only slowly filtered to the entire population, due to the rarity of the plant within its distribution range.

ACKNOWLEDGEMENTS

The original field and herbarium research that led to the present paper was supported by contract N01-NIH-DE-02425 with the National Institute of Dental Research, National Institutes of Health, Bethesda, Maryland (P.I.: A.D. Kinghorn).

The author thanks Ms. Jodi Slapcinsky, Research Assistant at the John G. Searle Herbarium, Field Museum, Chicago, who provided assistance in the search for certain data for the writing of the present paper, and Ms. Arietta Williams and Mary Lou Quinn, who generated the NA-PRALERT printouts, from which statistical and some biological data used in the paper have been derived.

The author also expresses his thanks to Dr Jacinto C. Regalado, Research Associate, University of Illinois at Chicago and the Field Museum, who provided additional ethnobotanical data on several species of *Stevia*, based on a search at the Oakes Ames Economic Herbarium of Harvard University. To the Director of the Oakes Ames Economic Herbarium, the author expresses his thanks for permission to consult the collection and for the use of data for this paper.

REFERENCES

Abbott, B. J., Leiter, J., Hartwell Jr., J. L., Caldwell, M. E., Beal, J. L., Perdue Jr., R. E. *et al.* (1966) Screening data from the Cancer Chemotherapy National Service Center Screening Laboratories. XXXIV. Plant extracts. *Cancer Research* **26**, 761–935.

Alvares, M., Bazzone, R. B., Godoy, G. L., Cury, R. and Botion, L. M. (1981) Hypoglycemic effect of *Stevia rebaudiana* Bertoni. *First Brazilian Seminar on* Stevia rebaudiana, *Instituto Technologico de Alimentos, Campinas, Brazil, June 25–26, 1981,* p. 13 (Abstracted in the NAPRALERT database, University of Illinois at Chicago, Chicago, IL).

Angelucci, E. (Coordinator) (1981) *I Seminario Brasileiro sobre* Stevia rebaudiana *Bertoni,* Campinas, Brazil, 25 e 26 de Junho de 1981. Governo do Estado de São Paulo, Secretaria de Agricultura e Abastecimiento, Coordinadoria da Pesquisa Agropecuaria, Instituto de Tecnología de Alimentos (Abstracts of Papers).

Anonymous (1893–1993) *Index Kewensis,* The Clarendon Press, Oxford, UK, vols. 1 and 2, and Supplements 1–20 (CD-ROM, 1996).

Anonymous (1929) *Catálogo de Plantas Reputadas Medicinales en la República de Guatemala* (2nd edn), Tipografía Nacional, Guatemala City, Guatemala, p. 32.

Anonymous (1980) Stevia components as sweetening agents and antibiotics. *Patent – Japan Kokai Tokkyo Koho* 80 92,323, 4 pp. (Abstracted in the NAPRALERT database, University of Illinois at Chicago, Chicago, IL).

Anonymous (1998) Paraguay: 'Bertoni Project'. *Helvetas – World Wide Web* (http://www.helvetas_projects/eparaguay_pa31.html).

Arizawa, M. (1994) Cell growth inhibition of KB cells by plant extracts. *Natural Medicine* **48**, 338–347.

Barreto, M. (1989) *El Maté: Su Historia y Cultura,* Biblioteca de Cultura Popular, Ediciones del Sol, Buenos Aires, Argentina, 123 pp.

Bejarano, R. C. (1960) *Canaí Vósú – Elementos Para El Estudio del Folklore Paraguayo,* Editorial Toledo, Asunción, Paraguay, 116 pp.

Bejarano, R. C. (1980) *Indígenas Paraguayos – Epoca Colonial,* Editorial Toledo, Asunción, Paraguay, 86 pp.

Bell, F. (1954) Stevioside: a unique sweetening agent. *Chemistry and Industry,* 17 July, pp. 897–898.

Bertoni, M. S. (1905) Le Kaá Hê-é – Sa nature et ses propriétés. *Anales Científicos Paraguayos, Serie I,* No. 5, 1–14.

Bertoni, M. S. (1914) *Plantas Usuales del Paraguay y Paises Limítrofes. Descripción Física y Económica del Paraguay* No. 31(1), Establecimiento Gráfico, M. Brossa, Asunción, Paraguay, p. 75.

Bertoni, M. S. (1918) La *Stevia rebaudiana* Bertoni. *Anales Científicos Paraguayos, Serie II,* No. 2, 129–134.

Boeckh, E. M. A. (1981) *Stevia rebaudiana* Bertoni – Cardio-circulatory effects of total aqueous extract in normal person and of stevioside in rats and frogs. *First Brazilian Seminar on* Stevia rebaudiana, *Instituto Technologico de Alimentos, Campinas, Brazil, June 25–26, 1981,* XI.1–XI.2. (Abstracted in the NAPRALERT database, University of Illinois at Chicago, Chicago, IL).

Bohlmann, F., Dutta, L. N., Dorner, W., King, R. M. and Robinson, H. (1979) Zwei neue Guajanolide sowie weitere Longipinester aus *Stevia*-Arten. *Phytochemistry* **18**, 673–675.

Bohlmann, F., Zdero, C., King, R. B. and Robinson, H. (1982) Sesquiterpenes, guaianolides and diterpenes from *Stevia myriadenia. Phytochemistry* **21**, 2021–2025.

Bracht, A., Alvares, M. and Bracht, A. M. K. (1981) Effect of water extract of *Stevia rebaudiana* Bertoni on the cellular and subcellular metabolism. *First Brazilian Seminar on* Stevia rebaudiana, *Instituto Technologico de Alimentos, Campinas, Brazil, June 25–26, 1981,* XIV.1–XIV.2. (Abstracted in the NAPRALERT database, University of Illinois at Chicago, Chicago, IL.)

Brondegaard, V. J. (1973) Contraceptive plant drugs. *Planta Medica* **23**, 167–172.

Bruneton, J. (1995) *Pharmacognosy, Phytochemistry, Medicinal Plants,* Lavoisier, Paris, pp. 422–423.

Cadogan, L. (1973) Some plants and animals in Guaraní and Guayakí Mythology. In *Paraguay – Ecological Essays,* J. R. Gorham (Ed.), Academy of the Arts and Sciences of the Americas, Miami, Florida, pp. 97–104.

Cardenas, M. (1943) *Notas Preliminares sobre La Materia Médica Boliviana. Sinopsis de la Flora Médica Boliviana,* Imprenta Universitaria, Cochabamba, Bolivia, p. 22.

Cardozo Ocampo, M. (1989) *Mundo Folklorico Paraguayo*, Editorial Cuadernos Republicanos, Asunción, Paraguay, 280 pp.

Carvalho Neto, P. (1961) *Folklore del Paraguay*, Editorial Universitaria, Quito, Ecuador.

Cocker, W. (1966) Some aspects of the chemistry of diterpene bitter principles. *Planta Medica* **14** (Suppl.), 78–85.

Cortés, S. (1919) *Flora de Colombia* (2nd edn), Libreria del Mensajero, Bogotá, pp. 153, 246, 260.

Dao, K. N. and Le, V. H. (1995) Biological properties of flavonoids from *Stevia rebaudiana* Bertoni. *Tap Chi Duoc Hoc* **2**(17), 18–21. (Abstracted in NAPRALERT database, University of Illinois at Chicago, Chicago, IL).

Del Pozo, E. C. (1979) Empiricism and magic in Aztec pharmacology. In *Ethnopharmacologic Search for Psychoactive Drugs*, D. H. Efron, B. Holmstedt, and N. S. Kline (Eds), Raven Press, New York, pp. 59–76.

Díaz, J. L. (Ed.) (1976a) *Indice y Sinonimia de las Plantas Medicinales de Mexico, Monografías Científicas I*, IME-PLAM, Mexico City, Mexico, p. 97.

Díaz, J. L. (1976b) *Usos de las Plantas Medicinales de Mexico, Monografías Científicas I*, IMEPLAM, Mexico City, Mexico, p. 97.

Farnsworth, N. R. (1996) NAPRALERT; A resource for research on medicinal plants. *ICAST 96 Proceedings (The 12th International Conference on Advanced Science and Technology)*, University of Illinois at Chicago, Chicago, 6–8 April, pp. 10–14.

Felippe, G. M. (1977) *Stevia rebaudiana* Bert.: Uma revisao. *Ciência e Cultura (São Paulo)*, **29**, 1240–1248.

Flores, L. (1907) Manual terapeutico de las plantas Mexicanas. Yerba del Borrego *Stevia eupatoria*. *Anales del Instituto Médico Nacional*, Mexico, **9**, 369–370.

González, J. N. and Ynsfrán, P. M. (1929) *El Paraguay Contemporáneo*, Editorial De Indias, Paris, France, Asunción, Paraguay, 203 pp.

Gorham, J. R. (1973) The history of natural history in Paraguay. In *Paraguay – Ecological Essays*, J. R. Gorham (Ed.), Academy of the Arts and Sciences of the Americas, Miami, Florida, pp. 1–8.

Gosling, C. (1901) Caá-êhê or azucá-caá. A letter from C. Gosling, British Consul to Asunción, Paraguay, to the Royal Botanic Gardens, Kew, United Kingdom. *Bulletin of Miscellaneous Information of the Royal Botanic Gardens, Kew* ('Kew Bulletin'), pp. 173–174.

Grashoff, J. E. (1972) *A Systematic Study of the North and Central American Species of Stevia*. Ph.D. Dissertation, University of Texas, Austin, TX.; University of Michigan Microfilm 73-7556, pp. 1–608.

Hanratty, D. M. and Meditz, S. W. (1990) *Paraguay – A Country Study* (2nd edn), The Superintendent of Documents, Washington, DC, 288 pp.

Harmaini, M. J. D. (1986) *Hypoglycemic Effect of* Stevia rebaudiana *Bertoni on Rabbits*. M.Sc. Thesis, Department of Chemistry, Faculty of Mathematics and Sciences, Institut Teknologi, Bandung, Indonesia (Abstracted in the NAPRALERT database, University of Illinois at Chicago, Chicago, IL).

Harvey, S. C. (1975) Topical drugs. In *Remington's Pharmaceutical Sciences* (15th edn), Mack Publishing Co., Easton, Pennsylvania, pp. 719–722.

Hegnauer, R. (1977) The chemistry of the Compositae. In *The Biology and Chemistry of the Compositae*, V. H. Heywood, J. B. Harborne and B. L. Turner (Eds), Academic Press, London, New York, **1**, 283–335.

Hemsley, W. B. (1906) Tabula 2816: *Stevia rebaudiana* Hemsley. In *Hooker's Icones Plantarum*, D. Prain (Ed.), Dulau and Co., London, UK (4th Ser.) **9**, pl. 2816.

Herken Krauer, J. C. (1984) *El Paraguay Rural Entre 1869 y 1913*, Centro Paraguayo de Estudios Sociológicos, Asunción, Paraguay, 224 pp.

Hernández, F. (1942) *Historia de las Plantas de Nueva España*, vol. 1. Instituto de Biología, UNAM, Imprenta Universitaria, Mexico City, Mexico.

Hernández, F. (1943) *Historia de las Plantas de Nueva España*, vol. 2. Instituto de Biología, UNAM, Imprenta Universitaria, Mexico City, Mexico.

Hernández, F. (1946) *Historia de las Plantas de Nueva España*, vol. 3. Instituto de Biología, UNAM, Imprenta Universitaria, Mexico City, Mexico.

Hernández, F. (1959) *Obras Completas Tomo II: Historia Natural de Nueva España*, vol. 1. Universidad Nacional de Mexico, Mexico City, Mexico.

Herrera, F. L. (1930) *Estudios sobre La Flora del Departamento de Cuzco*, Sanmartí y Compañia, Lima, Peru, pp. 187, 227.

Herrera, F. L. (1939) *Catálogo Alfabético de los Nombres Vulgares y Científicos de Plantas que Existen en el Peru*, Universidad Mayor de San Marcos, Lima, Peru.

Herz, W. (1977) Sesquiterpene lactones in the Compositae. In *The Biology and Chemistry of the Compositae*, V. H. Heywood, J. B. Harborne and B. L. Turner (Eds), Academic Press, London, New York, 1, 337–357.

Kinghorn, A. D., Soejarto, D. D., Nanayakkara, N. P. D., Compadre, C. M., Makapugay, H. C., Hovanec-Brown, J. M. *et al*. (1984) A phytochemical screening procedure for sweet *ent*-kaurene glycosides in the genus *Stevia* (Compositae). *Journal of Natural Products* **46**, 439–444.

Kinghorn, A. D. and Soejarto, D. D. (1985) Current status of stevioside as a sweetening agent for human use. In *Economic and Medicinal Plant Research*, H. Wagner, H. Hikino, and N. R. Farnsworth (Eds), Academic Press, New York, **1**, 1–52.

Kinghorn, A. D. and Soejarto, D. D. (1986) Sweetening agents of plant origin. *Critical Reviews in Plant Sciences* **4**, 79–120.

Kleinpenning, J. M. G. (1987) *Man and Land in Paraguay*, CEDLA, Amsterdam, The Netherlands.

Kleinpenning, J. M. G. (1992) *Rural Paraguay, 1870–1932*, CEDLA, Amsterdam, The Netherlands.

Latorre, D. L. and Latorre, F. A. (1977) Plants used by the Mexican Kickapoo Indians. *Economic Botany* **31**, 340–357.

Lewis, W. H. (1992) Early uses of *Stevia rebaudiana* (Asteraceae) leaves as a sweetener in Paraguay. *Economic Botany* **46**, 336–340.

Lipp, F. J. (1971) Ethnobotany of the Chinantec Indians, Oaxaca, Mexico. *Economic Botany* **25**, 234–244.

Mabberley, D. J. (1996) *The Plant-Book – A Portable Dictionary of the Higher Plants*, Cambridge University Press, Cambridge, UK, p. 291.

Martínez, M. (1969) *Las Plantas Medicinales de Mexico* (5th edn), Ediciones Botas, Mexico City, Mexico, pp. 163, 476.

Martínez, M. (1979) *Catálogo de Nombres Vulgares y Científicos de Plantas Mexicanas*, Fondo de Cultura Económica, Mexico City, Mexico.

Mata, R., Rodríguez, V., Pereda-Miranda, R., Bye, R. and Linares, E. (1991) A dammarane from *Stevia salicifolia*. *Phytochemistry* **30**, 3822–2823.

Mata, R., Rodríguez, V., Pereda-Miranda, R., Kaneda, N. and Kinghorn, A. D. (1992) Stevisalioside A, a new *ent*-atisene glycoside from the roots of *Stevia salicifolia*. *Journal of Natural Products* **55**, 660–666.

Melis, M. S. (1995) Chronic administration of aqueous extract of *Stevia rebaudiana* in rats: Renal effects. *Journal of Ethnopharmacology* **47**, 129–134.

Miquel, O. (1966) Un nuevo hipoglicemiante oral. *Revista Médica Paraguaya* **7**, 200.

Mors, W. B. and Rizzini, C. T. (1966) *Useful Plants of Brazil*, Holden-Day, Inc., San Francisco, CA, p. 93.

Morton, J. F. (1975) Current folk remedies of northern Venezuela. *Quarterly Journal of Crude Drug Research*. **13**, 97–121.

Oviédo, C. A. G., Fronciani, R. and Maas, L. C. (1970) Acción hipoglicemiante de la *Stevia rebaudiana* Bertoni (Kaá-Hê-ê). *Abstracts, Seventh Congress of the International Diabetes Federation, Buenos Aires, Argentina*, 23–24 August.

Pérez-Arbeláez, E. (1936) *Plantas Utiles de Colombia*, Imprenta Nacional, Bogotá, Colombia, Tomo I, p. 60.

Pérez-Arbeláez, E. (1937) *Plantas Medicinales y Venenosas de Colombia*, Published privately, Bogotá, Colombia, p. 280.

Pinheiro, C. E., Oliveira, S. S., Da Silva, S. M. B., Poletto, M. I. F. and Pinheiro, G. J. (1987) Effect of guaraná and *Stevia rebaudiana* Bertoni (leaves) extracts, and stevioside on the fermentation and synthesis of extracellular insoluble polysaccharides of dental plaque. *Revista Odontologica de la Universidade de São Paulo* **1**, 9–13.

Pio Corrêa, M. (1926) *Diccionario das Plantas Uteis do Brasil e das Exoticas Cultivadas*, Imprensa Nacional, Rio de Janeiro, Brazil, **1**, 348.

Pio Corrêa, M. and de Azeredo Pena, L. (1952) *Diccionario das Plantas Uteis do Brasil*, Ministerio de Agricultura, Rio de Janeiro, Brazil, **3**, 518.

Pittier, H. (1926) *Manual de las Plantas Usuales de Venezuela*, Litografía del Comercio, Caracas, Venezuela, pp. 209, 300.

Planas, G. M. and Kuc, S. (1968) Contraceptive properties of *Stevia rebaudiana*. *Science* **162**, 1007.

Randi, A. M. and Felippe, G. M. (1981) Substances promoting root growth from the achenes of *Stevia rebaudiana* Bertoni. *Revista Brasiliana de Botanica* **4**, 49–51.

Reed, R. K. (1995) *Prophets of Agroforestry: Guaraní Communities and Commercial Gathering*, University of Texas Press, Austin, TX, 251 pp.

Regalado, J. C. (1998) Personal communication to D. D. Soejarto, June 29, 1998. Dr J. C. Regalado, Botany Department, Field Museum, Roosevelt Road at Lake Shore Drive, Chicago, IL 60605, USA.

Ríos, T., Romo de Vivar, A. and Romo, J. (1967) Stevin, a new pseudoguaianolide isolated from *Stevia rhombifolia* HBK. *Tetrahedron* **23**, 4265–4269.

Rivarola Paoli, J. B. (1986) *La Economia Colonial*, J. Bautista, Asunción, Paraguay.

Robinson, B. L. (1930a) Observations on the genus *Stevia*. *Contributions from the Gray Herbarium, Harvard University* **90**, 36–58.

Robinson, B. L. (1930b) Stevias of Paraguay. *Contributions from the Gray Herbarium, Harvard University* **90**, 79–90.

Robinson, B. L. (1932) The Stevias of Bolivia. *Contributions from the Gray Herbarium, Harvard University* **100**, 37–69.

Rodríguez, E., Towers, G. H. N. and Mitchell, J. C. (1976) Biological activities of sesquiterpene lactones. *Phytochemistry* **15**, 1573–1580.

Rodríguez, E. (1977) Sesquiterpene lactones: chemotaxonomy, biological activity and isolation. *Revista Latinoamericana de Química* **8**, 56–62.

Roig y Mesa, J. T. (1953) *Diccionario Botánico de Nombres Vulgares Cubanos*, Seoane Fernández y Cia, Havana, Cuba, **1**, 381.

Sahagún, F. B. de (1961) *Florentine Codex: General History of the Things of New Spain*, A. J. D. Anderson and C. E. Dibble (Eds) (Transls.), Monograph No. 14, part 11, The School of American Research and the University of Utah, Santa Fé, New Mexico, p. 152.

Sahagún, F. B. de (1963) *Florentine Codex: General History of the Things of New Spain*, A. J. D. Anderson and C. E. Dibble (Eds) and (Transls.), Monograph No. 14, part 12, The School of American Research and the University of Utah, Santa Fé, New Mexico, pp. 167, 178, 197.

Sakaguchi, M. and Kan, T. (1982) As pesquisas Japonesas com *Stevia rebaudiana* (Bert.) Bertoni e o estevosídeo. *Ciencia é Cultura (São Paulo)* **34**, 235–248.

Salmón, M., Díaz, E. and Ortego, A. (1973) Christinine, a new epoxyguaianolide from *Stevia serrata* Cav. *Journal of Organic Chemistry* **38**, 1759–1760.

Salmón, M., Díaz, E. and Ortego, A. (1977) Epoxilactonas de *Stevia serrata* Cav. *Revista Latinoamericana de Química* **8**, 172–175.

Schultes, R. E. (1962) The role of ethnobotanist in the search for new medicinal plants. *Lloydia* **25**, 257–266.

Schuster, A. N. (1929) *Paraguay: Land, Volk, Geschichte, Wirtschaftsleben und Kolonisation*, Strecker und Schröder, Verlag, Stuttgart, Germany, 667 pp.

Sessé, M. and Mociño, J. M. (1893) *Plantae Novae Hispaniae*, Oficina Tipográfica de la Secretaria de Fomento, Mexico City, Mexico, p. 120.

Soejarto, D. D., Kinghorn, A. D. and Farnsworth, N. (1982) Potential sweetening agents of plant origin. III. Organoleptic evaluation of *Stevia* leaf herbarium samples for sweetness. *Journal of Natural Products* **45**, 590–599.

Soejarto, D. D., Compadre, C. M., Medon, P. J., Kamath, S. K. and Kinghorn, A. D. (1983a) Potential sweetening agents of plant origin II. Field search for sweet-tasting *Stevia* species. *Economic Botany* **37**, 71–78.

Soejarto, D. D., Compadre, C. M. and Kinghorn, A. D. (1983b) Ethnobotanical notes on *Stevia*. *Botanical Museum Leaflets, Harvard University* **29**, 1–25.

Soukup, J. (1970) *Vocabulario de los Nombres Vulgares de la Flora Peruana*, Colegio Salesiano, Lima, Peru, p. 329.

Spencer, C. F., Koniuszy, F. R., Rogers, E. F., Shavel, J. R., Easton, N. R., Kaczka, E. A. *et al.* (1947) Survey of plants for antimalarial activity. *Lloydia* **10**, 145–175.

Sticher, O. (1977) Plant mono-, di- and sesquiterpenoids with pharmacological or therapeutical activity. In *New Natural Products and Plant Drugs with Pharmacological, Biological or Therapeutical Activity*, H. Wagner and P. Wolff (Eds), Springer-Verlag, Berlin, New York, pp. 137–176.

Sumida, T. (1973) Reports on *Stevia rebaudiana* Bertoni introduced from Brazil as a new sweetness resource in Japan. *Miscellaneous Publications of the Hokkaido National Agricultural Experimental Station* No. 2, 69–83 (p. 81) (In Japanese, with a shorter English text).

Suzuki, H., Kasai, T., Sumihara, M. and Suginawa, H. (1977) Influence of the oral administration of stevioside on the levels of blood glucose and liver glycogen in intact rats. *Nogyo Kagaku Zasshi* **51**(3), 45. (Abstracted in the NAPRALERT database, University of Illinois at Chicago, Chicago, IL).

Takaki, M., de Campos Takaki, G. M., De Santana Diu, M. B., de Andrade, M. S. S. and Da Silva, E. C. (1985) Antimicrobial activity in leaf extract of *Stevia rebaudiana* Bertoni. *Revista de Instituto Antibiotico da Universidade Federal de Pernambuco Recife* **22**(1/2), 33–39.

Urbina, M. (1906) *Raices Comestibles entre Los Antiguos Mexicanos*, Imprenta del Museo Nacional, Mexico City, Mexico, p. 154.

Velasco-Neguerela, A., Pérez-Alonso, M. J. and Esenarro Abarca, G. (1995) Medicinal plants from Pampallakta: An Andean community in Cuzco (Peru). *Fitoterapia* **66**, 447–462.

Villanueva, A. (1967) *El Lenguaje del Mate*, Editorial Paídós, Buenos Aires, Argentina, pp. 1–119.

von Reis Altschul, S. (1973) *Drugs and Foods from Little Known Plants*, Harvard University Press, Cambridge, MA, pp. 298–299.

von Schmeling, G. M. (1967) Caá-Heê: Edulcorante natural nao calórico (*Stevia rebaudiana*). *Boletim do Sanatório Sao Lucas* **29**, 67–78.

von Schmeling, G. A., de Carvalho, F. V. and Domingos Espinosa, A. D. (1977) *Stevia rebaudiana* Bertoni. Avaliaçao do efeito hipoglicemiante em coelhos aloxanizados. *Ciencia é Cultura*, São Paulo, **29**, 599–601.

Washburn, C. A. (1871) *History of Paraguay*, vols. 1 and 2, Boston, Lee and Shepard, Publishers, Boston, MA.

Williams, J. H. (1979) *The Rise and Fall of the Paraguayan Republic 1800–1870*, Institute of Latin American Studies, University of Texas at Austin, Austin, TX.

Yasukawa, K., Yamaguchi, A., Arita, J., Sakurai, S., Ikeda, A. and Takido, M. (1993) Inhibitory effect of edible plant extracts on 12-*O*-tetradecanoylphorbol-13-acetate-induced ear oedema in mice. *Phytotherapy Research* **7**, 185–189.

Zamora-Martínez, M. C. and Pascual-Pola, C. N. (1992) Medicinal plants used in some rural populations of Oaxaca, Puebla and Veracruz, Mexico. *Journal of Ethnopharmacology* **35**, 229–257.

4 Sweet and non-sweet constituents of *Stevia rebaudiana*

Edward J. Kennelly

INTRODUCTION

Stevia rebaudiana has been studied in depth because this plant is the source of several well-known sweet-tasting compounds. Interest in using *S. rebaudiana* as a commercial sweetener, especially by the Japanese food industry, has led to extensive phytochemical investigations of the herb's constituents. To date more than 100 compounds have been identified from this species. The best known of these are the sweet-tasting *ent*-kaurene diterpenoid glycosides, particularly stevioside and rebaudioside A. Various review articles have appeared regarding the constituents of *S. rebaudiana*. Many review articles on the constituents of *S. rebaudiana* have been written in Japanese (Abe and Sonobe 1977; Morita 1977; Okazaki *et al*. 1977; Tanaka 1987; Yoshihira *et al*. 1987). Among the more comprehensive reviews in English of the *Stevia* constituents include those by Kinghorn and Soejarto (1985) and Hanson and De Oliveira (1993). A brief review of *Stevia* constituents by Crammer and Ikan (1986), and a more extensive treatment on the rebaudiosides (Crammer and Ikan 1987), have been published. Many other reviews of *S. rebaudiana* discuss the sweet-tasting constituents specifically, such as those by Tanaka (1980; 1982), Salvatore *et al*. (1984), Bakal and O'Brien Nabors (1986), Phillips (1987) and Kinghorn and Soejarto (1991). There are additional reviews written in Chinese (Wu 1987), German (Seidemann 1976), and Italian (Toffler and Orio 1981). Of historical interest are the early reviews of *Stevia* by Kobert (1915), Bell (1954), Fletcher (1955) and Jacobs (1955) that were written before the ultimate structural elucidation of stevioside.

In this chapter, the sweet and non-sweet chemical constituents of *S. rebaudiana*, as well as compounds obtained by tissue culture techniques, will be discussed. Summaries of the compounds obtained from *S. rebaudiana* are shown in Tables 4.1 and 4.2. Figures 4.1–4.5 show the structures of selected diterpenoid, triterpenoid, sterol, and flavonoid constituents of *S. rebaudiana*, and also certain derivatives of these compounds.

DITERPENOIDS

ent-Kaurene

Although a number of natural products have been isolated from *S. rebaudiana*, the best known are the diterpenoids, specifically the sweet-tasting *ent*-kaurene glycosides (Figure 4.1). Six naturally occurring *ent*-kaurene glycosides have been described from *S. rebaudiana*, comprising stevioside, rebaudiosides A and C–E, and dulcoside A (1–6), with stevioside (1) being the most abundant sweet-tasting compound in the leaves. Early phytochemical work on *S. rebaudiana*

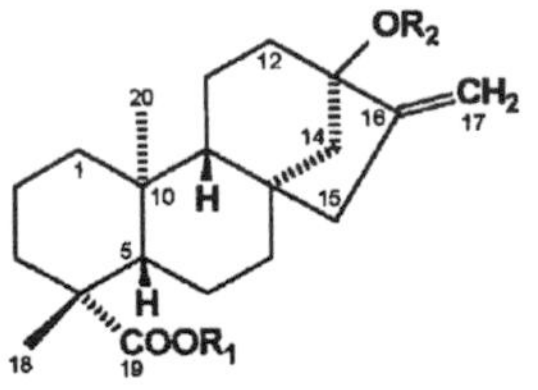

	R₁	R₂

(structure diagram with labels OR₂, CH₂, 17, 16, 14, 15, 12, 20, 1, 10, H, 5, H, 18, COOR₁, 19)

		R$_1$	R$_2$
1	Stevioside	Glc	Glc-Glc (2→1)
2	Rebaudioside A	Glc	Glc-Glc (2→1) | Glc (3→1)
3	Rebaudioside C (= Dulcoside B)	Glc	Glc-Rha (2→1) | Glc (3→1)
4	Rebaudioside D	Glc-Glc (2→1)	Glc-Glc (2→1) | Glc (3→1)
5	Rebaudioside E	Glc-Glc (2→1)	Glc-Glc (2→1)
6	Dulcoside A	Glc	Glc-Rha (2→1)

Glc = β-D-glucopyranosyl; rha = α-L-rhamnopyranosyl

Figure 4.1 Structures of sweet-tasting *ent*-kaurene diterpenes isolated from *Stevia rebaudiana*.

mainly focused on the isolation and structural determination of the sweet-tasting constituents. The structural elucidation of these sweet-tasting compounds, however, was not a straightforward process, and spanned more than 50 years in the chemical literature. Bertoni (1905) reported preliminary findings on the sweet-tasting constituents of *S. rebaudiana*. Dieterich (1908) isolated two sweet-tasting compounds, the crystalline eupatorin and the non-crystalline rebaudin, from an aqueous extraction of the powdered leaves and stems of *S. rebaudiana*. Through hydrolysis experiments, he found both of those compounds to be glycosides, with a sweetness of 'about 150–180 times greater than sucrose.' The French researchers Bridel and Lavieille (1931f) determined that the non-crystalline rebaudin was a mixture of eupatorin with various organic impurities. Eupatorin was later renamed stevioside because of the change in nomenclature of the plant from *Eupatorium rebaudianum* to *S. rebaudiana* (Bertoni 1918). Rasenack (1908) discussed the isolation of a sweet-tasting compound from *S. rebaudiana* with a probable molecular formula of $C_{42}H_{72}O_{21}$, and also noted the presence of uncharacterized tannins and resins in a crude extract.

In the early 1930s, Bridel and Lavieille (1931a–h) published an important series of papers on the chemical constituents of *S. rebaudiana*. They obtained the crystalline glycoside, stevioside (**1**), from an alcoholic extract of *S. rebaudiana*, and found it to be 300 times the sweetness of sucrose. Through hydrolysis experiments, these workers demonstrated stevioside to be a glycoside containing three sugar units and identified each as D-glucose, and proposed the following equation for the hydrolysis reaction (Bridel and Lavieille 1931e):

$$C_{38}H_{60}O_{18} + 3H_2O = C_{20}H_{30}O_3 + 3C_6H_{12}O_6$$

From their hydrolysis experiments, Bridel and Lavieille (1931d,e) showed that an enzyme-catalyzed hydrolysis of stevioside (1), using a gut extract of the vineyard snail *Helix pomatia*, yielded the aglycone steviol (7), whereas the acid-catalyzed hydrolysis of stevioside yielded isosteviol (8). Steviol (7) and isosteviol (8) were noted to be weakly acidic. Furthermore, Bridel and Lavieille found that alkaline solutions of steviol (7) and isosteviol (8) could be precipitated with carbon dioxide, which led them to hypothesize that stevioside (1) contained either a phenolic or acidic hydroxyl group. Since stevioside itself was not titratable, they concluded that the sugar unit was attached through the acidic functional group. They were not able, however, to determine fully the chemical structures of stevioside, steviol, or isosteviol. On the basis of Bridel and Lavieille's data, another researcher speculated erroneously that steviol and isosteviol were phytol derivatives (Bell 1954).

A research group at the National Institute of Arthritis and Metabolic Diseases, National Institutes of Health, Bethesda, Maryland published a series of papers that ultimately described the correct structure of stevioside (1) (Mosettig and Nes 1955; Wood *et al.* 1955; Vis and Fletcher 1956; Wood and Fletcher 1956; Dolder *et al.* 1960; Djerassi *et al.* 1961; Mosettig *et al.* 1961; Mosettig *et al.* 1963). Mosettig and Nes (1955) determined that steviol (7) and isosteviol (8) were diterpenoid acids with the former containing a 2,11-cyclopentanoperhydrophenanthrene skeleton, a hydroxy acid, and a terminal methylene group (Figure 4.2). They also found that isosteviol (8) could undergo Wolff-Kishner reduction to give isostevic acid, which in turn could be reduced to the hydrocarbon isostevane. Through examination of an analogue of stevioside, the ester-linked sugar was proposed to be a β-D-glucopyranosyl group (Wood and Fletcher 1956), whereas the other sugar units of stevioside (1) were shown to be β-sophorose [2-*O*-(β-D-glucopyranosyl)-D-glucose] (Vis and Fletcher 1956). Finally, with the absolute configuration of steviol (7) being determined by degradation reactions along with optical rotary dispersion and other physical measurements, the unambiguous structural elucidation of stevioside (1) was completed (Mosettig *et al.* 1963).

Stevioside (1) is the major sweet-tasting glycoside in *S. rebaudiana*, and has been reported to be 250–300 times sweeter than sucrose (Crammer and Ikan 1986). The yield of stevioside from dried leaves of *S. rebaudiana* can vary greatly, from about 5–22% of the weight of dry leaves, depending upon the cultivar and growing conditions (Kim and Dubois 1991). Stevioside has also been found in the flowers of *S. rebaudiana* at lower concentrations [0.9% (w/w)]

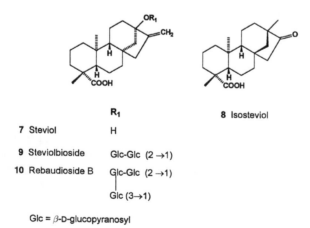

	R_1
7 Steviol	H
9 Steviolbioside	Glc-Glc (2 →1)
10 Rebaudioside B	Glc-Glc (2 →1)
	Glc (3→1)

8 Isosteviol

Glc = β-D-glucopyranosyl

Figure 4.2 Structures of derivatives of the *Stevia rebaudiana* sweet *ent*-kaurene diterpenoid constituents.

(Darise *et al*. 1983). Synonyms for stevioside (**1**) include eupatorin, rebaudin, stevin, and steviosin (Buckingham 1997). Isosteviol (**8**) has been shown to be a Wagner-Meerwein rearrangement product involving inversion of the D-ring of steviol (**7**), that can be acid catalyzed (Hanson and De Oliveira 1993). Saponification of stevioside (**1**) with strong base yields steviolbioside (**9**). Although steviolbioside has been identified in certain *S. rebaudiana* extracts, it is generally thought to be an artifact of extraction and/or isolation procedures rather than a naturally occurring glycoside (Kim and Dubois 1991).

In the 1970s further investigations of the sweet and non-sweet constituents of *S. rebaudiana* were conducted, with important contributions from certain Japanese laboratories, especially by Tanaka and his coworkers at Hiroshima University as summarized elsewhere in this volume. Eight sweet-tasting glycosides from *S. rebaudiana* contain a common aglycone called steviol (13-hydroxy-*ent*-kaur-16-en-19-oic acid) (**7**), and differ only in the glycosidic constituents attached at C-13 and/or C-19. Tanaka and co-workers reported the isolation of the novel *ent*-kaurenes rebaudioside A (**2**) and rebaudioside B (**10**) (Kohda *et al*. 1976). Rebaudioside A was obtained from a methanolic extraction of the leaves of *S. rebaudiana*, with a yield of 1.4% (w/w) (see Table 4.1). Hydrolysis of rebaudioside A with acid gave the known artifact isosteviol (**8**), while hydrolysis of rebaudioside A with hesperidinase yielded a partial hydrolysis product which was identified as rebaudioside B (**10**) (Kohda *et al*. 1976). Identification of steviol (**7**) as the genuine aglycone for both rebaudiosides A (**2**) and B (**10**) was achieved almost entirely by ^{13}C-NMR measurements (Kohda *et al*. 1976). The glycosidic portions of rebaudiosides A (**2**) and B (**10**) were determined through comparison of anomeric ^{13}C-NMR signals with those of stevioside (**1**) and steviolbioside (**9**), together with an alkaline saponification of rebaudioside A to rebaudioside B and MS analysis of the methylation product of rebaudioside B (Kohda *et al*. 1976). It is generally believed that rebaudioside B is formed from the partial hydrolysis of rebaudioside A during extraction (Kobayashi *et al*. 1977). Rebaudioside A (**2**) is the sweetest of the *ent*-kaurene glycosides isolated from *S. rebaudiana* to date, being approximately 350–450 times sweeter than sucrose, while rebaudioside B (**10**) is approximately 300–350 times sweeter than sucrose (Crammer and Ikan 1986). Rebaudioside A is the second most abundant *ent*-kaurene found in the leaves of *S. rebaudiana*, with yields approximately 25 to 54% the expected yield of stevioside from the dried leaves (Crammer and Ikan 1987). Rebaudioside A is more pleasant tasting and more water soluble than stevioside, and therefore it is better suited for use in food and beverages (Crammer and Ikan 1987). Rebaudioside A has also been identified in the flowers of *S. rebaudiana* at low concentrations, 0.15% (w/w) (Darise *et al*. 1983). Stevioside A$_3$ is a synonym for rebaudioside A, and stevioside A$_4$ is a synonym for rebaudioside B (Buckingham 1997).

From a methanol extraction of the leaves of *S. rebaudiana*, three minor *ent*-kaurene glycosides, rebaudiosides C–E (**3–5**) were isolated also by Tanaka and coworkers (Sakamoto *et al*. 1977a,b), having yields of 0.4%, 0.03%, and 0.03% w/w, respectively (see Table 4.1). The methanolic extract was recrystallized to remove stevioside (**1**), and the remaining extract subjected to repeated chromatography over silica gel to yield rebaudiosides C–E (**3–5**) (Sakamoto *et al*. 1977a). The structural elucidation of rebaudioside C (**3**) utilized ^{13}C-NMR techniques, including a partially relaxed Fourier transform method, to aid in the identification of the carbon resonances of individual sugar units (Sakamoto *et al*. 1977a). Chemical derivatizations of rebaudioside C (**3**), including enzymatic hydrolysis of the sugar units, saponification of the ester-linked sugars, and permethylation and methanolysis, were used to help in identification (Sakamoto *et al*. 1977a). The structural elucidation of rebaudiosides D (**4**) and E (**5**) also used ^{13}C-NMR experiments, as well as enzyme hydrolysis of the sugar units and alkaline saponification of the ester-linked sugars (Sakamoto *et al*. 1977b). The β-sophorosyl group [(2-*O*-(β-glucopyranosyl)-β-glucopyranosyl)] of rebaudioside D was determined by preparation of the β-sophorosyl ester

Table 4.1 Constituents of *Stevia rebaudiana*[a]

Compound Class (subclass)	Constituent (figure)	Plant part	% (w/w) Yield	References[b]
Diterpenoid				
ent-Kaurene	Dulcoside A (6)	Leaves	0.03	Kobayashi *et al.* 1977
	(−)-Kaurene[c]	Stem	ns[d]	Hanson and White 1968
	Rebaudioside A (2)	Leaves	1.43	Kohda *et al.* 1976
		Flowers	0.15	Darise *et al.* 1983
		Culture	ns	Hsing *et al.* 1983
	Rebaudioside B (10)[e]	Leaves	0.44	Kohda *et al.* 1976
	Rebaudioside C (3) (= dulcoside B)	Leaves	0.4	Sakamoto *et al.* 1977a
	Rebaudioside D (4)	Leaves	0.03	Sakamoto *et al.* 1977b
	Rebaudioside E (5)	Leaves	0.03	Sakamoto *et al.* 1977b
	Steviolbioside (9)[e]	Leaves	0.04	Kohda *et al.* 1976
	Stevioside (1)	Leaves	2.18	Kohda *et al.* 1976
		Flowers	0.92	Darise *et al.* 1983
		Culture	ns	Lee *et al.* 1982
Labdane	Austroinulin (12)	Leaves	0.06	Sholichin *et al.* 1980
		Flowers	0.21	Darise *et al.* 1983
	6-*O*-Acetylaustroinulin (13)	Leaves	0.15	Sholichin *et al.* 1980
		Flowers	0.18	Darise *et al.* 1983
	7-*O*-Acetylaustroinulin (14)	Flowers	0.08	Darise *et al.* 1983
	Jhanol (11)	Leaves	0.006	Sholichin *et al.* 1980
		Flowers	0.04	Darise *et al.* 1983
	Sterebin A (15)	Leaves	0.001	Oshima *et al.* 1986
	Sterebin B (16)	Leaves	0.0009	Oshima *et al.* 1986
	Sterebin C (17)	Leaves	0.0003	Oshima *et al.* 1986
	Sterebin D (18)	Leaves	0.0004	Oshima *et al.* 1988
	Sterebin E (19)	Leaves	0.002	Oshima *et al.* 1988
	Sterebin F (20)	Leaves	0.003	Oshima *et al.* 1988
	Sterebin G (21)	Leaves	0.0002	Oshima *et al.* 1988
	Sterebin H (22)	Leaves	0.0002	Oshima *et al.* 1988

Other	Gibberellin A$_{20}$	Leaves/Stems	0.0000001[f]	Alves and Ruddat 1979
Essential Oil	See Table 4.2			
Flavonoid	Apigenin 4'-O-glucoside (30)	Leaves	0.01	Rajbhandari and Roberts 1983
	Apigenin 7-O-glucoside (cosmosiin) (37)	Leaves	ns	Matsuo et al. 1986
	Kaempferol 3-O-rhamnoside (31)	Leaves	0.008	Rajbhandari and Roberts 1983
	Luteolin 7-O-glucoside (32)	Leaves	0.009	Rajbhandari and Roberts 1983
	Quercetin 3-O-arabinose (33)	Leaves	ns	Rajbhandari and Roberts 1983
	Quercetin 3-O-glucoside (34)	Leaves	ns	Rajbhandari and Roberts 1983
	Quercetin 3-O-rhamnoside (quercetrin) (35)	Culture	ns	Rajbhandari and Roberts 1983
	Quercetin 3-O rutinoside (rutin) (38)	Leaves	0.0007[f]	Suzuki et al. 1976
	5,7,3'-Trihydroxy 3,6,4'-trimethoxyflavone (centaureidin) (36)	Leaves	0.01	Rajbhandari and Roberts 1983
Sterol	Campesterol (27)	ns	ns	D'Agostino et al. 1984
	β-Sitosterol (25)	ns	ns	Sholichin et al.1980
	β-Sitosterol β-D-glucoside (28)	ns	ns	Matsuo et al. 1986
	Stigmasterol (26)	Leaves	ns	Nabeta et al. 1976
		Culture	0.001	Nabeta et al. 1976
	Stigmasterol β-D-glucoside (29)	Leaves	ns	Matsuo et al. 1986
Triterpenoid	β-Amyrin acetate (23)	Leaves	ns	Sholichin et al. 1980
	Lupeol	Leaves	ns	Sholichin et al. 1980
	Lupeol 3-palmitate (24)	ns	0.1	Yasukawa et al. 1993
	Lupeol esters[g]	Leaves	ns	Sholichin et al. 1980
Miscellaneous Organic	β-Carotene	ns	ns	Cheng and Chang 1983
	Chlorophyll A	ns	ns	Cheng and Chang 1983
	Chlorophyll B	ns	ns	Cheng and Chang 1983
	Citric acid	ns	0.535	Cheng and Chang 1983
	Formic acid	ns	0.871	Cheng and Chang 1983
	Glucose	Leaves	ns	Viana and Metivier 1980
	Indole-3-acetonitrile	Seeds	ns	Randi and Felippe 1981
	Lactic acid	ns	0.260	Cheng and Chang 1983
	Malic acid	ns	0.188	Cheng and Chang 1983
	Succinic acid	ns	0.400	Cheng and Chang 1983
	Sucrose	Leaves	ns	Viana and Metivier 1980
	Tannins[g]	Leaves	7.8	Rasenack 1908
	Tartaric acid	ns	1.715	Cheng and Chang 1983

Table 4.1 Continued

Compound Class (subclass)	Constituent (figure)	Plant part	% (w/w) Yield	References[b]
Inorganic	Calcium	ns	0.210	Cheng and Chang 1983
	Fluoride	Leaves	0.0012	Sakai et al. 1985
	Iron	ns	0.085	Cheng and Chang 1983
	Magnesium	ns	0.289	Cheng and Chang 1983
	Phosphorus	ns	0.098	Cheng and Chang 1983
	Potassium	ns	2.60	Cheng and Chang 1983
	Sodium	ns	0.031	Cheng and Chang 1983
	Zinc	ns	0.0026	Cheng and Chang 1983

Notes

a Table modified and expanded from Kinghorn and Soejarto (1985).
b In cases of multiple citations for a compound, the first reference containing the correct structure is given.
c Obtained from feeding cut stems of S. rebaudiana with DL-sodium mevalonate-2-^{14}C.
d Not stated in original publication.
e Thought to be an artifact of extraction.
f Weight/fresh weight.
g Chemical structures not determined.

of *ent*-kaur-16-en-19-oic acid, and subsequent comparison of the ^{13}C-NMR chemical shifts, and the coupling constant of the ester anomeric carbons (Sakamoto *et al.* 1977b). The structure of rebaudioside D was confirmed ultimately by its preparation from rebaudioside B (**10**), and rebaudioside E (**5**) was confirmed by its preparation from steviolbioside (**9**) (Sakamoto *et al.* 1977b).

In 1977 Kobayashi and coworkers at Hokkaido University reported the isolation of two minor *ent*-kaurene glycosides from *S. rebaudiana*, which they called dulcosides A (**6**) and B (**3**), and noted both to taste moderately sweet (Kobayashi *et al.* 1977). The structural elucidation of dulcoside A used spectroscopic methods, as well as chemical derivation, including methylation of dulcoside A to give its permethylate, that helped to establish the residue at C-13 as 2-*O*-rhamnosyl $(1 \rightarrow 2)$ glucose (also called neohesperidose) (Kobayashi *et al.* 1977). The structure of dulcoside B (**3**) was reported as 19-*O*-β-glucopyranosyl-13-*O*-[α-rhamnopyranosyl $(1 \rightarrow 2)$-β-glucopyranosyl $(1 \rightarrow 3)$]-β-glucopyranosylsteviol (Kobayashi *et al.* 1977), and is identical to the previously reported rebaudioside C (Sakamoto *et al.* 1977b). It has been noted that stevioside (**1**) and rebaudioside A (**2**) correspond to dulcoside A and rebaudioside C, respectively, with glucosyl groups of the former replaced in the latter by rhamnosyl groups (Kobayashi *et al.* 1977). This rhamnosyl substitution decreases the sweetness intensity of the steviol glycosides, with dulcoside A and rebaudioside C being only about one tenth as sweet as that of stevioside or rebaudioside A, respectively (Kobayashi *et al.* 1977).

The presence of the sweet-tasting *ent*-kaurene glycosides has been reported in the leaves, stems, and flowers of *S. rebaudiana*, but not in its roots (Tanaka 1982). Although other *Stevia* species contain *ent*-kaurene diterpenoids, only one other species, *S. phlebophylla* A. Gray has been found to contain sweet-tasting steviol glycosides (Kinghorn *et al.* 1984). This Mexican species, identified in a systematic screen of herbarium voucher specimens representing 110 *Stevia* species, is now thought to be extinct (Kinghorn *et al.* 1984). Other steviol glycosides, identified in the genus *Rubus* of the Rosaceae, are mentioned in Chapter 7. Due to the economic importance of the sweet-tasting *ent*-kaurene glycosides, considerable effort has been made to produce these compounds synthetically (see Chapter 6). Also, many studies have been conducted on *S. rebaudiana*, both *in vivo* and *in vitro*, in attempts to increase *ent*-kaurene glycoside yields. *Stevia rebaudiana* is a short-day perennial composite, which when grown in long-day conditions, has increased levels of stevioside (**1**) (Metivier and Viana 1979). Cheng *et al.* (1981) examined post-harvest levels of stevioside and rebaudioside A (**2**), and found that if *S. rebaudiana* is air dried for four to seven days to a moisture content of 15–20%, stevioside and rebaudioside content decrease significantly. The optimum temperature for the production of stevioside in greenhouse-cultivated *S. rebaudiana* plants has been reported to be 25°/20°C (day/night) (Mizukami *et al.* 1983). Seed propagation of *S. rebaudiana* plants results in a great variation in the levels of steviol glycosides (Tamura *et al.* 1984a). This phenomenon has been attributed to the fact that *S. rebaudiana* is a self-incompatible plant, with steviol glycoside variation due to gene segregation (Tamura *et al.* 1984a). Therefore, attempts have been made to produce homogeneous populations of *S. rebaudiana* by propagation from shoot tip cultures (Tamura *et al.* 1984a,b) and leaf explants (Ferreira and Handro 1987). Some success in producing uniform yields of stevioside (**1**) and rebaudioside A (**2**) with clonal propagation of tip culture have been reported (Tamura *et al.* 1984a). Early work with *in vitro* callus cultures of *Stevia* did not result in the production of measurable levels of the sweet-tasting glycosides (Handro *et al.* 1977; Wada *et al.* 1981). Within several years, however, investigators reported the accumulation of stevioside and rebaudioside A in callus culture of *S. rebaudiana* (Lee *et al.* 1982; Hsing *et al.* 1983). Other groups have reported the production of steviol glycosides in organ culture systems, including shoot tip cultures (Yamazaki *et al.* 1991; Swanson *et al.* 1992). Studies on the biosynthetic pathway of stevioside led to the isolation of $(-)$-kaurene in *S. rebaudiana* stems fed D,L-sodium mevalonate (Hanson and White 1968). This compound and (−)-kaur-16-en-19-oic acid are considered biosynthetic precursors of stevioside (**1**) (Hanson and White 1968). In another biosynthetic study, Alves and Ruddat (1979) isolated gibberellin A_{20} from the stems and leaves of *S. rebaudiana* and proposed that steviol (**7**) may be a precursor for C-13 hydroxy-gibberellins.

Labdane

In addition to *ent*-kaurene diterpenes, a number of labdane-type diterpenes have been identified from *S. rebaudiana* (Figure 4.3). Two known labdane diterpenes, jhanol (**11**) and austroinulin (**12**), and one new diterpene, 6-*O*-acetylaustroinulin (**13**), were isolated from a methanolic extract of *S. rebaudiana* leaves (Sholichin *et al.* 1980). The novel labdane was identified using ^{1}H NMR, ^{13}C NMR, UV, and by saponification of the compound to yield austroinulin. The diterpenoid constituents from *S. rebaudiana* flowers have been examined, and jhanol (**11**), austroinulin (**12**), and 6-*O*-acetylaustroinulin (**13**) identified, along with a new labdane, 7-*O*-acetylaustroinulin (**14**) with yields of 0.04, 0.21, 0.18, and 0.08% w/w, respectively (see Table 4.1) (Darise *et al.* 1983). The novel compound, 7-*O*-acetylaustroinulin, was shown not to be an artifact formed from acyl migration of 6-*O*-acetylaustroinulin since the two compounds could be

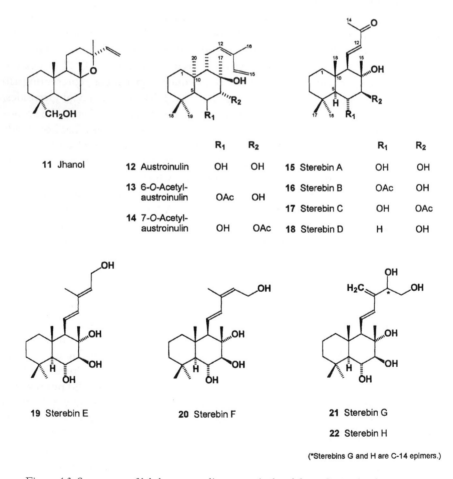

	R_1	R_2		R_1	R_2
11 Jhanol					
12 Austroinulin	OH	OH	**15** Sterebin A	OH	OH
13 6-*O*-Acetyl-austroinulin	OAc	OH	**16** Sterebin B	OAc	OH
			17 Sterebin C	OH	OAc
14 7-*O*-Acetyl-austroinulin	OH	OAc	**18** Sterebin D	H	OH

19 Sterebin E

20 Sterebin F

21 Sterebin G

22 Sterebin H

(*Sterebins G and H are C-14 epimers.)

Figure 4.3 Structures of labdane-type diterpenes isolated from *Stevia rebaudiana*.

identified by TLC in the crude methanolic extract and the ether-soluble fraction (Darise *et al*. 1983). Furthermore, 7-*O*-acetylaustroinulin could not be found in a crude methanolic extract of the leaves which contained high levels of 6-*O*-acetylaustroinulin (Darise *et al*. 1983).

From the leaves of *S. rebaudiana*, a series of eight novel labdane-type diterpenoids, sterebins A–H (**15–22**), have been identified (Oshima *et al*. 1986; 1988). Analysis of the spectroscopic data for sterebin A (**15**) revealed its general structural relationship to austroinulin (**12**), both having a *trans* double bond, two secondary hydroxyls, and one tertiary hydroxyl (Oshima *et al*. 1986). The structure of sterebin A (**15**) was verified by the preparation of its benzoyl derivative using benzoyl chloride in pyridine in the presence of *p*-dimethylaminopyridine to produce the dibenzoate that was then hydrogenated over 5% palladized charcoal catalyst to produce the corresponding saturated ketone, which confirmed the absolute stereochemistry of C-6 and C-7 as R and S, respectively (Oshima *et al*. 1986). Sterebin B (**16**) was saponified with 1 *N* potassium hydroxide to yield sterebin A (**15**), indicating it to be a monoacetate of sterebin A (Oshima *et al*. 1986). Sterebin C (**17**) was found to have the same molecular formula as sterebin B (**16**), and similar ^{1}H-NMR spectral data. Careful analysis of the ^{1}H-NMR and ^{13}C-NMR data revealed that sterebin C (**17**) has an acetate functional group at C-7, rather than at C-6, as found in its structural isomer sterebin B (**16**) (Oshima *et al*. 1986). Sterebin D (**18**), unlike sterebins A–C (**15–17**), displayed only one carbinyl hydrogen signal at about 3.5 ppm in the ^{1}H-NMR spectrum. From analysis of the *J* values of this proton, it was determined that sterebin D (**18**) contains an equatorially oriented hydroxyl group either at C-1, C-3, or C-7 (Oshima *et al*. 1986). The position of the hydroxyl functional group was determined to be at C-7 based upon comparison of its ^{13}C-NMR chemical shifts with those of model compounds, and by comparison with calculated ^{13}C-NMR values (Oshima *et al*. 1986). A series of NOE experiments conducted on the monoacetate of sterebin D helped to confirm the position of the β-hydroxyl at C-7 (Oshima *et al*. 1986). Sterebins A–D (**15–18**) occur in low yields in the leaves of *S. rebaudiana*, namely, 0.001%, 0.0009%, 0.0003%, and 0.0004% w/w, respectively (see Table 4.1). These four compounds are bisnorditerpenoids, that is, they are C_{18} diterpenoids, lacking C-14 and C-15 of the typical labdane skeleton. Bisnorditerpenoids are unusual compounds, and sterebins A–D (**15–18**) were the first to be reported with such highly-oxidized B-rings (Oshima *et al*. 1986).

The four additional novel labdane-type diterpenoids from *S. rebaudiana*, sterebins E–H (**19–22**), are not bisnorditerpenoids (Oshima *et al*. 1988). NMR data measured for sterebin E (**19**) indicated the absence of a carbonyl group, as found in sterebins A–D (**15–18**), and the presence of a trisubstituted double bond, a disubstituted double bond, and a 5-hydroxy-3-methyl-1,3-pentadienyl side chain at C-9 (Oshima *et al*. 1988). Sterebin F (6α, 7β, 8α, 11E, 13Z) (**20**) was found to be a C-13 geometrical isomer of sterebin E (6α, 7β, 8α, 11E, 13E) (**19**) (Oshima *et al*. 1988). Although the A–B rings of sterebin G (**21**) and H (**22**) were determined to be identical to those of sterebins E (**19**) and F (**20**), sterebin G was found to contain a 4,5-dihydroxy-3-methylene-1-pentenyl group at C-9 (Oshima *et al*. 1988). Sterebins G and H were identified as C-14 positional isomers (Oshima *et al*. 1988). Low yields were reported for sterebins E–H, 0.002%, 0.003%, 0.0002%, and 0.0002% (w/w), respectively (see Table 4.1) (Oshima *et al*. 1988).

TRITERPENOIDS AND STEROIDS

The known triterpenoid β-amyrin acetate (**23**) and three unidentified esters of lupeol were obtained from the ether-soluble portion of a methanolic extract of *S. rebaudiana* leaves

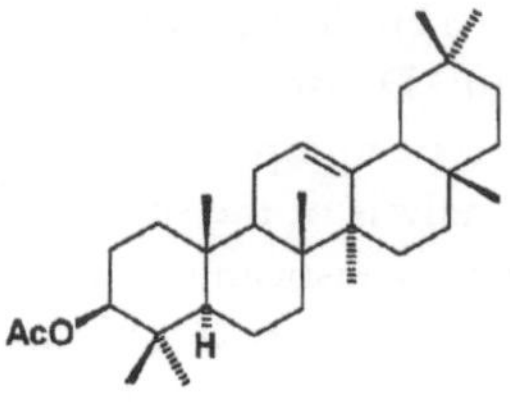

23 β-Amyrin acetate

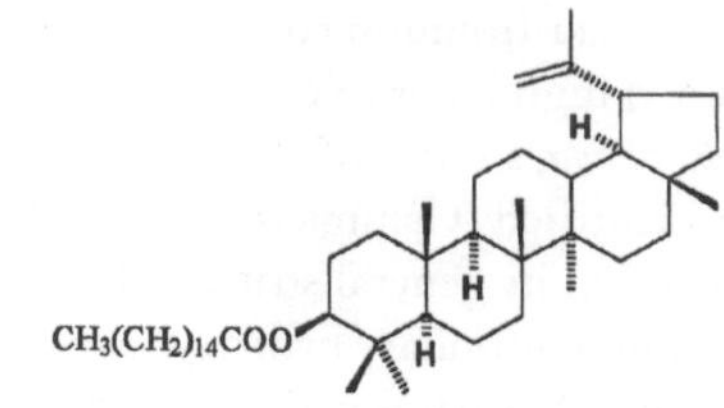

24 Lupeol 3-palmitate

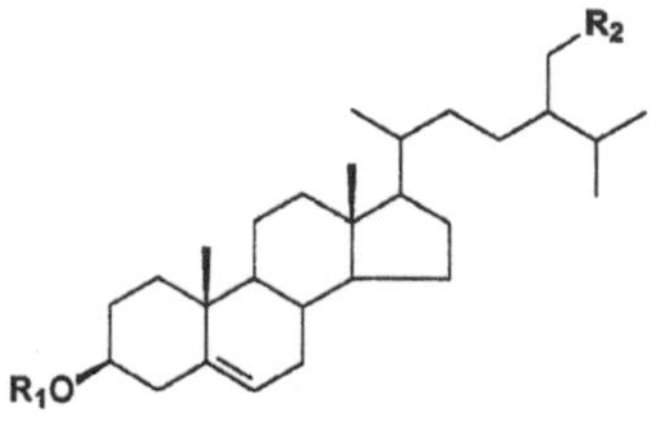

	R₁	R₂
25 β-Sitosterol	H	CH₃
27 Campesterol	H	H
28 β-Sitosterol-β-D-glucoside	Glc	CH₃

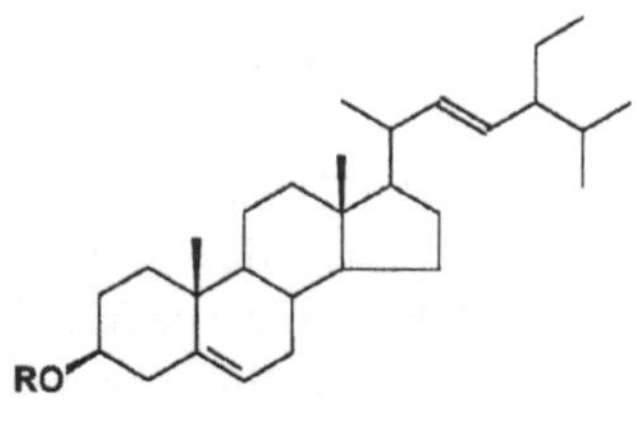

	R
26 Stigmasterol	H
29 Stigmasterol β-D-glucoside	Glc

Glc = β-D-glucopyranosyl

Figure 4.4 Structures of triterpenes and sterols isolated from *Stevia rebaudiana*.

(Sholichin *et al*. 1980). More recently, a lupeol ester, lupeol 3-palmitate (**24**), was identified from a methanolic extraction of *S. rebaudiana* by GC-MS (Yasukawa *et al*. 1993). The ubiquitous plant sterols, β-sitosterol (**25**) and stigmasterol (**26**), have been identified in *S. rebaudiana*, and were respectively found to comprise 39.4% and 45.8% (w/w) of the total sterol fraction (D'Agostino *et al*. 1984). Campesterol (**27**) was identified from *S. rebaudiana* by GC retention time and MS of the acetylated compound and found to constitute about 13.1% of the sterol fraction of the plant extract (D'Agostino *et al*. 1984). The callus culture from the leaves of *S. rebaudiana* has also been shown to produce stigmasterol (**26**) (Nabeta *et al*. 1976). In their investigation of the non-sweet glycosides of *S. rebaudiana*, Matsuo *et al*. (1986) isolated two steroid glycosides, β-sitosterol-β-D-glucoside (**28**) and stigmasterol-β-D-glucoside (**29**), with their structures being established by NMR and MS (Figure 4.4).

FLAVONOIDS

Examination of the flavonoid content of *S. rebaudiana* leaves (Rajbhandari and Roberts 1983) resulted in the identification of six flavonoid glycosides in the ethyl acetate fraction, comprising apigenin 4′-O-glucoside (**30**), kaempferol 3-O-rhamnoside (**31**), luteolin 7-O-glucoside (**32**), quercetin 3-O-arabinoside (**33**), quercetin 3-O-glucoside (**34**), and quercetin 3-O-rhamnoside

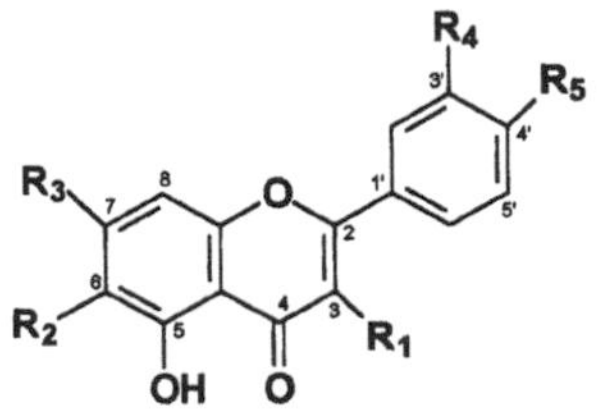

| Compound | R₁ | R₂ | R₃ | R₄ | R₅ |

	Compound	R_1	R_2	R_3	R_4	R_5
30	Apigenin 4'-O-glucoside	H	H	OH	H	Glc
31	Kaempferol 3-O-rhamnoside	Rha	H	OH	H	OH
32	Luteolin 7-O-glucoside	H	H	Glc	OH	OH
33	Quercetin 3-O-arabinoside	Ara	H	OH	OH	OH
34	Quercetin 3-O-glucoside	Glc	H	OH	OH	OH
35	Quercetin 3-O-rhamnoside	Rha	H	OH	OH	OH
36	Centaureidin	OMe	OMe	OH	OH	OMe
37	Apigenin 7-O-glucoside	H	H	Glc	H	OH
38	Quercetin 3-O-rutinoside	Rut	H	OH	OH	OH

Glc = O-β-D-glucopyranosyl; rha = O-α-L-rhamnopyranosyl; ara = O-α-L-arabinopyranosyl;
rut = 6-O-α-L-rhamnopyranosyl-D-glucopyranosyl

Figure 4.5 Structures of flavonoids isolated from *Stevia rebaudiana*.

(quercitrin) (**35**) (Figure 4.5). These compounds were identified using standard methods of UV and ¹H-NMR spectroscopy, and mass spectrometry. From the chloroform extract, a methoxylated flavonoid, 5,7,3' trihydroxy-3,6,4' trimethoxyflavone (centaureidin) (**36**), was also isolated by Rajbhandari and Roberts (1983). In their investigation of the non-sweet glycosides of *S. rebaudiana*, Matsuo *et al.* (1986) identified apigenin 7-O-glucoside (cosmosiin) (**37**) from the leaves. In cell culture, *S. rebaudiana* exhibits a noticeable yellowish color, with one of the pigments having been identified as the flavonoid glycoside rutin (quercetin 3-β-rutinoside) (**38**), with a yield of 0.0007% of the fresh weight of the culture cells (see Table 4.1) (Suzuki *et al.* 1976).

VOLATILE OIL COMPONENTS

The volatile oils from the leaves and infloresences of *S. rebaudiana* have been studied, and a number of monoterpenes, sesquiterpenes, alkanols, aldehydes, and aromatic alcohols have been identified (Fujita *et al.* 1977; Martelli *et al.* 1985), as summarized in Table 4.2. Using GC-MS, Fujita *et al.* (1977) were able to identify 32 components of the essential oil of *S. rebaudiana*, including the sesquiterpenes β-caryophyllene, *trans*-β-farnesene, α-humulene, δ-cadinene, caryophyllene oxide, nerolidol, and among the monoterpenes, linalool, terpinen-4-ol, and α-terpineol were found as major constituents. An additional 22 peaks were noted by GC, MS, or IR, but could not be identified (Fujita *et al.* 1977). In another study, the dried leaves of *S. rebaudiana* were extracted by steam distillation, and the two major essential oil components in the extract, caryophyllene oxide and spathulenol, were removed by column chromatography (Martelli *et al.* 1985). Analysis of the remaining minor constituents by GC-MS showed over 100 peaks, of which 54 were identified (Martelli *et al.* 1985). Martelli *et al.* (1985) also examined the volatile oils present in a fresh specimen of *S. rebaudiana* (see Table 4.2).

Table 4.2 The essential oil components identified from *Stevia rebaudiana*[a]

Compound Class	Compound	Yield (% w/w) dried herb[b]; [dried inflorescence][c]	Yield (% w/w) dried leaves[d]; [fresh leaves and stem][c]
Monoterpene	Borneol	−[f] [−]	+[g] [−]
	Camphor	0.0017 [0.0017]	−[−]
	Carvacrol	−[−]	+[+]
	1,8-Cineole	0.0008 [0.0009]	+[−]
	p-Cymene	0.0008 [0.002]	+[−]
	Geraniol	0.0016 [0.0009]	−[−]
	Limonene	0.0001 [0.003]	+[−]
	Linalool	0.0067 [0.0129]	+[+]
	cis-Linalool oxide	0.0026 [0.0004]	+[−]
	trans-Linalool oxide	0.0029 [0.0004]	+[−]
	Myrcene	−[−]	+[−]
	Myrtenal	−[−]	+[−]
	Myrtenol	−[−]	+[−]
	α-Pinene	0.0005 [0.0004]	+[+]
	β-Pinene	0.002 [0.007]	+[+]
	Pinocarveol	−[−]	+[−]
	Pinocarvone	−[−]	+[−]
	Sabinene	−[−]	+[−]
	γ-Terpinene	0.0002 [0.0009]	+[−]
	Terpinene-4-ol	0.0012 [0.0172]	−[−]
	α-Terpineol	0.0054 [0.0034]	+[+]
	Terpinolene	−[−]	+[−]
	trans-Verbenol	−[−]	+[−]
Sesquiterpene	α-Bergamotene	−[−]	+[+]
	Bisabolene	−[−]	+[−]
	β-Bourbonene	−[−]	+[+]
	γ-Cadinene	0.0034 [0.0082]	+[+]
	δ-Cadinene	0.0012 [0.0133]	+[+]
	α-Cadinol	0.0017 [0.0065]	+[+]
	tert- Cadinol	0.0028 [0.0060]	−[−]
	α-Calacorene	0.0024 [0.0004]	+[−]
	Calamenene	0.0018 [0.0004]	+[−]
	β-Caryophyllene	0.0013 [0.0149]	+[+]
	Caryophyllene oxide	0.0188 [0.0267]	0.0051 [+]
	α-Copaene	0.0001 [0.0026]	+[−]
	α-Cubebene	0.0001 [0.0017]	+[−]

	β-Cubebene	−[−]	+[+]
	β-Elemene	0.0006 [0.0108]	+[+]
	γ-Elemene	−[−]	+[+]
	trans-β-Farnesene	0.0008 [0.0344]	+[+]
	Germacrene D	−[−]	+[−]
	α-Humulene	0.0029 [0.0262]	0.0007 [+]
	β-Ionone	−[−]	+[−]
	Epoxy-β-ionone	−[−]	+[−]
	Nerolidol	0.0088 [0.0331]	0.0013 [+]
	β-Selinene	0.0029 [0.0043]	−[−]
	Spathulenol	†[h]	0.0057 [+]
	Torreyol	−[−]	+[+]
Miscellaneous	Anethole	−[−]	+[−]
	Benzyl alcohol	0.0012 [0.0004]	−[−]
	Cuminic aldehyde	−[−]	+[−]
	Eugenol	−[−]	+[+]
	n-Hexanal	0.001 [0.0004]	−[−]
	cis-Hex-3-en-1-ol	−[−]	−[+]
	trans-Hex-2-en-1-ol	−[−]	−[+]
	Hex-3-enyl acetate	−[−]	−[+]
	Hex-3-enyl-2-methyl-butanoate	−[−]	+[−]
	Hexan-1-ol	0.9	−[−]
	Methyleugenol	−[−]	+[−]
	Octa-2,3-dione	−[−]	+[−]
	Octan-3-ol	0.0004 [0.0009]	+[−]
	Oct-1-en-3-ol	0.0008 [0.0146]	+[+]
	Pentanoic acid	−[−]	+[−]
	2-Pentylfuran	−[−]	+[−]
	Phenylethyl valerate	−[−]	+[−]

Notes

a Information compiled from Fujita *et al.* 1977; Martelli *et al.* 1985; Kinghorn and Soejarto 1985; 22 compounds were not identified in the study by Fujita *et al.* (1977), and about 50 trace compounds remain unidentified from the study of Martelli *et al.* (1985).

b The percentage (w/w) yield of oil in the herb was 0.12 (Fujita *et al.* 1977).

c The percentage (w/w) yield of oil in the inflorescence was 0.43 (Fujita *et al.* 1977).

d The percentage (w/w) yield of the oil in the dried leaves was 0.025 (Martelli *et al.* 1985).

e The percentage (w/w) yield of the oil in the fresh plant material was not reported (Martelli *et al.* 1985).

f The compound was not detected.

g The compound was detected, but not quantified.

h Possibly the unidentified sesquiterpene alcohol, molecular weight 220, reported by Fujita *et al.* (1977) from the herb and flowers of *S. rebaudiana*.

MISCELLANEOUS CONSTITUENTS

A number of common phytochemicals have been identified in *S. rebaudiana*, such as the pigments chlorophylls A and B and β-carotene (see Table 4.1) (Cheng and Chang 1983). The same investigators also noted that various gums comprised about 7–15% of the total extracted solids from *S. rebaudiana*. Tartaric acid was the major organic acid in an extract of *S. rebaudiana*, with citric, formic, lactic, malic, and succinic acids also being identified (Cheng and Chang 1983). The phytohormone indole-3-acetonitrile has been reported from the seeds of *S. rebaudiana*, and identified by retention time and color tests (Randi and Felippe 1981). Unspecified tannins have also been reported in *S. rebaudiana* (Chung and Lee 1978). Inorganic compounds comprise approximately 13% of the total extractables from *S. rebaudiana*, with potassium being the major inorganic constituent of the *Stevia* extract (Cheng and Chang 1983). Other inorganic substances identified from *S. rebaudiana* include calcium, iron, magnesium, phosphorus, sodium, and zinc (Cheng and Chang 1983). In a screening procedure on fluoride levels in plants used in the People's Republic of China, the leaves of *S. rebaudiana* were found to contain 12.2 ppm of fluoride (Sakai *et al.* 1985). Some of the inorganic compounds identified may be from contamination, such as fluorine from air pollution on a plant surface (Jacobson and McCune 1969).

SUMMARY AND CONCLUSIONS

The phytochemistry of *S. rebaudiana* has been studied intensely for decades now, and many sweet and non-sweet compounds have been identified. Six new, naturally occurring, sweet-tasting steviol glycosides have been identified from *S. rebaudiana*, with stevioside (**1**) being the most abundant, and rebaudioside A (**2**) being the second most abundant. The remaining four natural product *ent*-kaurene glycosides, rebaudiosides C–E (**3–5**) and dulcoside A (**6**), are present in lesser amounts. Due to their economic importance, attempts have been made to increase the yield of the steviol glycosides *in vivo* and also to produce these compounds *in vitro* using various culture systems. In addition to the *ent*-kaurene glycosides, *S. rebaudiana* leaves contain a series of novel non-sweet labdane diterpenes, sterebins A–H (**15–22**). *S. rebaudiana* also elaborates a complex mixture of known sterols, triterpenoids, essential oils, flavonoids, and other compounds.

ACKNOWLEDGEMENTS

The author wishes to thank Dr Aiko Ito, Hiroshima University, Hiroshima, Japan, for translating certain Japanese articles, and Mr Nam-Cheol Kim, University of Illinois at Chicago, for kindly obtaining a number of references. Drs William Horwitz and Paul M. Kuznesof, Center for Food Safety and Applied Nutrition, US Food and Drug Administration, Washington, DC, are also thanked for their helpful comments on this manuscript.

REFERENCES

Abe, K. and Sonobe, M. (1977) Use of stevioside in the food industry. *New Food Industry* **19**, 67–72.

Alves, L. M. and Ruddat, M. (1979) The presence of gibberellins A_{20} in *Stevia rebaudiana* and its significance of the biological activity of steviol. *Plant Cell Physiology* **20**, 123–130.

Bakal, A. I. and O'Brien Nabors, L. (1986) Stevioside, In *Alternative Sweeteners*, L. O'Brien Nabors and R. C. Gelardi (Eds), Marcel Dekker, New York, pp. 295–307.

Bell, F. (1954) Stevioside: a unique sweetening agent. *Chemistry and Industry*, London, 17 July, pp. 897–898.

Bertoni, M. S. (1905) Le Kaá hê-é: sa nature et ses propriétés. *Anales Científicos Paraguayos Serie I* **5**, 1–14.

Bertoni, M. S. (1918) La *Stevia rebaudiana* Bertoni: la estevina y la rebaudina, nuevas substancias edulcorantes. *Anales Científicos Paraguayos Serie II* **2**, 129–134.

Bridel, M. and Lavieille, R. (1931a) Le principe à saveur sucrée du Kaà-hê-é (*Stevia rebaudiana* Bertoni). *Journal de Pharmacie et de Chimie* **14**, 99–113.

Bridel, M. and Lavieille, R. (1931b) Le principe à saveur sucré du Kaà-hê-é (*Stevia rebaudiana* Bertoni) III. Propriétés du stévioside. *Journal de Pharmacie et de Chimie* **14**, 154–161.

Bridel, M. and Lavieille, R. (1931c) La rebaudine de Dieterich est du stévioside impur. *Journal de Pharmacie et de Chimie* **14**, 161–163.

Bridel, M. and Lavieille, R. (1931d) Sur le principe sucré du Kaà-hê-é *(Stevia rebaudiana* Bertoni) II. L'hydrolyse diastasique du stévioside. III. Le stéviol de l'hydrolyse diastasique et l'isostéviol de l'hydrolyse acide. *Journal de Pharmacie et de Chimie* **14**, 321–328.

Bridel, M. and Lavieille, R. (1931e) Sur le principe sucré du Kaà-hê-é *(Stevia rebaudiana* Bertoni) II. L'hydrolyse diastasique du stévioside. III. Le stéviol de l'hydrolyse diastasique et l'isostéviol de l'hydrolyse acide. *Journal de Pharmacie et de Chimie* **14**, 369–379.

Bridel, M. and Lavieille, R. (1931f) La rebaudine de Dieterich et du stévioside impur. *Bulletin de la Societe de Chimie Biologique* **13**, 656–657. [*Biological Abstracts* (1932) **6**, 21498].

Bridel, M. and Lavieille, R. (1931g) Sur le principe sucré des feuilles de Kaà-hê-é. *Comptes Rendus Hebdomadaires des Seances de l'Académie des Sciences* **192**, 1123–1125.

Bridel, M. and Lavieille, R. (1931h) Sur le principe sucré du Kaà-hê-é *(Stevia rebaudiana* Bertoni): II. Les produits d'hydrolyse diastasique du stévioside; glucose et stéviol. *Comptes Rendus Hebdomadaires des Seances de l'Académie des Sciences* **193**, 72–74.

Buckingham, J. (Ed) (1997) *Dictionary of Natural Products*, Chapman & Hall, New York, CD-ROM, release 6:1.

Cheng, T.-F. and Chang, W.-H. (1983) Studies on the non-stevioside components of *Stevia* extracts. *National Science Council Monthly, Taipei* **11**, 96–108.

Cheng, T.-F., Chang, W.-H. and Chang, T. R. (1981) A study on the post-harvest changes in steviosides contents of *Stevia* leaves and stems. *National Science Council Monthly, Taipei* **9**, 775–782.

Chung, M. H. and Lee, M. Y. (1978) Studies on the development of hydrangea and *Stevia* as natural sweetening products. *Saengyak Hakhoechi* **9**, 149–156.

Crammer, B. and Ikan, R. (1986) Sweet glycosides from the *Stevia* plant. *Chemistry in Britain* **22**, 915–916, and 918.

Crammer, B. and Ikan, R. (1987) Progress in the chemistry and properties of the rebaudiosides. In *Developments in Sweeteners – 3*, T. H. Grenby (Ed), Elsevier Applied Science, London, pp. 45–64.

D'Agostino, M., De Simone, F., Pizza, C. and Aquino, R. (1984) Sterols from *Stevia rebaudiana* Bertoni. *Bollettino-Societa Italiana di Biologia Sperimentale* **60**, 2237–2240. [*Chemical Abstracts* (1985) **102**, 109851d].

Darise, M., Kohda, H., Mizutani, K., Kasai, R. and Tanaka, O. (1983). Chemical constituents of flowers of *Stevia rebaudiana* Bertoni. *Agricultural and Biological Chemistry* **47**, 133–135.

Dieterich, K. (1908) The constituents of *Eupatorium rebaudianum*, 'Kaa-he-e' and their pharmaceutical value. *Pharmazeutische Zentralhalle* **50**, 435–458. [*Chemical Abstracts* (1909) **3**, 2485–2491].

Djerassi, C., Quitt, P., Mosettig, E., Cambie, R. C., Rutledge, P. S. and Briggs, L. H. (1961) Optical rotatory dispersion studies. LVIII. The complete absolute configurations of steviol, kaurene and the diterpene alkaloids of the garryfoline and atisine groups. *Journal of the American Chemical Society* **83**, 3720–3722.

Dolder, F., Lichti, H., Mosettig, E. and Quitt, P. (1960) The structure and stereochemistry of steviol and isosteviol. *Journal of the American Chemical Society* **82**, 246–247.

Ferreira, C. M. and Handro, W. (1988) Micropropagtion of *Stevia rebaudiana* through leaf explants from adult plants. *Planta Medica* **54**, 157–160.

Fletcher, H. G., Jr (1955) The sweet herb of Paraguay. *Chemurgic Digest* **14**, 7, 18–19.

Fujita, S.-I., Taka, K. and Fujita, Y. (1977) Miscellaneous contributions to the essential oils of plants from various territories. XLI. On the components of the essential oil of *Stevia rebaudiana* Bertoni. *Yakugaku Zasshi* **97**, 692–694.

Handro, W., Hell, K. G. and Kerbauy, G. B. (1977) Tissue culture of *Stevia rebaudiana*, a sweetening plant. *Planta Medica* **32**, 115–117.

Hanson, J. R. and White, A. F. (1968) Studies in terpenoid biosynthesis – II. The biosynthesis of steviol. *Phytochemistry* **7**, 595–597.

Hanson, J. R. and De Oliveira, B. H. (1993) Stevioside and related sweet diterpenoid glycosides. *Natural Product Reports* **10**, 301–309.

Hsing, Y. I., Su, W. F. and Chang, W. C. (1983) Accumulation of stevioside and rebadioside A in callus culture of *Stevia rebaudiana* Bertoni. *Botanical Bulletin of Academia Sinica* **24**, 115–119.

Jacobs, M. B. (1955) The sweetening power of stevioside. *The American Perfumer*, December, pp. 45–46.

Jacobson, J. S. and McCune, D. C. (1969) Interlaboratory study of analytical techniques for fluorine in vegetation. *Journal of the Association of Official Analytical Chemists* **52**, 894–899.

Kim, S.-H. and Dubois, G. E. (1991) Natural high potency sweeteners. In *Handbook of Sweeteners*, S. Marie and J. R. Piggott (Eds), Avi, New York, pp. 116–185.

Kinghorn, A. D. and Soejarto, D. D. (1985) Current status of stevioside as a sweetening agent for human use. In *Economic and Medicinal Plant Research*, H. Wagner, H. Hikino and N. R. Farnsworth (Eds), Academic Press, New York, pp. 1–52.

Kinghorn, A. D. and Soejarto, D. D. (1991) Stevioside. In *Alternative Sweeteners* (2nd edn, Revised and Expanded), L. O'Brien Nabors and R. C. Gelardi (Eds), Marcel Dekker, New York, pp. 157–171.

Kinghorn, A. D., Soejarto, D. D., Nanayakkara, N. P. D., Compadre, C. M., Makapugay, H. C., Hovanec-Brown, J. M. *et al.* (1984) A phytochemical screening procedure for sweet *ent*-kaurene glycosides in the genus *Stevia. Journal of Natural Products* **47**, 439–444.

Kobayashi, M., Horikawa, S., Degrandi, I. H., Ueno, J. and Mitsuhashi, H. (1977) Dulcosides A and B, new diterpene glycosides from *Stevia rebaudiana. Phytochemistry* **16**, 1405–1408.

Kobert, R. (1915) Two sweet tasting drugs. *Berichte der Deutschen Pharmazeutisch en Gesellschaft* **25**, 162–185.

Kohda, H., Kasai, R., Yamasaki, K., Murakami, K. and Tanaka, O. (1976) New sweet diterpene glucosides from *Stevia rebaudiana. Phytochemistry* **15**, 981–983.

Lee, K. R., Park, J. R., Choi, B. S., Han, J. S., Oh, S. L. and Yamada, Y. (1982) Studies on the callus culture of *Stevia* as a new sweetening source and the formation of stevioside. *Hanguk Sikp'um Kwahakhoe Chi* **14**, 179–183. [*Chemical Abstracts* (1982) **97**, 71003s].

Martelli, A., Frattini, C. and Chialva, F. (1985) Unusual essential oils with aromatic properties. I. Volatile components of *Stevia rebaudiana* Bertoni. *Flavour and Fragrance Journal* **1**, 3–7.

Matsuo, T., Kanamori, H. and Sakamoto, I. (1986) Nonsweet glucosides in the leaves of *Stevia rebaudiana. Hiroshima-ken Eisei Kenkyusho Kenkyu Hokoku* **33**, 25–29. [*Chemical Abstracts* (1987) **107**, 93559o].

Metivier, J. and Viana, A. M. (1979) The effect of long and short day length upon the growth of whole plants and the level of soluble proteins, sugars, and stevioside in leaves of *Stevia rebaudiana* Bert. *Journal of Experimental Botany* **30**, 1211–1222.

Mizukami, H., Shiba, K., Satoshi, I. and Ohashi, H. (1983) Effect of temperature on growth and stevioside formation of *Stevia rebaudiana* Bertoni. *Shoyakugaku Zasshi* **37**, 175–179.

Morita, E. (1977) Outlook for *Stevia* natural sweetening agents. *Shokuhin to Kagaku* **19**, 83–87.

Mosettig, E. and Nes, W. R. (1955) Stevioside. II. The structure of the aglucon. *Journal of Organic Chemistry* **20**, 884–899.

Mosettig, E., Quitt, P., Beglinger, U., Waters, J. A., Vorbrueggen, H. and Djerassi, C. (1961) A direct correlation of the diterpene and hydrocarbons of the phyllocladene group. Interconversion of garryfoline and steviol. *Journal of the American Chemical Society* **83**, 3163–3164.

Mosettig, E., Beglinger, U., Dolder, F., Lichti, H., Quitt, P. and Waters, J. A. (1963) The absolute configuration of steviol and isosteviol. *Journal of the American Chemical Society* **85**, 2305–2309.

Nabeta, K., Kasai, T. and Sugisawa, H. (1976) Phytosterol from the callus of *Stevia rebaudiana* Bertoni. *Agricultural and Biological Chemistry* **40**, 2103–2104.

Okazaki, K., Nakayama, Y. and Owada, K. (1977) *Stevia* – new natural sweetener. *Seikatsu Eisei* **21**, 181–185.

Oshima, Y., Saito, J.-I. and Hikino, H. (1986) Sterebins A, B, C and D, bisnorditerpenoids of *Stevia rebaudiana* leaves. *Tetrahedron* **42**, 6443–6446.

Oshima, Y., Saito, J.-I. and Hikino, H. (1988) Sterebins E, F, G and H, diterpenoids of *Stevia rebaudiana* leaves. *Phytochemistry* **27**, 624–626.

Phillips, K. C. (1987) *Stevia*: steps in developing a new sweetener. In *Developments in Sweeteners – 3*, T. H. Grenby (Ed), Elsevier Applied Science, London, pp. 1–43.

Rajbhandari, A. and Roberts, M. F. (1983) The flavonoids of *Stevia rebaudiana. Journal of Natural Products* **46**, 194–195.

Randi, A. M. and Felippe, G. M. (1981) Substances promoting root growth from the achenes of *Stevia rebaudiana. Revista Brasileira Botanica* **4**, 49–51. [*Chemical Abstracts* (1982) **97**, 52510p].

Rasenack, P. (1908) Sweet substances of *Eupatorium rebaudianum* and of licorice. *Arbeiten aus der Biologischen Abteilung fuer Land- und Forstwirtschaft am Kaiserliche Gesundheitsamte* **28**, 420–423. [*Chemical Abstracts* (1909) **3**, 688–692].

Sakai, T., Kobashi, K., Tsunezuka, M., Hattori, M. and Namba, T. (1985) Studies on dental caries prevention by traditional Chinese medicines (Part VI). On the fluoride contents in crude drugs. *Shoyakugaku Zasshi* **39**, 165–169.

Sakamoto, I., Yamasaki, K. and Tanaka, O. (1977a) Application of ^{13}C NMR spectroscopy to chemistry of natural glycosides: rebaudioside-C, a new sweet diterpene glycoside of *Stevia rebaudiana*. *Chemical and Pharmaceutical Bulletin* **25**, 844–846.

Sakamoto, I., Yamasaki, K. and Tanaka, O. (1977b) Application of ^{13}C NMR spectroscopy to chemistry of plant glycosides: rebaudiosides-D and -E, new sweet diterpene-glucosides of *Stevia rebaudiana* Bertoni. *Chemical and Pharmaceutical Bulletin* **25**, 3437–3439.

Salvatore, G., Paganuzzi, A. S., Silano, V. and Aureli, A. D. (1984) Stevioside: An updating of its chemical and biological properties. *Reports of the Higher Institute of Health, Rome* **84**(13), pp. 1–34.

Seidemann, J. (1976) Steviosid, ein interessantes, natürliches Süßungsmittel. *Nahrung* **20**, 675–679.

Sholichin, M., Yamasaki, K., Miyama, R., Yahara, S. and Tanaka, O. (1980) Labdane-type diterpenes from *Stevia rebaudiana*. *Phytochemistry* **19**, 326–327.

Suzuki, H., Ikeda, T., Matsumoto, T. and Noguchi, M. (1976) Isolation and identification of rutin from cultured cells of *Stevia rebaudiana* Bertoni. *Agricultural and Biological Chemistry* **40**, 819–820.

Swanson, S. M., Mahady, G. B. and Beecher, C.W.W. (1992) Stevioside biosynthesis by callus, root, shoot and rooted-shoot cultures *in vitro*. *Plant Cell, Tissue and Organ Culture* **28**, 151–157.

Tamura, Y., Nakamura, S., Fukui, H. and Tabata, M. (1984a) Comparison of *Stevia* plants grown from seeds, cuttings and stem-tip cultures for growth and sweet diterpene glucosides. *Plant Cell Reports* **3**, 180–182.

Tamura, Y., Nakamura, S., Fukui, H. and Tabata, M. (1984b) Clonal propagation of *Stevia rebaudiana* Bertoni by stem-tip culture. *Plant Cell Reports* **3**, 183–185.

Tanaka, O. (1980) Chemistry of *Stevia rebaudiana* Bertoni: new source of natural sweeteners. *Saengyak Hakhoechi* **11**, 219–227.

Tanaka, O. (1982) Steviol-glycosides: new natural sweeteners. *Trends in Analytical Chemistry* **1**, 246–248.

Tanaka, O. (1987) Natural sweet principles. Recent advances. *Kagaku to Kogyo, Osaka* **61**, 404–410. [*Chemical Abstracts* (1987) **107**, 233038m].

Toffler, F. and Orio, O. A. (1981) Accenni sulla pianta topicale 'kaá-hé-é' o 'erba dulce.' *Rivista della Societa Italiana di Scienza dell'Alimentazione* **10**, 225–230.

Viana, A. M. and Metivier, J. (1980) Changes in the levels of total soluble proteins and sugars during leaf ontogeny in *Stevia rebaudiana* Bert. *Annals of Botany* **45**, 469–474.

Vis, E. and Fletcher, H. G., Jr (1956) Stevioside. IV. Evidence that stevioside is a sophoroside. *Journal of the American Chemical Society* **75**, 4709–4710.

Wada, Y., Tamura, T., Kodama, T., Yamaki, T. and Uchida, Y. (1981) Callus cultures and morphogenesis of *Stevia rebaudiana* Bertoni. *Yukagaku* **30**, 215–219. [*Chemical Abstracts* (1981) **94**, 203745k].

Wood, H. B., Jr, Allerton, R., Diehl, H.W. and Fletcher, H. G., Jr (1955) Stevioside. I: the structure of the glucose moieties. *Journal of Organic Chemistry* **20**, 875–883.

Wood, H. B., Jr and Fletcher, H. G., Jr (1956) Stevioside. III. The anomeric 2,3,4,6-tetra-O-acetyl-1-O-acetyl-1-O-mesitoyl-D-glucopyranoses and their behavior with alkali. *Journal of the American Chemical Society* **78**, 207–210.

Wu, Y.-P. (1987) Extraction and utilization of steviosides. *Yiyao Gongye* **18**, 228–231.

Yamazaki, T., Flores, H. E., Shimomura, K. and Yoshihira, K. (1991) Examination of steviol glucosides production by hairy root and shoot culture of *Stevia rebaudiana*. *Journal of Natural Products* **54**, 986–992.

Yasukawa, K., Yamaguchi, A., Arita, J., Sakurai, S., Ikeda, A. and Takido, M. (1993) Inhibitory effect of edible plant extracts on 12-O-tetradecanoyphorbol-13-acetate-induced ear oedema in mice. *Phytotherapy Research* **7**, 185–189.

Yoshihira, K., Matsui, M. and Ishidate, M. (1987) Chemical characteristics and biological safty [sic] of a glycosidic sweetener, stevioside. *Tokishikoroji Foramu* **10**, 281–289.

5   The phytochemistry of *Stevia*: a general survey

Carlos M. Cerda-García-Rojas and
Rogelio Pereda-Miranda

INTRODUCTION

Stevia, a New World genus of the tribe Eupatorieae in the Asteraceae family, extends from the southwestern United States to northern Argentina, through Mexico, Central America, the South American Andes and the Brazilian highlands in ecosystems located at altitudes 1000 m above sea level. Records indicate that *Stevia* is not represented in the Bahamas, the Antilles, and Amazonia. Estimates on the number of species within the genus range from 150 to 300 (Grashoff 1972), of which more than 80 species are known to occur in North America, with at least 70 of these being native to Mexico. The South American species do not appear to have been studied recently from a taxonomic point of view and perhaps 90 species occur in the area bounded by Bolivia, southern Brazil, and northern Argentina (Grashoff 1972).

The aim of this review is to outline the chemistry of *Stevia*, giving emphasis to the most significant components of the secondary metabolism in the form of a concise and general survey. This report covers all the relevant literature published up to May 1998, excluding from this compilation references related to the sweet and non-sweet constituents of *S. rebaudiana*, which are reviewed in Chapter 4 of this volume.

CHEMICAL CONSTITUENTS

Despite the fact that *Stevia* is one of the largest and most easily recognized genus in the tribe Eupatorieae (Robinson and King 1977), this morphologically well-delineated taxon in the Piqueria group is surprisingly diverse in its chemical composition. However, the chemical profile for *Stevia* does not differ widely from that described for other members of the Eupatorieae (Domínguez 1977). To our knowledge, phytochemical information is available to date on only 58 species, with some of the Mexican and Argentinean members having received particular attention, e.g. *S. salicifolia* and *S. satureiaefolia*.

Sesquiterpenoids are by far the major and most typical constituents of the aerial parts and roots of *Stevia*. The overwhelming majority of these compounds belong to the guaiane, longipinane, and germacrane groups. The *ent*-kaurene glycosides have recently attracted much attention due to their economic importance as potential low-calorie sweeteners (Kinghorn and Soejarto 1985). Positive alkaloid tests have been reported for the seeds of *S. serrata* var. *arguta*, *S. serrata* var. *linoides* and *S. uniaristata* (Willaman and Li 1970). No attempts to isolate and characterize the alkaloids of *Stevia* have been performed. However, pyrrolizidine alkaloids are present in leaves, flowers and roots of some *Eupatorium* species (Domínguez 1977), the proto-type genus of the tribe Eupatorieae. In the present review, the chemical constituents of *Stevia* are classified according to their structural types and some representative examples of the structures of each group of natural products are presented.

ESSENTIAL OILS AND OTHER VOLATILE CONSTITUENTS

Although the flowers and leaves of *Stevia* species could be rich sources of essential oils, as expected for any member of the Asteraceae (Domínguez 1977), only three studies have been carried out to fully analyze the lower terpenoids and volatile substances by modern techniques. The first study involved the gas chromatographic analysis of the essential oil of *S. satureiaefolia*, collected in Argentina, which yielded the monoterpenes borneol (**1**), cineole (**2**), pulegone (**3**), geraniol (**4**), nerol (**5**), linalool acetate (**6**), limonene (**7**), and the phenylpropanoid derivative eugenol, as the main constituents (Montes 1969). A second analytical study was conducted on the essential oil obtained from the flowers of *S. porphyrea* collected in Mexico at the beginning and the end of the flowering season in September and November, respectively. The composition of both these essential oils, which were analyzed by gas chromatography coupled to mass spectroscopy, turned out to be markedly different since the essential oil obtained in September mainly afforded thymol (**8**), α-pinene (**10**) and 7-ethyl-2,4-dimethylazulene, while that obtained in November yielded essentially nerol (**5**), α-phellandrene (**9**), and the sesquiterpenes α-humulene (α-caryophyllene) (**12**) and farnesol (**20**) (Pérez *et al*. 1995). Recently, the volatile

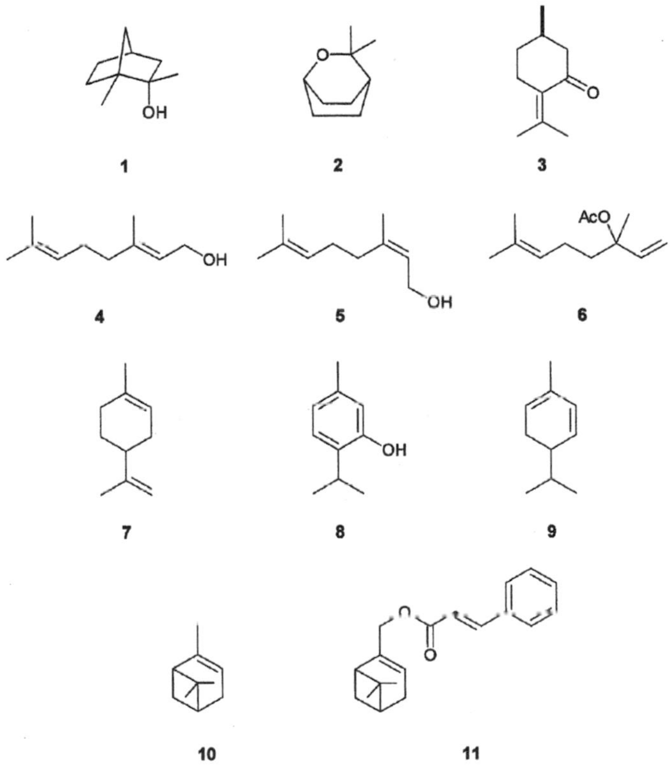

Figure 5.1 Selected monoterpenoids (**1–11**) from the essential oils of *Stevia* species.

constituents of *S. achalensis* were characterized. The main components of the essential oil were β-selinene, β-caryophyllene, and α-muurolene (Zygadlo *et al.* 1997).

In addition to these studies, the flowers of *S. serrata* collected in Mexico afforded, on steam distillation, good yields of the sesquiterpene chamazulene (**15**) (Román *et al.* 1990), while α-pinene (**10**) and myrtenyl cinnamate (**11**) were found in the lipophilic extracts obtained from the aerial parts of *S. salicifolia* (Bohlmann and Zdero 1985). Figure 5.1 illustrates the structure of some monoterpenes present in the essential oils of *Stevia* species.

A variety of volatile sesquiterpenes have been also isolated from *Stevia*. This group comprises hydrocarbons and mono-oxygenated substances. Selected examples are included in Figure 5.2. Most of these compounds have been isolated from the lipophilic fractions obtained by column chromatography of the total crude extracts prepared from either aerial parts or roots of the analyzed plant materials. Thus, α-humulene (**12**) has been found in the aerial parts of *S. sarensis* and *S. yaconensis* (Zdero *et al.* 1988) collected in Bolivia. The roots of the North American species *S. polycephala*, *S. serrata* and *S. ovata* (Bohlmann *et al.* 1977) afforded isohumulene (**13**), while the aerial parts of *S. salicifolia* (Bohlmann and Zdero 1985) yielded γ-humulene (**14**). This compound (**14**) was also isolated from the roots of *S. aristata* collected in Paraguay (Zdero *et al.* 1987), while himachalol (**16**) was isolated from the aerial parts of *S. berlandieri* collected in Mexico (Bohlmann and Zdero 1985). The roots of *S. myriadenia* collected in northeastern Brazil

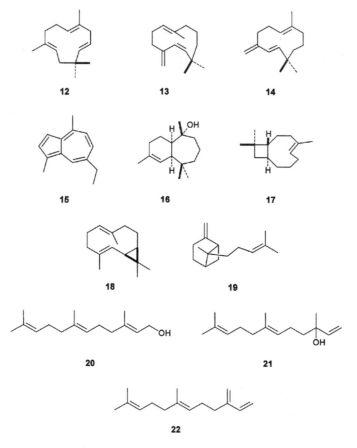

Figure 5.2 Selected volatile sesquiterpenoids (**12–22**) from *Stevia* species.

(Bohlmann *et al.* 1982) yielded caryophyllene (**17**), while *S. berlandieri*, *S. salicifolia* (Bohlmann and Zdero 1985) and *S. amambayensis* (Schmeda-Hirschmann *et al.* 1986) afforded caryophyllene epoxide. The 4α,5β-epoxy-8-hydroxy derivative of compound **17** was isolated from *S. triflora* (Amaro-Luis and Adrián 1997). Bicyclogermacrene (**18**) has been found in *S. achalensis* (Bohlmann *et al.* 1986), *S. amambayensis* (Schmeda-Hirschmann *et al.* 1986), *S. myriadenia* (Bohlmann *et al.* 1982) and *S. polyphylla* (Zdero *et al.* 1988). β-Bergamotene (**19**) was isolated from the roots of *S. amambayensis* (Schmeda-Hirschmann *et al.* 1986). The aerial parts of *S. ovata* (Bohlmann *et al.* 1977) afforded the acyclic sesquiterpenoids nerolidol (**21**) and β-farnesane (**22**), together with the polyacetylene pentaynene. Nerolidol (**21**) has been also obtained from the aerial parts of *S. salicifolia* (Bohlmann and Zdero 1985). Volatile bisabolane, longipinane, and germacrane derivatives are present as constituents of the essential oils of *Stevia*. However, these chemical types of sesquiterpenes will be reviewed in the following sections. It is worthy of mention that no functionalized humulane, bicyclogermacrane, or bergamotane derivatives have been so far isolated from this genus.

ACETOPHENONE, BENZOFURAN, AND CHROMENE DERIVATIVES

Figure 5.3 shows the structures of acetophenone, benzofuran, and chromene derivatives so far isolated from *Stevia* species. *p*-Hydroxyacetophenone (**23**) was isolated from the aerial parts of *S. setifera* from Bolivia, together with the benzofuran **25** (Bohlmann *et al.* 1979). The aerial parts of *S. hyssopifolia*, collected in Chile, yielded an acetophenone derivative (**24**) and ageratochromene (**35**) (Zdero *et al.* 1991). The roots of a Mexican collection of *S. lemmonia* afforded the benzofuran **26** and the chromene **36**, while the aerial parts yielded the benzofuran **30** (Bohlmann and Zdero 1985). This substance (**30**) has been isolated from the roots of *S. breviaristata* (Hernández *et al.* 1994), the aerial parts of *S. salicifolia* (Bohlmann and Zdero 1985), and the roots of *S. mercedensis* (Bohlmann *et al.* 1986). In addition, the aerial parts of *S. salicifolia* yielded the benzofurans **27–29** and the chromene **36** (Bohlmann and Zdero 1985), while the roots of *S. mercedensis* afforded the chromene **31** (Bohlmann *et al.* 1986). Methylripariochromene (**32**) was isolated from the leaves of a collection of *S. serrata*, which was cultivated at an experimental station in Japan (Kohda *et al.* 1976b). The aerial parts of *S. monardaefolia* afforded chromenes **33** and **34** (Quijano *et al.* 1982), while the aerial parts of *S. purpurea* yielded chromone **37** (Bohlmann *et al.* 1976). These kinds of benzofuran and chromene derivatives have been suggested to represent chemical markers of the tribe Eupatorieae (Domínguez 1977).

FUNCTIONALIZED SESQUITERPENOIDS

Bisabolanes

Only ten bisabolane derivatives have been isolated from the genus *Stevia*, and representative structures of this class of compound are summarized in Figure 5.4. The absolute configuration of bisabolen-1-one (**38**) isolated from *S. purpurea* (Bohlmann *et al.* 1976) was established by synthesis of its enantiomer starting from (R)-(+)-citronellal (Ghisalberti *et al.* 1979). In addition, *S. purpurea* yielded the 2α,3α-epoxide of **38**, which has been also isolated from *S. ovata* (Bohlmann *et al.* 1977) and *S. samaipatensis* (Zdero *et al.* 1988), while its 2α,3α-epoxide has been found in *S. berlandieri*, *S. salicifolia* (Bohlmann and Zdero 1985), and *S. amambayensis* (Schmeda-

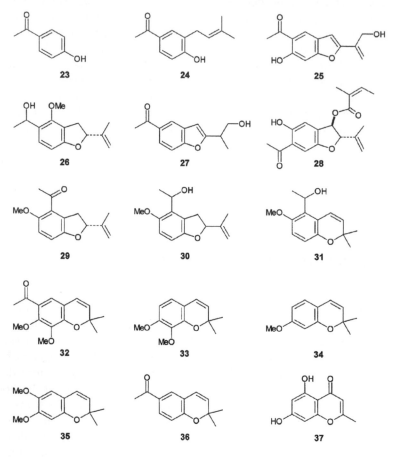

Figure 5.3 Acetophenone, benzofuran and chromene derivatives (**23–37**) from *Stevia* species.

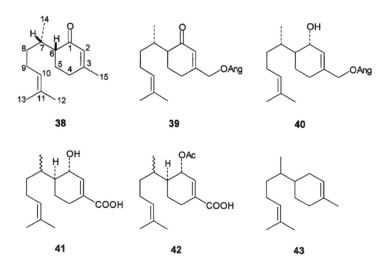

Figure 5.4 Selected bisabolanes (**38–43**) from *Stevia* species.

Hirschmann *et al.* 1986). The 15-hydroxy and 15-acetyloxy derivatives of **38** were isolated from the roots of *S. ovata* (Bohlmann *et al.* 1977), while 15-angeloyloxybisabolen-1-one (**39**) was isolated from *S. amambayensis* (Schmeda-Hirschmann *et al.* 1986) together with its $2\alpha,3\alpha$-epoxide and the alcohol **40**. This alcohol was previously isolated from the roots of *S. myriadenia* (Bohlmann *et al.* 1982). Bisabolene acid derivatives **41** and **42** were isolated as their methyl esters from the aerial parts of *S. salicifolia* (Calderón *et al.* 1984). Treatment of **41** with Jones reagent afforded the corresponding keto derivative, which is a positional isomer of dehydrojuvabione. The bisabolene **43** has been isolated from the aerial parts of *S. hyssopifolia* (Zdero *et al.* 1991). After reviewing the literature on the bisabolane derivatives found in *Stevia*, it may be pointed that additional chemical work is necessary to fully determine the relative and absolute stereochemistry of several members of this group. Finally, a very short synthesis of racemic **38** has been described using tandem phenylation-reduction as the key reaction (Singh *et al.* 1995).

Longipinanes

Longipinane derivatives are one of the most representative classes of compounds present in *Stevia*. This group comprises 61 compounds, 43 longipinenes and 18 longipinanes, which can be classified systematically into 18 basic skeletons (**44–61**) according to their degree of oxygenation and the position of the functional groups. As can be seen in Figure 5.5, the characteristic functional groups present in these compounds are a carbonyl group at C-1 and hydroxyl or ester groups at C-7, C-8, C-9, and/or C-15. Longipinen-2-one (**46**), isolated from *S. salicifolia* (Bohlmann and Zdero 1985), represents an exception to this behavior. In saturated compounds (**55–61**), the methyl group at C-3 occurs always with an α configuration while the hydroxyl or ester groups at C-7, C-8 and C-9 are always β, α, and α oriented, respectively.

It is worth mentioning that the structure and stereochemistry of derivatives with a longipinanetriolone skeleton (**59**) has been revised in light of the X-ray diffraction analysis of rastevione (**59a**, $R^1 = R^2 = $ Ang, $R^3 = $ H), isolated in high yields from the roots of *S. serrata* and *S. rhombifolia* (Román *et al.* 1981). Further chemical correlation of rastevione with unsaturated compounds of type **49** (longipinenediolones) and **52** (longipinenetriolones) also allowed the correction of the structure and stereochemistry of this kind of compound published prior to 1985 (Bohlmann *et al.* 1976; 1977; 1979), in which inversion in the relative stereochemistry of chiral centers at C-7 and C-9, and in the case of saturated compounds, the configuration at C-3, as well as the interchange of ester substituents at position C-7 and C-9 must be taken into consideration (Román *et al.* 1985). The absolute configuration of longipinene derivatives from this genus has been assigned by chemical correlation with (+)-longipinene (Bohlmann and Zdero 1985) and by circular dichroism studies (Joseph-Nathan *et al.* 1986). Conformational analysis of these substances has revealed that the seven-membered ring exists in a twist-chair conformation whereas the C-7 and C-8 substituents remain with a *pseudo*-equatorial orientation and the C-9 substituent is *pseudo*-axial (Joseph-Nathan *et al.* 1986). Longipinene (**44**) has been found in *S. boliviensis* and *S. mandonii* (Bohlmann *et al.* 1979), while longipin-2-en-1-one (**45**) was isolated from the aerial parts of *S. salicifolia* (Bohlmann and Zdero 1985). Longipinen-7β-ol-1-one esters (**47**) were isolated from *S. lemmonia*, *S. salicifolia* (Bohlmann and Zdero 1985), *S. mercedensis* (Bohlmann *et al.* 1986), and *S. serrata* (Sánchez-Arreola *et al.* 1995), while the alcohol **47a** (R = H) was found in the roots of *S. polycephala* (Bohlmann *et al.* 1977). Only two longipinen-7β,8α-diol-1-one diesters (**48**) have been isolated from *Stevia*. They were found in *S. serrata* (Bohlmann *et al.* 1977; Sánchez-Arreola *et al.* 1995) and *S. lemmonia* (Bohlmann and Zdero 1985).

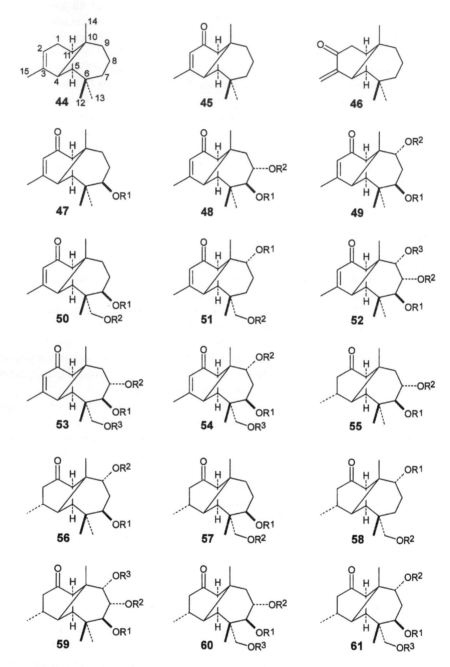

Figure 5.5 Different types of longipinanes (**44–61**) from *Stevia* species.

In contrast, more than twenty longipinene-7β,9α-diol-1-one diesters (**49**) have been detected in the genus. Each of the analyzed species generally produces several of these substances generating complex mixtures that are difficult to separate. The most common ester residues are angelate (Ang), tiglate (Tigl), senecioate (Sen), methacrylate (Meacr), and acetate (Ac), although epoxyangelate (Epang), isovalerate (*i*-Val) and isobutyrate (*i*-Bu) residues have been also found. Such diesters have been isolated from *S. boliviensis*, *S. mandonii* (Bohlmann *et al*. 1979), *S. lucida* (Amaro *et al*. 1988), *S. jaliscensis* (Bohlmann *et al*. 1977), *S. salicifolia* (Bohlmann and Zdero 1985), *S. serrata*, *S. polycephala* (Bohlmann *et al*. 1977), and *S. subpubescens* (Román *et al*. 1989). In particular, the diangelate derivative of skeleton **49** ($R^1 = R^2 = $ Ang) has been found as the most widely distributed longipinene diester, since it has been found in *S. aristata* (Zdero *et al*. 1987), *S. mercedensis* (Bohlmann *et al*. 1986), *S. berlandieri*, *S. lemmonia* (Bohlmann and Zdero 1985), and *S. potrerensis* (Gil *et al*. 1987). A procedure for analyzing the complex mixtures of these lipophilic substances has been developed. The analytical sequence consists essentially of the separation of small quantities of each component by reversed-phase HPLC, determination of the ^{1}H-NMR spectrum of the isolated fractions, and comparison of the spectral data with a systematic series of previously prepared longipinene-$7\beta,9\alpha$-diol-1-one diesters synthesized from the parent diol **49a** ($R^1 = R^2 = $ H) (Joseph-Nathan *et al*. 1989). This methodology was successfully applied to the mixture present in the roots of *S. subpubescens* var. *intermedia* from which twelve diesters were identified (Joseph-Nathan *et al*. 1991). Two new substances of type **49** were isolated using this analytical approach from the roots of *S. lucida* (Guerra-Ramírez *et al*. 1998).

The aerial parts of two Argentinean species, *S. achalensis* and *S. potrerensis*, have also afforded several longipinanes. Diol **50a** ($R^1 = R^2 = $ H) and diester **50b** ($R^1 = $ Ac, $R^2 = $ Ang) were found in the first one (Bohlmann *et al*. 1986), while the second species gave angelate esters with the skeleton types **50**, **54**, **56**, and **61** (Gil *et al*. 1987). Longipinanes belonging to classes **50** and **54** have also been found in *S. eupatoria* (Zdero *et al*. 1991). A collection of *S. lemmonia* (Bohlmann and Zdero 1985) afforded diangelate and triangelate esters of types **50**, **52**, **53**, **59**, and **60**, while the roots of *S. origanoides* yielded derivatives with basic skeletons **50** and **54**, bearing angelate, senecioate and acetate ester residues (Cerda-García-Rojas *et al*. 1993). Only one compound of type **51** ($R^1 = R^2 = $ Ang) has been obtained from the roots of *S. elatior* collected in Guatemala (Bohlmann *et al*. 1977). Extensive studies on the roots of *S. serrata* (Bohlmann *et al*. 1977; Román *et al*. 1981; Román *et al*. 1993; Sánchez-Arreola *et al*. 1995) have yielded compounds of types **52** and **55**, together with the two positional isomers of rastevione, thus completing the isomeric series **59a** ($R^1 = R^2 = $ Ang, $R^3 = $ H), **59b** ($R^1 = R^3 = $ Ang, $R^2 = $ H) and **59c** ($R^1 = $ H, $R^2 = R^3 = $ Ang). Additionally, triflorestevione (**59d**, $R^1 = R^2 = $ Tigl, $R^3 = $ H) was isolated from the aerial parts of *S. triflora* collected in Venezuela (Amaro *et al*. 1988). The roots of *S. aristata* (Zdero *et al*. 1987) and *S. boliviensis* (Bohlmann *et al*. 1979) produced longipinanes of type **56**, while the roots of *S. viscida* (Román *et al*. 1995a) afforded the only two derivatives of type **58**. Finally, a collection of *S. yaconensis* yielded derivatives of type **57** and **61** (Zdero *et al*. 1988).

A large number of tricyclic sesquiterpene hydrocarbons and some of their alcohol derivatives are used in the perfume industry as scents. The interest in the generation of new hydrocarbon skeletons with potential use as fragrances has encouraged a group of Mexican researchers to carry out the study of molecular rearrangements of longipinanes from *Stevia* (Román *et al*. 1991; 1992; 1995a; 1996; Joseph-Nathan and Cerda-García-Rojas 1994). These tricyclic strained substances have produced several novel skeletons since bond migrations can be promoted easily to release the four-membered ring strain. Some rearrangement products of longipinane derivatives are illustrated in Figure 5.6. Acid treatment of rastevione (**59a**, $R^1 = R^2 = $ Ang, $R^3 = $ H) afforded, after alkaline hydrolysis of the ester groups, the moreliane derivative **62** by a Wagner–Meerwein rearrangement. Likewise, diacetate **59e** ($R^1 = R^2 = $ Ac, $R^3 = $ H), its C-3 epimer, and diacetate **52a** ($R^1 = R^2 = $ Ac, $R^3 = $ H) yielded, after hydrolysis, compounds **62**, **63**, and **64**, respectively (Román *et al*. 1991). Longipinan-$7\beta,9\alpha$-diol-1-one (**56a**, $R^1 = R^2 = $ H) gave, in addition to alcohol **65**, ketone **66** which was formed from **65** itself by a transannular hydride migration (Román *et al*. 1992). On the other hand, alkaline treatment

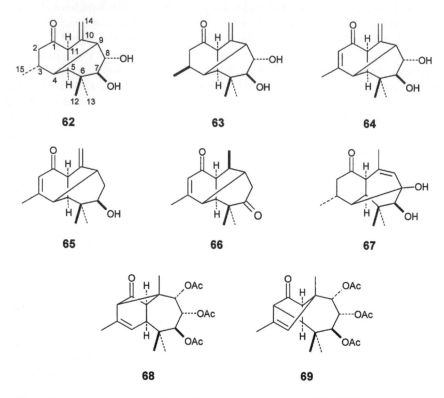

Figure 5.6 Rearrangement products of longipinane derivatives (**62–69**) from *Stevia* species.

of rastevione mesylate (**59f**, $R^1 = R^2 = Ang$, $R^3 = Ms$) mainly afforded the arteagane derivative **67** (Román *et al*. 1995c). The mechanistic pathway for this transformation involved a 1,3-bond migration (Román *et al*. 1996). All these new structures have been studied by X-diffraction analysis. In addition, photochemical rearrangement of triester derivative **52b** ($R^1 = R^2 = R^3 = Ac$) produced, *via* [1,3] sigmatropic shifts, the vulgarone A derivative **68** and the pingilonene derivative **69** (Joseph-Nathan *et al*. 1996).

Germacranes

This group comprises 58 substances among three germacrenes, 34 germacranolides, three melampolides, 16 heliangolides and two *cis,cis*-germacranolides. The structures drawn in Figure 5.7 can be used to summarize the *trans,trans*-germacrenes present in this genus. Thus, germacrene D (**70**) is a very common component present in the genus *Stevia*, since it has been isolated from at least 22 species. The alcohol **71** was found in the roots of *S. ovata* (Bohlmann *et al*. 1977). The roots of *S. myriadenia* yielded small amounts of the 15-acetyloxy derivative of **71** (Bohlmann *et al*. 1982). Costunolide (**72**) was isolated from the aerial parts of *S. yaconensis* and *S. chamaedrys* (Zdero *et al*. 1988). The latter species also afforded the 8β-(4′-hydroxytigloyl)oxy derivative of costunolide 20-desoxyeupatoriopicrin (Zdero *et al*. 1988). Other costunolide derivatives have been isolated from the aerial parts of *S. lemmonia* (Bohlmann and Zdero 1985), *S. mercedensis* (Bohlmann *et al*. 1986), and *S. sanguinea* (De Hernández *et al*. 1997). Carmeline (11α, 13-dihydro-8β,3α-diacetyloxycostunolide) was isolated from *S. serrata* collected in the southern

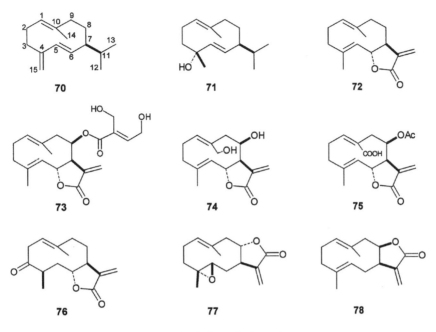

Figure 5.7 Selected germacranes (**70–78**) from *Stevia* species.

region of Mexico City (Salmón *et al.* 1975). The aerial parts of *S. sarensis* (Zdero *et al.* 1988) yielded 20-desoxyeupatoriopicrin, eupatoriopicrin (**73**), and 20-tigloyloxyeupatoriopicrin. Compound **73** was also found in *S. procumbens* (Sosa *et al.* 1985) and *S. maimarensis* (Hernández *et al.* 1996b). In addition, the latter plant material afforded six germacranolides which are derived from 8β,14-dihydroxycostunolide (**74**). These substances possess a tiglate or a 4-hydroxytiglate moiety at position C 8, a tiglate at C-14 and/or a hydroxyl or acetyloxy group at C-3 (Hernández *et al.* 1996b). Four different ester derivatives of 8β,14-dihydroxycostunolide (**74**) have been isolated from the aerial parts of *S. vaga*. Other 8β,14-dihydroxycostunolide derivatives have been isolated from *S. aristata* (Zdero *et al.* 1987) and *S. breviaristata* (Hernández *et al.* 1994). The germacranolide carboxylic acid **75** was isolated from the aerial parts of *S. potrerensis* collected in Argentina (Gil *et al.* 1987). Three germacranolide analogs of compound **75** bearing tiglate, 4-hydroxytiglate, or 4-acetyloxytiglate esters instead of the acetate group at C-8 have been isolated from *S. aristata* (Zdero *et al.* 1987) and *S. amambayensis* (Schmeda-Hirschmann *et al.* 1986). The aerial parts of *S. grisebachiana* collected in Argentina afforded 4-*epi*-tansanin (**76**) whose structure was confirmed by X-ray analysis (Sigstad *et al.* 1991). The 8β-sarracenyl and 8β-acetyl-oxysarracenyl derivatives of **76** have been found in the aerial parts of *S. jujuyensis*, together with 8β-acetylsarracenyloxy-3β-hydroxycostunolide (Gil *et al.* 1992; De Gutiérrez *et al.* 1992). Leaves and flowers of *S. ovata* yielded the 12,8α-germacranolide 4α,5β-epoxy-8-*epi*-inunolide (**77**) whose structure was confirmed by chemical correlation with 11,13-dehydroeriolin studied by X-ray diffraction (Calderón *et al.* 1987b). The first 12,8β-germacrolide **78** was obtained from the aerial parts of *S. polyphylla* from Bolivia, and its structure was elucidated by NMR spectroscopy and chemical transformations (Zdero *et al.* 1988).

Figure 5.8 shows the structure of selected heliangolides isolated from *Stevia*. Eucannabino-lide (**79**) was isolated from the aerial parts of *S. origanoides* collected in Mexico (Calderón *et al.*

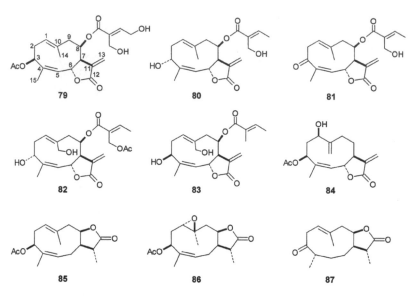

Figure 5.8 Selected heliangolides (**79–87**) from *Stevia* species.

1987a), together with eucannabinolide 19-*O*-acetate. Eucannabinolide (**79**) was also obtained from the aerial parts of *S. sarensis* from Bolivia (Zdero *et al.* 1988). The aerial parts of *S. monardaefolia* collected in Mexico afforded 11,13-dihydroeucannabinolide which represented the first heliangolide isolated from the genus (Gómez *et al.* 1983). Its structure was determined from spectroscopic and chemical data. Four heliangolides were isolated from the aerial parts of *S. jujuyensis* collected in Argentina. Their structures corresponded to lactone **80**, its 3-*O*-acetyl derivative (Gil *et al.* 1992), lactone **81**, and its 20-*O*-acetyl derivative (De Gutiérrez *et al.* 1992). The aerial parts of *S. vaga* from Argentina yielded three heliangolides, **82, 83**, and its 20-acetyloxy derivative (Hernández *et al.* 1996a). The aerial parts of *S. alpina* collected in Argentina (De Heluani *et al.* 1989) afforded 3-acetylpulverolide (**84**) together with two known heliangolides. The aerial parts of *S. yaconensis* var. *aristifera*, from Argentina, yielded three heliangolides **85–87** with an unusual *cis*-fused lactone ring closed to C-8 (Gil *et al.* 1990).

Figure 5.9 shows the melampolides and *cis,cis*-germacranolides isolated from *Stevia* species. Sesquiterpene lactones of the melampolide type have been only found in South American specimens of this genus. The aerial parts of *S. amambayensis* collected in Paraguay yielded the novel melampolide derivative **88** and its 4Z-isomer, the *cis,cis*-germacranolide **89** (Schmeda-Hirschmann *et al.* 1986). A year later, the corresponding desacetyl derivatives of both substances **90** and **91**, together with **88**, were found in a collection of *S. aristata* from Paraguay (Zdero *et al.* 1987). Compound **53** was also isolated from the aerial parts of *S. breviaristata* from Argentina (Hernández *et al.* 1994). Finally, the aerial parts of *S. vaga* afforded the melampolide **92**, whose structure has been recently determined by comparison of its NMR data with appropriate previously reported models (Hernández *et al.* 1996a).

Elemanes

Only five elemane derivatives have been found in *Stevia*, whose structures are illustrated in Figure 5.10. Elemanolides **93–95** were isolated from the aerial parts of *S. achalensis* collected

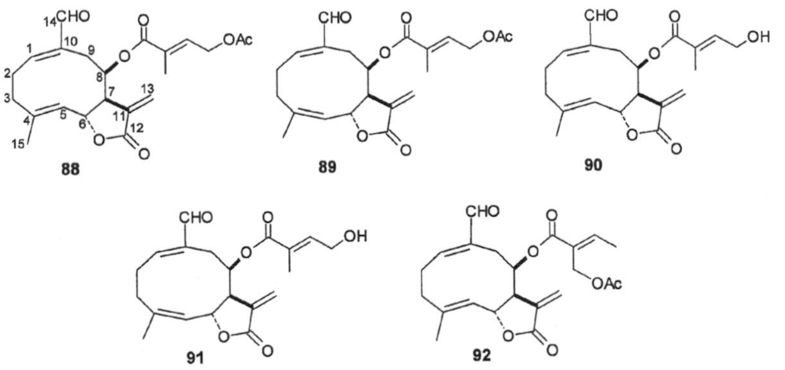

Figure 5.9 Selected melampolides and *cis,cis*-germacranolides (**88–92**) from *Stevia* species.

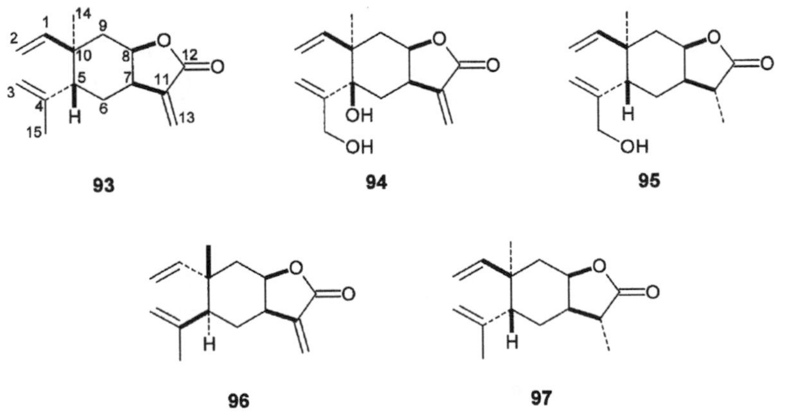

Figure 5.10 Selected elemanolides (**93–97**) from *Stevia* species.

in Argentina (Bohlmann *et al*. 1986). Compound **93** was also found in the aerial parts of *S. poly-phylla* from Bolivia, together with its 5,10-*epi* derivative **96** (Zdero *et al*. 1988). The aerial parts of *S. yaconensis* var. *aristifera* yielded the elemadienolide callitrin **97** (Gil *et al*. 1990).

Eudesmanes

The structures included in Figure 5.11 can be used to summarize the 24 eudesmane derivatives so far isolated from plants in the genus *Stevia*. The aerial parts of *S. chamaedrys* collected in Chile afforded costic acid (**98**) and its Δ^3-isomer (**99**), together with the eudesmanolides reynosin (**101**) and santamarin (**102**) (Zdero *et al*. 1991). The aerial parts of *S. achalensis* have turned out to be a rich source of eudesmane derivatives (Bohlmann *et al*. 1986), since they yielded the Δ^4-isomer of costic acid (**100**), the 2α-hydroxy and the 2-keto derivatives of **99**, the 3β-hydroxy and the 3-keto derivatives of **100**, the 4α,11β,13,15-tetrahydro-3α-hydroxy derivative of **105**, the 3-keto derivative of **106**, named isotelekin, eudesmanolide **107**, the 3-keto derivative of **107**, the 1α-hydroxy derivative of **108**, and onoseriolide (**109**). The aerial parts of *S. polyphylla* from Bolivia yielded the eudesmanolide **108**, together with costic acid (**98**), isoalantolactone

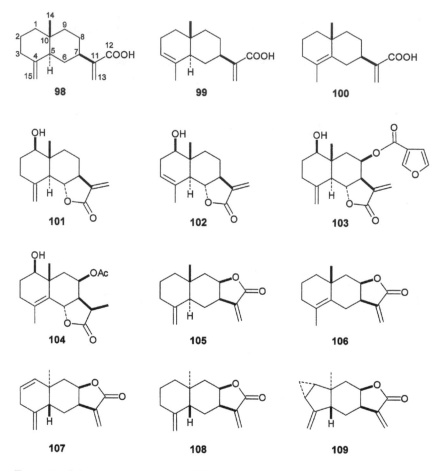

Figure 5.11 Selected eudesmanes (**98–109**) from *Stevia* species.

(**105**), and its Δ^4-isomer (**106**). Since eudesmanolide **108** was obtained from germacranolide **78** on standing in chloroform containing a trace of hydrochloric acid, it cannot be excluded that **108** represents an artifact of extraction (Zdero *et al.* 1988). The aerial parts of *S. breviaristata* from Argentina afforded an eudesmane derivative whose structure corresponded to 8β-(4-hydroxy-tigloyl)oxysantamarin (Hernández *et al.* 1994). This substance has also been isolated from the aerial parts of *S. maimarensis*, from Argentina (Hernández *et al.* 1996b), together with compounds **103** and 8β-(4-hydroxytigloyl)oxyreynosin. The aerial parts of *S.* aff. *tomentosum* collected in Mexico afforded three 11,13-dihydroeudesmanolides. Their structures, which corresponded to **104** and its Δ^3- and Δ^4-isomers, were elucidated by chemical correlation and spectroscopic methodologies (Martínez-Vázquez *et al.* 1990).

Eremophilanes

Literature investigation has revealed that only two *Stevia* species have yielded eremophilane derivatives. These compounds comprise a group of 12 substances whose parent skeletons are represented in Figure 5.12. The aerial parts of *S. achalensis* contained eremophilenic acid (**110**),

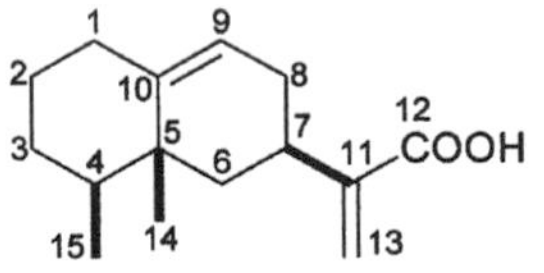

Figure 5.12 Basic structures of eremophilane derivatives (**110–113**) from *Stevia* species.

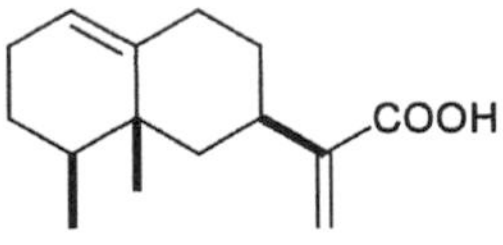

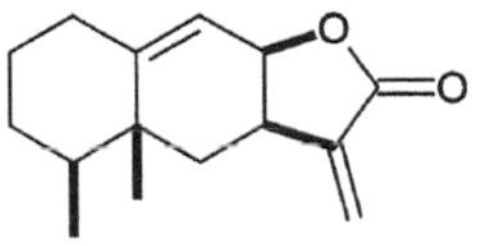

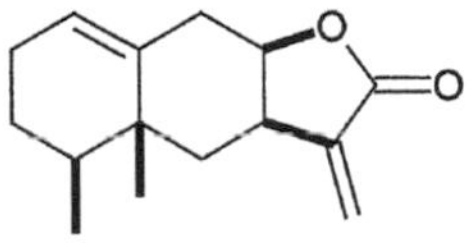

the 2-keto derivative of **111**, eremophilanolides **112** and **113**, together with eight substances whose structures corresponded to the 2β-hydroxy, the 3-keto, and the 3β-hydroxy derivatives of **110**; the 3β-hydroxy, and the 3-keto derivatives of **112**; and the 3-keto, the 3β-hydroxy, and the 3α-hydroxy derivatives of **113** (Bohlmann *et al*. 1986). The aerial parts of *S. polyphylla* yielded the eremophilane derivatives **110–113** (Zdero *et al*. 1988).

Guaianes

Guaianolides are the largest group of compounds isolated from 25 of the phytochemically analyzed *Stevia* species. This group comprises 91 substances whose structures are represented by a selection of 30 structural formulae in Figure 5.13. The aerial parts of *S. alpina* afforded two epoxy-estafiatins, compound **114** and its 10-*epi*-derivative, together with achillin (**115**), its 11,13-dehydro-derivative, 2-oxo-8-deoxyligustrin (**116**), and estafiatin (**117**). The stereochemistry of the new substances was verified by X-ray analysis of natural product **114** (De Heluani *et al*. 1989). Estafiatin (**117**) was also isolated from the aerial parts of *S. boliviensis* (Bohlmann *et al*. 1979). The aerial parts of *S. breviaristata* yielded breviarolide (**118**) (Oberti *et al*. 1986). Reinvestigation of the minor components of this species afforded 10-*epi*-breviarolide, 14-*O*-acetylbreviarolide, 19-*O*-acetylbreviarolide, and 19-deoxybreviarolide (Hernández *et al*. 1994). The aerial parts of *S. chamaedrys* gave six known substances: eupahakonenin B (**119**) and five derivatives, while the aerial parts of *S. eupatoria* yielded the guaianolide (4,5-dihydroxytigloyl)-eupahakonenin B, together with two known derivatives (Zdero *et al*. 1991). The aerial parts of *S. gilliesii*, collected in Argentina, afforded four known sesquiterpene lactones and the guaianolide **120** (Hernández *et al*. 1995). The aerial parts of *S. grisebachiana* afforded a series of seven hydroxylated guaianolides, compounds **121–123**, together with estafiatin (**117**), its 11α,13-dihydro-derivative, the 11,13-dehydro-derivative of **115**, and lactone **137** (Sigstad *et al*. 1991). The aerial parts of *S. lemmonia*, collected in Mexico, yielded two guaianolides, **124** and its 2α,3α-epoxide (Bohlmann and Zdero 1985). The 20-deoxo-derivative of eupahakonenin B (**119**) was isolated from the aerial parts of *S. mercedensis* collected in Argentina. Also, a mixture of 2,5-dihydro-5-hydroxy-3-furylcarboxylates (**126**) and a mixture of its Δ^{10}-isomers were obtained from this species, together with eupahakonenin B (**119**) and lactone **125** (Bohlmann *et al*. 1986). The aerial parts of *S. myriadenia* yielded two sesquiterpene lactones, whose structures corresponded to **127** and its 4,5-dihydroxytigloyl derivative (Bohlmann *et al*. 1982). The aerial parts of *S. pilosa*, collected in

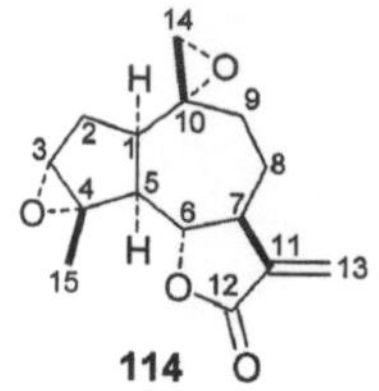

114

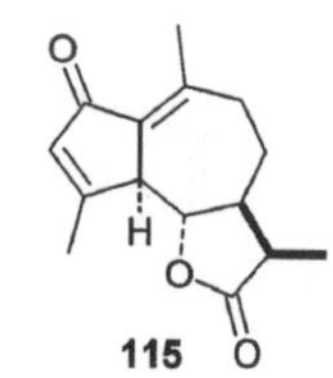

115

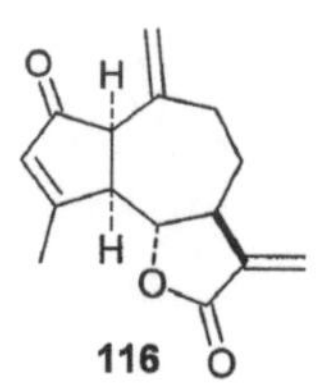

116

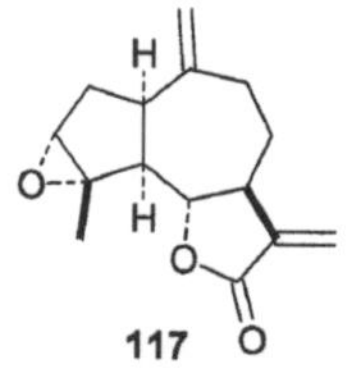

117

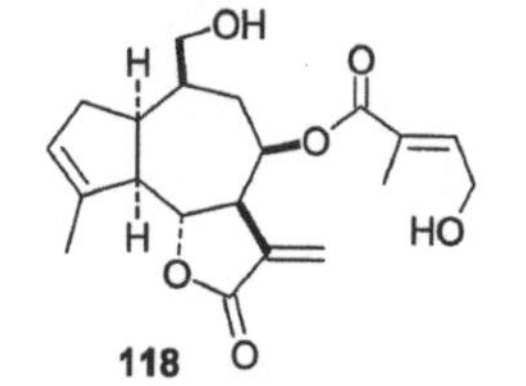

118

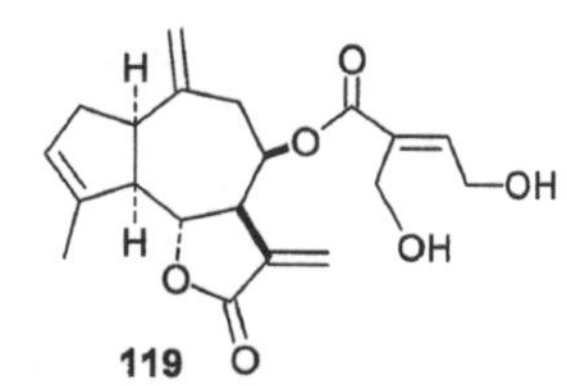

119

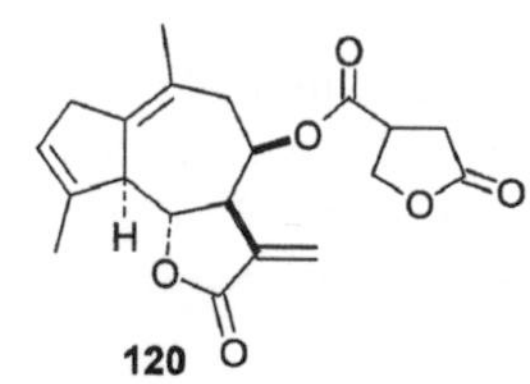

120

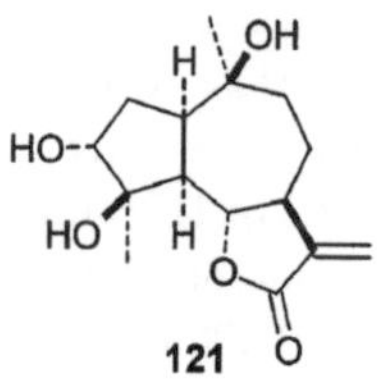

121

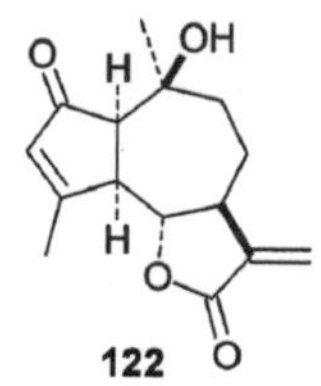

122

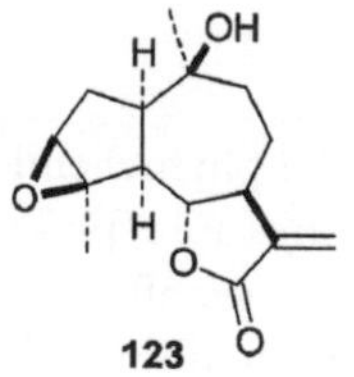

123

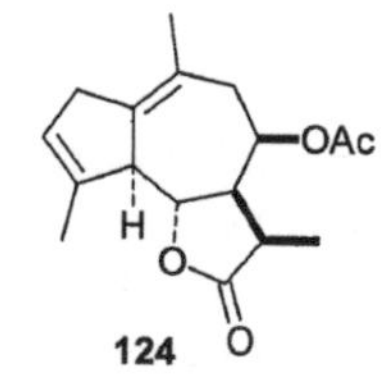

124

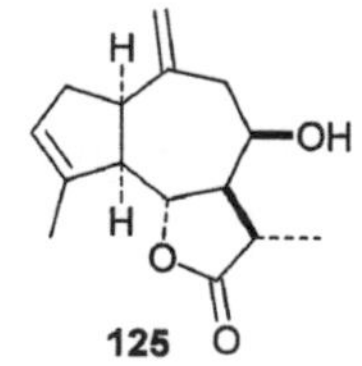

125

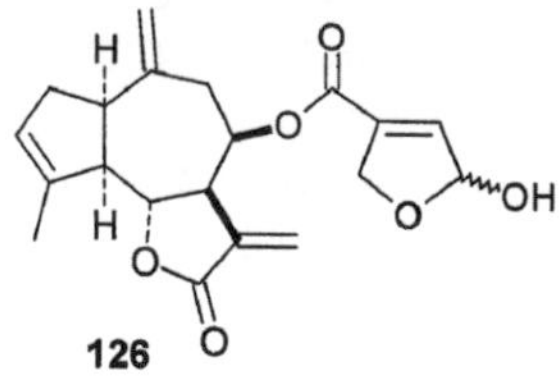

126

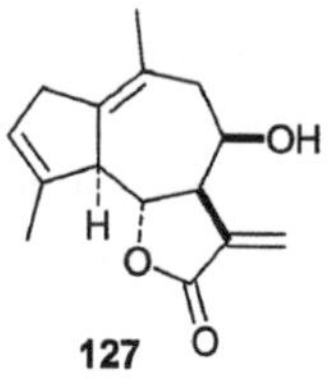

127

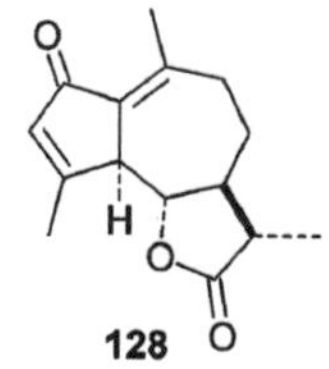

128

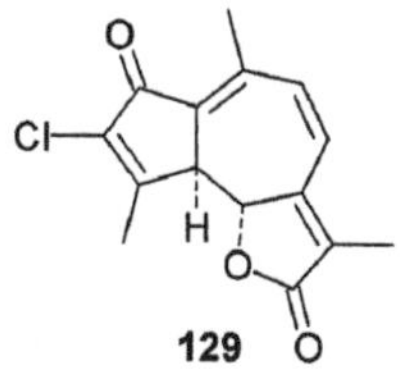

129

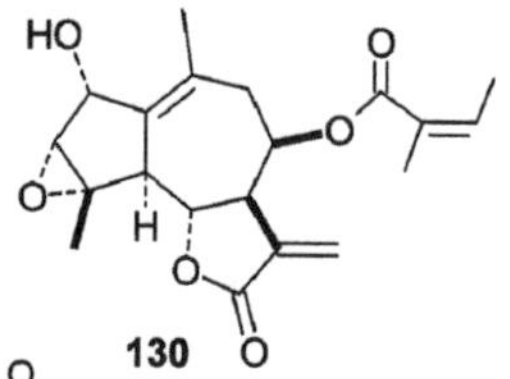

130

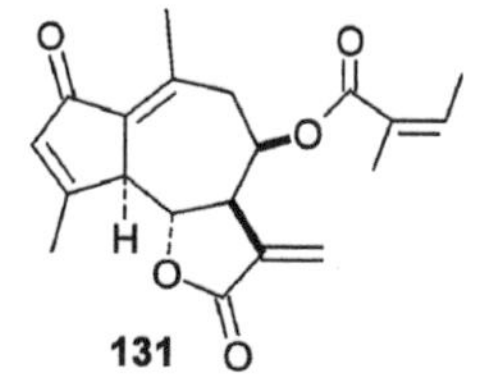

131

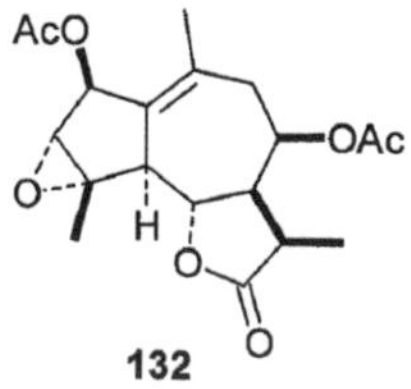

132

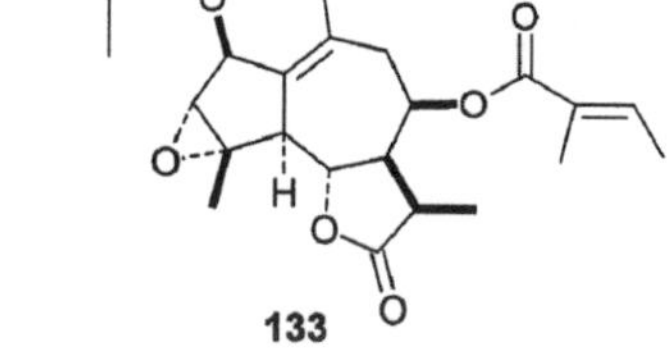

133

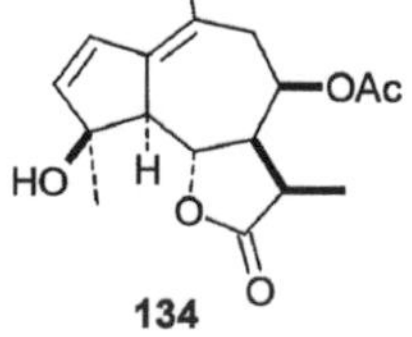

134

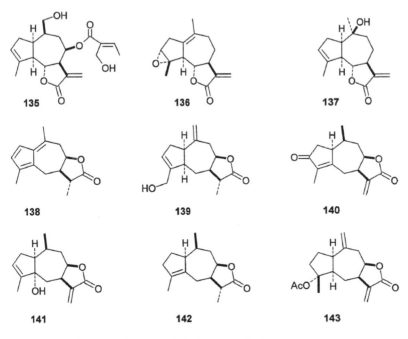

Figure 5.13 Selected guaianolides (**114–143**) from *Stevia* species.

Mexico, afforded leukodin (**128**), whose conformation and that of achillin (**115**) were studied by NMR and X-ray diffraction analysis (Martínez *et al.* 1988). Four guaianolide derivatives were isolated from the aerial parts of *S. sanguinea* collected in Argentina. Their structures were determined by NMR and mass spectroscopy as the chloro-derivative **129**, the epoxylactone **130**, its 2-epimer, and angelate **131** (Gil *et al.* 1989). A recent reinvestigation of the aerial parts of this species afforded twelve new and five known guaianolides. This study revised the structure of compound **130** (De Hernández *et al.* 1997). The aerial parts of *S. satureiafolia* and *S. procumbens* afforded eupahakonenin B (**119**) (Sosa *et al.* 1984; 1985), while the aerial parts of *S. sarensis* afforded eupahakonenin B (**119**), together with its Δ^{10}-isomer (Zdero *et al.* 1988). The aerial parts of *S. setifera* from Bolivia also yielded eupahakonenin B (**119**), guaianolide **125**, and its 11,13-dehydro-derivative (Bohlmann *et al.* 1979). The stereochemistry of the chiral center C-1 in these substances was corrected by Zdero and coworkers (1988; 1991). Therefore, the guaianolide derivatives from *S. setifera* correspond to the 1αH-guaianolides and not to the 1βH series as stated in the original paper (Bohlmann *et al.* 1979). Three epoxyguaianolides: christinine (**132**), christinine II (**133**), and the 8β-acetate-2β-isovalerate, named christinine III, were isolated from *S. serrata*. The relative stereochemistry and the conformation of christinine (**132**) were deduced mainly from the ^{1}H-NMR data measured in the presence of a chemical shift reagent (Salmón *et al.* 1973, 1977). Two prochamazulene lactones, steviserrolide A (**134**) and its C-4 epimer, steviserrolide B, were isolated from the leaves of *S. serrata*. Their structures were established by NMR (Calderón *et al.* 1989). The aerial parts of *S. vaga* yielded four guaianolides whose structures were elucidated as **135**, its 20-O-acetate, its C-10 epimer, and the corresponding 20-O-acetate (Hernández *et al.* 1996a). The aerial parts of *S. yaconensis* collected in Bolivia afforded 2-oxo-8-deoxyligustrin (**116**) and estafiatin (**117**) (Zdero *et al.* 1988). The aerial parts of *S. yaconensis* var. *subeglandulosa* afforded, in addition to the known guaianolides ludartin (**136**) and the 11,13-dehydro-derivative of leucodin (**128**), 10-*epi*-8-deoxycumambrin B (**137**), whose relative and absolute configuration was determined by X-ray crystallography and circular dichroism measurements (Sosa *et al*. 1989). The aerial parts of *S. samaipatensis* and *S. yaconensis* var. *aristifera* gave several guaianolides including stevisamolide (**138**), its 2,5α-dihydro-derivative, and alcohol **139** and its acetate (Zdero *et al*. 1988; Gil *et al*. 1990). Two guaianolides, achalensolide (**140**) and its 11β,13-dihydro-derivative, were isolated from the aerial parts of *S. achalensis*. Their structures were determined by NMR, chemical transformations, and X-ray diffraction analysis of **140** (Oberti *et al*. 1983). Both substances have also been isolated from *S. sarensis* and *S. polyphylla* (Zdero *et al*. 1988). In addition, the aerial parts of *S. achalensis* yielded the 7α-hydroxy and the 1α-hydroxy-11β,13-dihydro-derivatives of **140**, as well as the 10,11,13,14-tetrahydro-Δ^{4,15}-derivative of **142** (Bohlmann *et al*. 1986), while the aerial parts of *S. polyphylla* afforded the guaianolides **141**, **142** and seven derivatives (Zdero *et al*. 1988). Finally, the aerial parts of *S. ovata* yielded 4-acetyl-8-*epi*-inuviscolide (**143**), together with inuviscolide (Calderón *et al*. 1987b).

Pseudoguaianes

In contrast to the large number of structurally diverse guaianolides found in the genus *Stevia*, only four pseudoguaianolides have been isolated from two Mexican species. Figure 5.14 shows their structures. The aerial parts of *S. rhombifolia* afforded the first pseudoguaianolide isolated from *Stevia*, stevin (**144**) whose structure was deduced by NMR and chemical transformations (Ríos *et al*. 1967), while the aerial parts of *S. isomeca* yielded pseudoguaianolides **145–147** (Bohlmann *et al*. 1985).

Miscellaneous sesquiterpenoids

The structures of sesquiterpenoids with a novel carbon framework isolated from *Stevia* species are illustrated in Figure 5.15. The jujuyane derivatives **148** and **149** are two rearranged germacranolides isolated from the aerial parts of *S. jujuyensis* from Argentina (Gil *et al*. 1992; De Gutiérrez *et al*. 1992). This new skeleton could be formed by biological rearrangement of

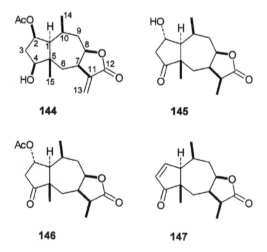

144   **145**

146   **147**

Figure 5.14 Pseudoguaianolides (**144–147**) from *Stevia* species.

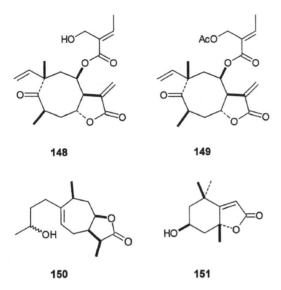

148   **149**

150   **151**

Figure 5.15 Miscellaneous sesquiterpenoids (**148–151**) from *Stevia* species.

a 4-*epi*-tansanin (**76**) derivative (Gil *et al.* 1992). The aerial parts of *S. isomeca* yielded the xanthan-olide **150** formed by ring opening of an 8,12-guaianolide, which also may be the precursor of the pseudoguaianolides isolated from this species (Bohlmann *et al.* 1985). Finally, loliolide (**151**), a degraded sesquiterpene, was isolated from the aerial parts of *S. grisebachiana* collected in Argentina (Sigstad *et al.* 1991).

DITERPENOIDS

Tetracyclic diterpenoids are the most typical C_{20} constituents of *Stevia*. The greater part of these belong to the *ent*-kaurene group, but labdane and clerodane diterpenoids have also been found in this genus. A review of diterpenoids isolated from Mexican *Stevia* species was published by Zdero and Bohlmann (1986).

Acyclic and bicyclic diterpenoids

The aerial parts of *S. myriadenia*, the only species in the genus for which acyclic diterpenoids have been reported, were shown to contain geranyl-linalol (**152**) and 20-hydroxygeranylnerol (**153**), as well as several kaurene derivatives and bicyclic diterpenoids (Bohlmann *et al.* 1982). Examples of both enantiomeric series of the labdane skeleton, i.e. the 10-Meα-labdane and the enantiomeric 10-Meβ-labdane, have been described in *Stevia*. The normal labdane type has been found in the isolated diterpenes from the aerial parts of *S. monardaefolia* (Quijano *et al.* 1982), *S. sarensis* (Zdero *et al.* 1988), and *S. lucida* (Amaro-Luis and Hung 1988; Salmón *et al.* 1983), e.g. abienol (**154**, $R^1 = R^2 = H$) and its derivatives **154a** ($R^1 = H$; $R^2 = OH$), **154b** ($R^1 = OAng$; $R^2 = OH$) and **154c** ($R^1 = R^2 = OH$). To this stereotype series also belong the 2β,15-dihydroxylabd-7-ene (**155**, $R = OH$; $X = H_2$) and the 3-oxolabd-7-en-15-ol (**155**, $R = H$;

X = O) derivatives from the aerial parts of *S. samaipatensis* (Zdero *et al.* 1988), and 2α,7α-dihydroxy-8(17),13Z-labdadien-15-oic acid (**156**) from *S. jujuyensis* (De Gutiérrez *et al.* 1992). Figure 5.16 illustrates the structures of acyclic compounds and selected examples of bicyclic diterpenes with the normal labdane stereochemistry.

The aerial parts of *S. aristata* afforded as the main constituent austroinulin 7-*O*-acetate (**157**, R¹ = OH; R² = Ac), an *ent*-labdane diterpene. A series of related derivatives of the *E/Z*-isomers of austroinulin (**157**, R¹ = OH; R² = H) and 6-deoxyaustroinulin (**157**, R¹ = R² = H) has been obtained from the aerial parts of *S. andina*, *S. aristata* (Zdero *et al.* 1987), *S. berlandieri* (Bohlmann and Zdero 1985), and *S. salicifolia* (Ortega *et al.* 1980), as well as from the leaves and flowers of *S. rebaudiana* (Sholichin *et al.* 1980; Darise *et al.* 1983). A study of *S. seleriana* afforded *ent*-labd-7-en-3-oxo-15-oic acid (**158**) and two *seco*-labdenes, compounds **159** and **159a** (Δ^{13Z}) (Escamilla and Ortega 1991). From the methanolic extract of *S. polycephala*, a clerodane-type diterpene, stephalic acid (**160**, R¹ = β-Me; R² = OH), was isolated (Angeles *et al.* 1982). Its structure and relative stereochemistry were determined by single-crystal X-ray analysis to be similar to the previously reported 2β-acetoxy-13,14Z-kolavenoic acid (**160**, R¹ = αMe; R² = H) from *S. myriadenia* (Bohlmann *et al.* 1982), with the exception of the configuration at the chiral center C-8. Molecular structures of selected *ent*-labdane and clerodane diterpenoids from *Stevia* are shown in Figure 5.17.

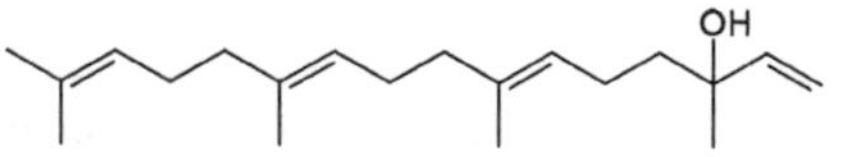

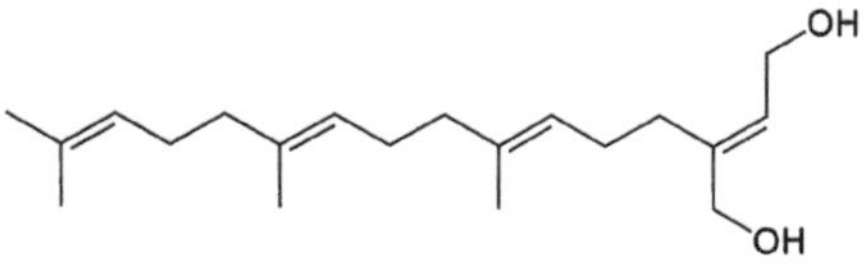

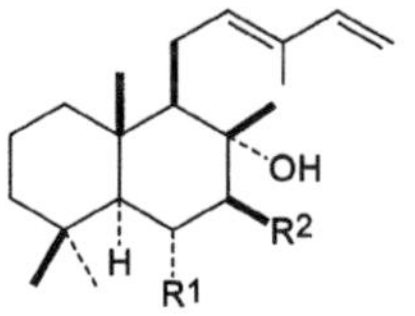

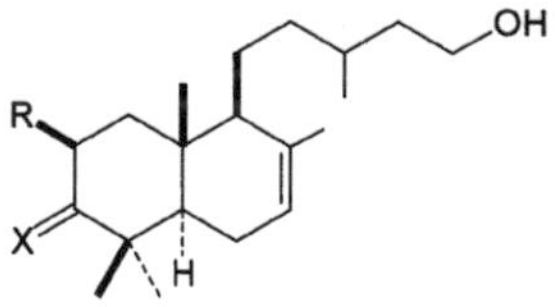

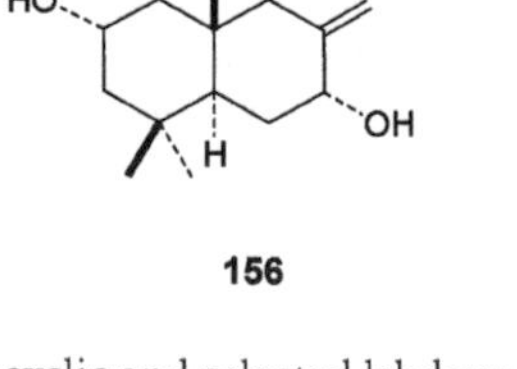

Figure 5.16 Acyclic and selected labdane diterpenoids (**152–156**) from *Stevia* species.

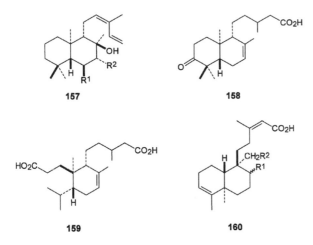

Figure 5.17 Selected *ent*-labdane and clerodane diterpenoids (**157–160**) from *Stevia* species.

Tetracyclic diterpenoids

The genus *Stevia* is well-known as the source of eight important sweet-tasting *ent*-kaurene glycosides (Figure 5.18), namely, stevioside (**161**), steviolbioside (**162**), rebaudiosides A–E (**163–167**), and dulcoside A (**168**), which are extracted commercially from the South American plant, *S. rebaudiana* (Ishikawa 1991). However, such sweet glycosides appear to have a very limited distribution in *Stevia*, since aside from *S. rebaudiana*, they were only detected in *S. phlebophylla*, among 108 other *Stevia* species examined by a phytochemical screening procedure, using a combination of TLC and HPLC, followed by GC-MS. Stevioside (**168**) was found in leaf herbarium samples of *S. rebaudiana*, collected in Paraguay in 1919, and *S. phlebophylla*, collected in Mexico in 1889 (Kinghorn *et al*. 1984). The additional sweet compounds, rebaudiosides A (**163**) and C (**165**), were detected in the *S. rebaudiana* sample. These results were correlated with those of a preliminary organoleptic evaluation of the same herbarium specimens for sweetness, where no other sample exhibited an intensity of sweetness equivalent to that of *S. rebaudiana* (Soejarto *et al*. 1982). Interestingly, the *S. phlebophylla* sample exhibited a slightly sweet taste due to the persistence of detectable amounts of stevioside (**161**). These results attested to the stability of this diterpene glycoside and validated a phytochemical screening procedure by modern techniques for the identification of potentially sweet compounds from plants (Kinghorn and Kim 1993).

It has been found that structural analogs exhibiting bitterness occur for all of the known naturally occurring diterpene sweeteners, embracing *ent*-kaurene glycosides from *S. rebaudiana* (Kinghorn and Soejarto 1985) and *Rubus suavissimus* (Rosaceae) (Hirono *et al*. 1990), and labdane glycosides from *Phlomis betonicoides* (Labiatae) (Tanaka *et al*. 1985) and *Baccharis gaudichaudiana* (Asteraceae) (Fullas *et al*. 1991). So far, five non-sweet *ent*-kaurene glycosides, named paniculosides I–V (**169–173**), have been reported to occur in the leaves of cultivated samples of the medicinal species *S. paniculata* (Kohda *et al*. 1976a) and *S. ovata* (Kaneda *et al*. 1978). Some studies on the application of ^{13}C-NMR spectroscopy to the structural elucidation of *Stevia* diterpene glycosides with aglycones unstable to acid hydrolysis were described during the structural elucidation of compounds **169–173** (Yamasaki *et al*. 1976; 1977). The molecular structures of paniculosides I–V are shown in Figure 5.19.

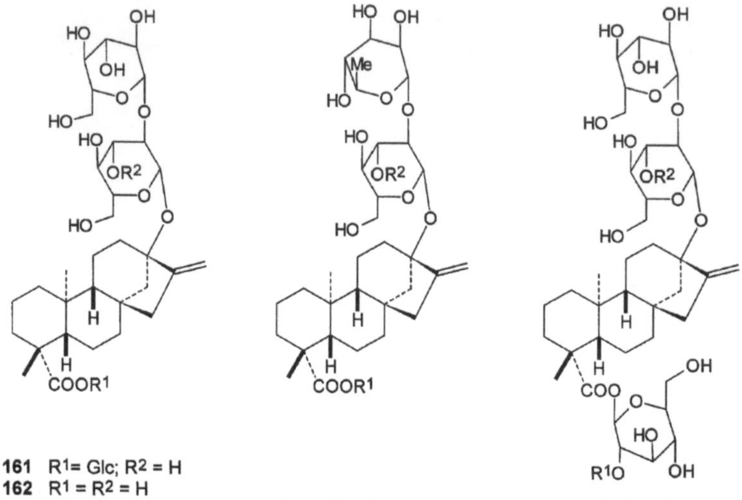

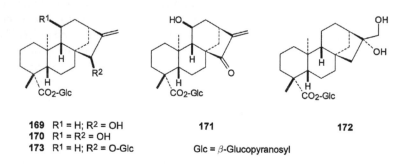

161 R1= Glc; R2 = H
162 R1 = R2 = H
163 R1 = R2 = Glc
164 R1 = H; R2 = Glc

165 R1 = R2 = Glc
168 R1 = Glc; R2 = H

166 R1 = H; R2 = Glc
167 R1 = Glc; R2 = H

Glc = β-Glucopyranosyl

Figure 5.18 Structures of sweet *ent*-kaurene glycosides (**161–168**) from *Stevia rebaudiana*.

169 R1 = H; R2 = OH
170 R1 = R2 = OH
173 R1 = H; R2 = O-Glc

171

172

Glc = β-Glucopyranosyl

Figure 5.19 Structures of paniculosides I–V (**169–173**) from *Stevia* species.

The leaves of *S. subpubescens* afforded a new non-sweet *ent*-kaurane glycoside, named sub-pubescensoside (**174**). Acid hydrolysis of the bitter natural product gave D-glucose and the aglycone, which was characterized as 11β,16-oxo-*ent*-kauran-19-oic acid (**174**, R = H). The X-ray crystal analysis of the methyl ester prepared from the aglycone was reported to verify its structure. Use of 2-D homonuclear NMR techniques (COSY) on the peracetylated and permethylated derivatives of the natural product made possible the assignment of the inter-glycosidic connectivities and, therefore, the determination of the saccharide substitution (Román *et al.* 1995b). Figure 5.20 shows the structure of subpubescensoside (**174**) and the preferred conformation of its aglycone.

Recently, it was rationalized that the bitter taste of infusions and decoctions prepared with the roots of *S. salicifolia*, a medicinal plant among the Tarahumara Indians in northern Mexico (Bye 1985; Heinrich 1996), might conceivably be due to either *ent*-kaurene or labdane glycosides. The phytochemical investigation of this plant material revealed that the major

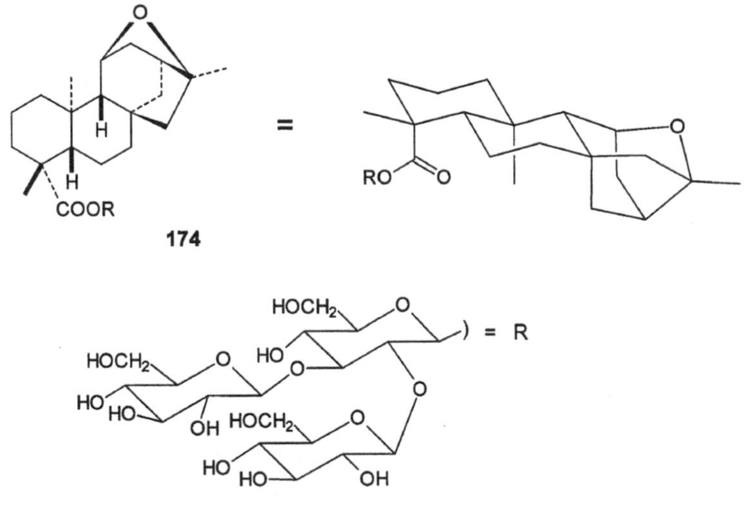

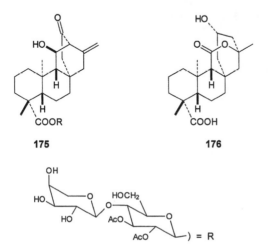

Figure 5.20 Structure of subpubescensoside (**174**) and the conformation of its aglycone.

bitter-tasting principle was an acetylated *ent*-atisene diterpene glycoside, stevisalioside A (**175**). The structure was established by the interpretation of spectral data, with the NMR assignments of this compound being based on ¹H-¹H COSY, ¹H-¹³C HETCOR, and selective INEPT experiments. While acidic hydrolysis led to a complex mixture of decomposition products, alkaline treatment afforded a major rearranged lactone derivative (**176**) which supported the structure of the natural product (Mata *et al*. 1992). Figure 5.21 illustrates the structure of stevisalioside A (**175**) and its enzymatic (**175**, R = H) and alkaline (**176**) hydrolysis products. In spite of extensive phytochemical studies on *Stevia* species, no other atisane derivative has yet been described from this genus. Furthermore, only two *ent*-atisane glycosides, from *Pteris purpureorachis* (Pteridaceae), have been reported previously in nature (Murakami *et al*. 1983).

Figure 5.21 Structure of stevisalioside A and its enzymatic and alkaline hydrolysis products (**175** and **176**).

The absolute configuration of the aglycone moieties of subpubescensoside (**174**) and stevi-salioside (**175**) was determined as corresponding to the enantiomeric 10-Meα-stereotype skeleton by optical rotatory dispersion, because of the intense levo-rotations at 589–365 nm demonstrated by their corresponding genins obtained from enzymatic hydrolysis, as has been consistently observed for other *ent*-kaur-16-en-19-oic and *ent*-atis-16-en-19-oic acid derivatives (Mata *et al*. 1992; Román *et al*. 1995b).

A number of hydroxylated kaurenoic acid derivatives (Figure 5.22), analogous to those obtained by the enzymatic hydrolysis of the diterpene glucoside mixtures of *S. paniculata* (Kohda *et al*. 1976), have been reported from the lipophilic fractions of crude extracts of several *Stevia* species. The investigation of the roots of *S. myriadenia* from Brazil afforded *ent*-kaur-15-en-19-oic acid, *ent*-kaur-16-en-19-oic acid (**177**, R = H), 15β-senecioyloxy-*ent*-kaur-16-en-19-oic acid (**177**, R = OSen), and 15β-tigloyloxy-*ent*-kaur-16-en-19-oic acid (**177**, R = OTigl) (Bohlmann *et al*. 1982). The aerial parts of a Venezuelan collection of *S. lucida* yielded, besides compound **177**, *ent*-kaur-16-en-19-ol, 16β,17-dihydroxy-*ent*-kauran-19-oic acid and a kaurenolide, named stevionolide (**178**). The structure of this minor constituent was established by spectral methods as 7-oxo-16β-acetyloxy-*ent*-kaur-5-en-19,6-olide (Amaro-Luis 1993). Some additional kauranes, including compound **177**, have been isolated from *S. amambayensis* (Schmeda-Hirschmann *et al*. 1986), *S. andina* (Zdero *et al*. 1987), *S. eupatoria* (Ortega *et al*. 1985), *S. paniculata* (Yamasaki *et al*. 1976), *S. setifera* (Bohlmann *et al*. 1979), *S. triflora* (Amaro-Luis and Adrián 1988), and *S. yaconensis* (Zdero *et al*. 1988). Examination of the aerial parts of *S. aristata* afforded beyerenic acid and two hydroxy derivatives, 7β-hydroxybeyerenic acid (**179**) and 12α-hydroxybeyerenic acid (**180**) (Zdero *et al*. 1987) (see Figure 5.22).

TRITERPENOIDS AND STEROLS

Besides the widespread sterols, β-sitosterol, stigmasterol, and campesterol, *Stevia* species also produce pentacyclic triterpenoids with the oleanane, lupane, taraxastane and friedelane

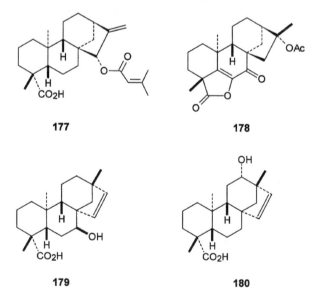

Figure 5.22 Selected kaurenoic and beyerenic acids derivatives (**177–180**) from *Stevia* species.

skeletons (Figure 5.23). β-Amyrin (**181**, $R^1 = H$; $R^2 = CH_3$), oleanolic acid (**181**, $R^1 = H$, $R^2 = CO_2H$) and their acetyl derivatives have been isolated from *S. alpina* (Sosa *et al.* 1989), *S. mandonii* (Bohlmann *et al.* 1979), *S. seleriana* (Escamilla and Ortega 1991), and *S. jujuyensis* (De Gutiérrez *et al.* 1992). Lupeol (**182**) and its acetate are present in *S. amambayensis* (Schmeda-Hirschmann *et al.* 1986), *S. andina* (Zdero *et al.* 1987), *S. myriadenia* (Bohlmann *et al.* 1982), and *S. setifera* (Bohlmann *et al.* 1979). Taraxasterol (**183**) and its acetate are the triterpenoids found in *S. berlandieri* (Domínguez *et al.* 1974), *S. mercedensis* (Bohlmann *et al.* 1986), and *S. paniculata* (Kohda *et al.* 1976a). Friedelin (**184**, R = O), friedelan-3β-ol (**184**, R = βOH, αH) and its C-3 epimer (**184**, R = αOH, βH) were isolated from *S. triflora*, *S. origanoides* (Amaro-Luis and Adrián 1988), and *S. subpubescens* var. *intermedia* (Joseph-Nathan *et al.* 1991). The *D:C*-friedours-7-en-3β-ol acetate (**185**) was also isolated from the latter species. Two dammaranes have been isolated from *Stevia*. Dammarenediol acetate was isolated from *S. boliviensis* (Bohlmann *et al.* 1979). More recently, a new dammarane was obtained from the hexane extract of the roots of *S. salicifolia*. This tetracyclic triterpenoid was characterized by spectroscopic and chemical methods as (20 *S*)-dammar-13(17),24-diene-3β-yl acetate (**186**) (Mata *et al.* 1991).

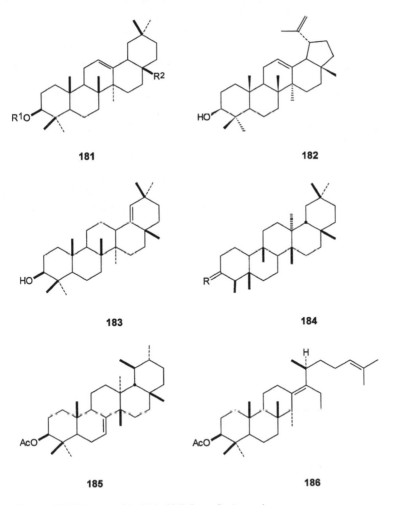

Figure 5.23 Triterpenoids (**181–186**) from *Stevia* species.

FLAVONOIDS

Members of the Asteraceae have been found to be a rich source of flavonoid aglycones which accumulate externally on the plant surfaces. Similar substitution patterns are found among flavones and flavonols. The flavones apigenin (5,7,4'-OH) and luteolin (5,7,3',4'-OH) are common, as are the corresponding flavonols, kaempferol (3,5,7,4'-OH) and quercetin (3,5,7,3',4'-OH). In both cases polymethoxylated types including compounds with 6- and, somewhat often less, 8-hydroxyl and methoxyl substitutions are widely present (Wollenweber and Valant-Vetschera 1996). This substitution pattern seems to represent a basic trend in flavonoids of plants in the Asteraceae and is considered an advanced biosynthetic character of the family (Giannasi 1988).

Stevia embraces some species with the richest production of exude flavonoids among the Asteraceae (Wollenweber *et al.* 1989), e.g. the amount of the leaf exudate from *S. berlandierei* corresponded to 2.9% of dry weight and for *S. subpubescens* to 3.0% and, as for other members of the Eupatorieae (Seligmann 1996), such exudates are rich in lipophilic polymethoxylated flavones and flavonols in the A-ring and their glycosides (Domínguez 1977). From the published data, it is apparent that the additional hydroxylation and O-methylation of the apigenin and quercetin nucleus at the C-6 position represents an infrageneric feature of possible systematic significance. Structures of some representative flavonoids from *Stevia* species are shown in Figure 5.24. It is of interest to note that all the North and Central American species so far investigated contained a greater diversity of methoxylated aglycones, and flavonoids glycosides of luteolin and quercetin than the South American species (Rajbhandari and Roberts 1985a). For example, *S. microchaeta*, which is generally considered to be one of the more primitive North American members of the genus, was notable for its lack of methoxylated flavonoid aglycones. However, the dried leaves of *S. origanoides* showed a higher variety of methoxylated aglycones, including hispidulin (**187**), eupatorin (**188**), 5,7,4'-trihydroxy 3,6-dimethoxyflavone (**189**), santin (**190**) and centaureidin (**191**), and was particularly rich in the common flavonoid quercetin 3-O-β-D-galactoside, hyperoside (**192**). The flavonoid pattern of *S. nepetifolia* and *S. monardifolia* proved to be similar to that of the previously reported South

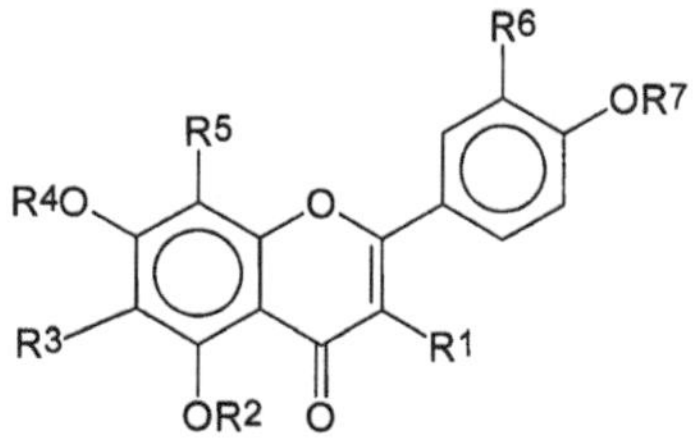

187	R1 = R2 = R4 = R5 = R6 = R7 = H; R3 = OMe
188	R1 = R2 = R5 = H; R3 = OMe; R4 = R7 = Me; R6 = OH
189	R1 = R3 = OMe; R2 = R4 = R5 = R6 = R7 = H
190	R1 = R3 = OMe; R2 = R4 = R5 = R6 = H; R7 = Me
191	R1 = R3 = OMe; R2 = R4 = R5 = H; R6 = OH; R7 = Me
193	R1 = R2 = R3 = R5 = R6 = R7 = H; R4 = Me
194	R1 = OH; R2 = R3 = R5 = R6 = R7 = H; R4 = Me
195	R1 = R2 = R3 = R4 = R5 = R6 = H; R7 = Me
196	R1 = OMe; R2 = R3 = R4 = R5 = R6 = H; R7 = Me
197	R1 = R3 = OMe; R2 = R5 = H; R6 = OH; R4 = R7 = Me

Figure 5.24 Some representative flavonoids (**187–191, 193–197**) from *Stevia* species.

American species, e.g. *S. galeopsidifolia* and *S. rebaudiana* (Rajbhandari and Roberts 1983; 1985b), in that the methoxylated flavonid centaureidin (**191**), and flavonoid glycosides of quercetin were found in the four species (Rajbhandari and Roberts 1984). The South American species *S. cuzcoensis* (Rajbhandari and Roberts 1985b), *S. procumbens* and *S. satureiaefolia* yielded a major aglycone, eupatorin (**188**) (Sosa *et al*. 1985). These results would support the view that the genus has spread into South America from origins to be found in the southern states of Mexico (Rajbhandari and Roberts 1985b).

The coexistence of methoxylated flavones with their corresponding flavonols is exempli-fied with the apigenin and kaempferol derivatives detected in *S. subpubescens* (Wollenweber *et al*. 1989), e.g. genkwanin (**193**), rhamnocitrin (**194**), acacetin (**195**), and ermanin (**196**), as well as with the C-6 substituted lutcolin and quercetin derivatives of *S. breviaristata* (Hernández *et al*. 1994), i.e. eupatorin (**188**) and casticin (**197**). Additional polymethoxylated aglycones have been isolated from *S. jujuyensis* (De Gutiérrez *et al*. 1992) and *S. maimarensis* (Hernández *et al*. 1996b).

Glycosylation may occur at the 3-, 7- or 4'-positions in the luteolin and quercetin nucleus and this substitution has been found to be similar in *S. cuzcoensis, S. galeopsidifolia, S. monardifolia, S. nepetifolia, S. origanoides, S. serrata, S. soratensis,* and *S. triflora* (Amaro-Luis *et al*. 1997; Rajbhandari and Roberts 1985b). The sugar component of these glycosides is usually glucose or galact-ose. The isolated monosides from *S. microchaeta* illustrate the glycosidic patterns for the genus (Rajbhandari and Roberts 1985a) and were as follows: luteolin 7-*O*-β-D-glucoside (**198**), luteolin-4'-*O*-β-D-glucoside (**199**), quercetin 3-*O*-β-D-galactoside (**192**) and quercetin 3-*O*-β-D-glucoside (**200**) (Figure 5.25).

BIOLOGICAL ACTIVITY

As far as a systematic study on the biological activity of compounds isolated from *Stevia* is concerned, more research has to be conducted in order for definitive conclusions to be made. The biological potential of a very small number of isolated substances has been reported, in contrast to all the extensive pharmacological and toxicological studies of the sweet diterpene glycosides of *S. rebaudiana*, which are reviewed in Chapter 8 of this volume. Furthermore, it have been demonstrated that several compounds present in *Stevia* species, but obtained from different plant sources, possess important biological and/or pharmacological activities.

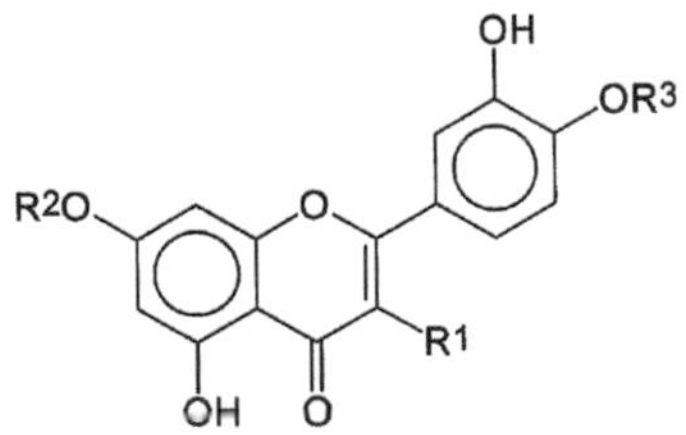

192	R¹ = *O*-β-D-galactoside; R² = R³ = H
198	R¹ = R³ = H; R² = β-D-glucoside
199	R¹ = R² = H; R³ = β-D-glucoside
200	R¹ = *O*-β-D-glucoside; R² = R³ = H

Figure 5.25 Some representative flavonoid glycosides (**192, 198–200**) from *Stevia* species.

Sesquiterpene lactones constitute a large and diverse group of biologically active plant constituents of the Asteraceae, and as expected, some of the medicinal properties of *Stevia* could be a consequence of the presence in high yields of these compounds. Previous major reviews have discussed the chemistry, taxonomic significance and biological activities of this group of active principles (Marles *et al.* 1995; Robles *et al.* 1995). Costunolide (**72**), an active constituent of several medicinal herbs, including the aerial parts of *S. yaconensis* and *S. chamaedrys* (Zdero *et al.* 1988), suppresses hepatitis B virus surface antigen gene expression in human hepatoma cells (Chen *et al.* 1995). Eupatoriopicrin (**73**), isolated from *S. sarensis* (Zdero *et al.* 1988), *S. procumbens* (Sosa *et al.* 1985), and *S. maimarensis* (Hernández *et al.* 1996b), as well as leucodin derivatives, isolated from *S. yaconensis* var. *subeglandulosa* (Sosa *et al.* 1989), have been identified as important constituents of plants consumed by chimpanzees for presumed therapeutic properties (Robles *et al.* 1995). Eupatoriopicrin (**73**) has shown strong anti-tumor activity in the Lewis lung tumor system in mice, while the 11,13-dehydro-derivative of leucodin (**128**) displayed anti-ulcer activity (Robles *et al.* 1995). This substance has also been identified as the gastric cytoprotective agent of the leaves of *Artemisia douglasiana* (Asteraceae: Anthemideae) which are used in folk medicine as an anti-ulcer agent (Giordano *et al.* 1990). Ludartin (**136**), isolated from *S. yaconensis* var. *subeglandulosa* (Sosa *et al.* 1989), displayed a stronger cytoprotective effect than dehydroleucodin (Giordano *et al.* 1990).

Crude extracts of *Stevia* have shown anti-microbial and anti-fungal activities which in some cases have been associated with the presence of either sesquiterpene lactones and/or flavonoids (Montanaro *et al.* 1996). Casticin (**197**), found in *S. breviaristata* (Hernández *et al.* 1994), was responsible for the anti-fungal activity displayed by extracts of *Psiadia trinervia* (Asteraceae: Astereae) against *Cladosporium cucumerinum*, which was evaluated by bioautographic assays (Wang *et al.* 1989). The flavonoid eupatorin (**188**), found in *S. origanoides* (Rajbhandari and Roberts 1985a), *S. satureiaefolia*, *S. procumbens* (Sosa *et al.* 1989), and *S. breviaristata* (Hernández *et al.* 1994), displayed inhibitory activity against mitochondrial succinoxidase and NADH-oxidase, demonstrating a primary site of inhibition in the complex I (NADH-coenzyme Q reductase) portion of the respiratory chain (Hodnick *et al.* 1994). The inhibition of agriculturally important fungal pathogens growth by selected sesquiterpene lactones has been studied. The most active compounds were the isoalantolactone (**105**) and their isomers (Picman and Schneider 1993; Calera *et al.* 1995), which are important constituents of *S. polyphylla* (Zdero *et al.* 1988).

Methylripariochromene (**32**), isolated from *Eupatorium riparium* (Asteraceae: Eupatorieae), displayed anti-fungal activity against five fungal species including *Colletotrichum gloeosporioides*, a tropical pathogen species (Bandara *et al.* 1992). As previously mentioned, this substance is a known constituent of *S. serrata* (Kohda *et al.* 1976b). Bisabolene epoxides related to **38**, which was isolated from *S. purpurea* (Bohlmann *et al.* 1976), were found as the main components of the male sex pheromone in the insect *Nezaria viridula* (Heteroptera: Pentatomidae) (Brezot *et al.* 1994). A sample of loliolide **151**, a degraded sesquiterpene isolated from *Eucommia ulmoides* (Eucommiaceae) and also present in *S. grisebachiana* (Sigstad *et al.* 1991), showed a moderate immunosuppressive activity. It was suggested that this substance preferentially perturbs T-lymphocyte proliferation (Okada *et al.* 1994).

CONCLUSIONS

In summarizing the status of phytochemical research on *Stevia* species, we can conclude that at the present time all screening studies to determine the infrageneric distribution of the known constituents have been limited to the search for sweet *ent*-kaurene glycoside principles and, therefore, an extensive and rational approach is needed to fully understand the chemo-taxonomic significance of the secondary metabolism in this genus. Phytochemical information is available on only 58 species of this large genus, and sesquiterpenoids of the guaiane, longipinane, and germacrane groups represent the most characteristic constituents of their aerial parts and roots. Nevertheless, the distribution of *ent*-kaurene glycosides in *Stevia* has received considerable attention during the last two decades due to their commercial importance as a major class of so-called 'high-intensity sweeteners' of natural origin.

Despite the fact that a definitive chemical link between the *Stevia* constituents described above and the abundance of ethnomedical literature has not been established, many of these forgoing results offer an appealing insight into how such a link might be found. The results of future studies are awaited with considerable interest.

REFERENCES

Amaro, J. M., Adrián, M., Cerda, C. M. and Joseph-Nathan, P. (1988) Longipinene derivatives from *Stevia lucida* and *S. triflora*. *Phytochemistry* **27**, 1409–1412.

Amaro-Luis, J. M. (1993) An *ent*-kaurenolide from *Stevia lucida*. *Phytochemistry* **32**, 1611–1613.

Amaro-Luis, J. M. and Adrián, R. M. (1988) Terpenoids and steroids from *Stevia triflora*. *Fitoterapia* **59**, 512.

Amaro-Luis, J. M. and Adrián, R. M. (1997) A new caryophyllane derivative from *Stevia triflora*. *Pharmazie* **52**, 162–163.

Amaro-Luis, J. M. and Hung, P. M. (1988) Phytochemical studies on the Venezuela Andean flora. IV. Structure of labd-13(Z)-en-8α-ol-15-oic acid, a component from *Stevia lucida* Lagasca. *Acta Científica Venezolana* **39**, 21–24.

Amaro-Luis, J. M., Adrián, M. and Díaz, C. (1997) Isolation, identification and antimicrobial activity of ombuoside from *Stevia triflora*. *Annales Pharmaceutiques Francaises* **55**, 262–268.

Angeles, E., Folting, K., Grieco, P. A., Huffman, J. C., Miranda, R. and Salmón, M. (1982) Isolation and structure of stephalic acid, a new clerodane diterpene from *Stevia polycephala*. *Phytochemistry* **21**, 1804–1806.

Bandara, B. M. R., Hewage, C. M., Karunaratne, V., Wannigama, G. P. and Adikaram, N. K. B. (1992) An antifungal chromene from *Eupatorium riparium*. *Phytochemistry* **31**, 1982–1985.

Bohlmann, F. and Zdero, C. (1985) Stevisalicinon, ein neuer Diterpentyp, sowie weitere Inhaltsstoffe aus *Stevia*-Arten. *Liebigs Annalen der Chemie*, 1764–1783.

Bohlmann, F., Zdero, C. and Schöneweiss, S. (1976) Über die Inhaltsstoffe aus *Stevia*-Arten. *Chemische Berichte* **109**, 3366–3370.

Bohlmann, F., Suwita, A., Natu, A. A., Czerson, H. and Suwita, A. (1977) Über weitere α-Longipinen-Derivate aus *Compositen*. *Chemische Berichte* **110**, 3572–3581.

Bohlmann, F., Dutta, L. N., Dorner, W., King, R. M. and Robinson, H. (1979) Zwei neue Guajanolide sowie weitere Longipinenester aus *Stevia*-Arten. *Phytochemistry* **18**, 673–675.

Bohlmann, F., Zdero, C., King, R. M. and Robinson, H. (1982) Sesquiterpenes, guaianolides and diterpenes from *Stevia myriadenia*. *Phytochemistry* **21**, 2021–2025.

Bohlmann, F., Umemoto, K. and Jakupovic, J. (1985) Pseudoguaianolides related to confertin from *Stevia isomeca*. *Phytochemistry* **24**, 1017–1019.

Bohlmann, F., Zdero, C., King, R. M. and Robinson, H. (1986) Neue Sesquiterpenlactone und andere Inhaltsstoffe aus *Stevia mercedensis* und *Stevia achalensis*. *Liebigs Annalen der Chemie*, 799–813.

Brezot, P., Malosse, C., Mori, K. and Renou, M. (1994) Bisabolene epoxides in sex pheromone in *Nezara viridula* (L.) (Heteroptera, Pentatomidae): Role of cis isomer and relation to specificity of pheromone. *Journal of Chemical Ecology* **20**, 3133–3147.

Bye, R. (1985) Medicinal plants of the Tarahuamara Indians from Chihuahua, Mexico. In *Two Mummies from Chihuahua: A Multidisciplinary Study*. R. A. Tyson and D. V. Elerick (Eds), San Diego Museum Papers, San Diego, No. 19, pp. 77–104.

Calderón, J. S., Angeles, E., Salmón, M. and García de la Mora, G. A. (1984) Bisabolene derivatives from *Stevia salicifolia. Phytochemistry* **23**, 186–188.

Calderón, J. S., Quijano, L., Gómez, F. and Ríos, T. (1987a) Heliangolides from *Stevia origanoides. Journal of Natural Products* **50**, 522–522.

Calderón, J. S., Quijano, L., Gómez-Garibay, F., Sanchez, D. M., Ríos, T. and Fronczek, F. R. (1987b) Sesquiterpene lactones from *Stevia ovata* and crystal structure of 11,13-dehydroeriolin. *Phytochemistry* **26**, 1747–1750.

Calderón, J. S., Quijano, L., Gómez, F. and Ríos, T. (1989) Prochamazulene sesquiterpene lactones from *Stevia serrata. Phytochemistry* **28**, 3526–3527.

Calera, M. R., Soto, F., Sánchez, P., Bye, R., Hernández-Bautista, B., Anaya, A. L., Lotina-Hennsen, B. and Mata, R. (1995) Biochemically active sesquiterpene lactones from *Ratibida mexicana. Phytochemistry* **40**, 419–425.

Cerda-García-Rojas, C. M., Sánchez-Arreola, E., Joseph-Nathan, P., Román, L. U. and Hernández, J. D. (1993) Longipinene derivatives from *Stevia origanoides. Phytochemistry* **32**, 1219–1223.

Chen, H. C., Chou, C. K., Lee, S. D., Wang, J. C. and Yeh, S. F. (1995) Active compounds from *Saussurea lappa* Clark that suppress hepatitis B virus surface antigen gene expression in human hepatoma cells. *Antiviral Research* **27**, 99–109.

Darise, M., Kohda, H., Mitzutani, K., Kasai, R. and Tanaka, O. (1983) Chemical constituents of flowers of *Stevia rebaudiana* Bertoni. *Agricultural and Biological Chemistry* **47**, 133–135.

De Gutiérrez, A. N., Catalán, C. A. N., Díaz, J. G. and Herz, W. (1992) Sesquiterpene lactones and other constituents of *Stevia jujuyensis. Phytochemistry* **31**, 1818–1820.

De Heluani, C. S., De Lampasona, M. P., Catalán, C. A. N., Goedken, V. L., Gutiérrez, A. B. and Herz, W. (1989) Guaianolides, heliangolides and other constituents from *Stevia alpina. Phytochemistry* **28**, 1931–1935.

De Hernández, Z. N. J., Hernández, L. R., Catalán, C. A. N., Gedris, T. E. and Herz, W. (1997) Guaianolides and germacranolides from *Stevia sanguinea. Phytochemistry* **46**, 721–727.

Domínguez, X. A. (1977) Eupatorieae – chemical review. In *The Biology and Chemistry of the Compositae.* V. H. Heywood, J. B. Harborne, and B. L. Turner (Eds), Academic Press, London, **1**, 487–502.

Domínguez, X. A., González, A., Zamudio, M. A. and Garza, A. (1974) Taraxasterol from *Stevia berlandieri* and *Cirsium texanum. Phytochemical Reports* **13**, 2001.

Escamilla, E. M. and Ortega, A. (1991) Labdane diterpenoids from *Stevia seleriana. Phytochemistry* **30**, 599–602.

Fullas, K., Hussain, R. A., Bordas, E., Pezzuto, J. M., Soejarto, D. D. and Kinghorn, A. D. (1991) Gaudichaudiosides A–E, five novel diterpene glycoside constituents from the sweet-tasting plant, *Baccharis gaudichaudiana. Tetrahedron* **47**, 8515–8522.

Ghisalberti, E. L., Jefferies, P. R. and Stuart, A. D. (1979) The absolute configuration of a sesquiterpene enone from *Stevia purpurea* and a phenol from *Lasianthaea podocephala. Australian Journal of Chemistry* **32**, 1627–1630.

Giannasi, D. E. (1988) Flavonoids and evolution in the dicotyledons. In *The Flavonoids. Advances in Research since 1980.* J. B. Harborne (Ed.), Chapman and Hall, London, pp. 479–504.

Gil, R. R., Oberti, J. C., Sosa, V. E. and Herz, W. (1987) Longipinanes and a germacranolide carboxylic acid from *Stevia potrerensis. Phytochemistry* **26**, 1459–1461.

Gil, R. R., Pastoriza, J. A., Oberti, J. C., Gutiérrez, A. B. and Herz, W. (1989) Guaianolides from *Stevia sanguinea. Phytochemistry* **28**, 2841–2843.

Gil, R. R., Oberti, J. C., Gutiérrez, A. B. and Herz, W. (1990) Heliangolides from *Stevia yaconensis* var. *aristifera. Phytochemistry* **29**, 3881–3884.

Gil, R. R., Pacciaroni, A. D. V., Oberti, J. C., Díaz, J. G. and Herz, W. (1992) A rearranged germacranolide and other sesquiterpene lactones from *Stevia jujuyensis. Phytochemistry* **31**, 593–596.

Giordano, O. S., Guerreiro, E., Pestchanker, M. J., Guzmán, J., Pastor, D. and Guardia, T. (1990) The gastric cytoprotective effect of several sesquiterpene lactones. *Journal of Natural Products* **53**, 803–809.

Gómez, G. F., Quijano, L., Calderón, J. S., Perales, A. and Ríos, T. (1983) 11,13-Dihydroeucannabinolide, a heliangolide from *Stevia monardaefolia. Phytochemistry* **22**, 197–199.

Grashoff, J. L. (1972) *A Systematic Study of the North and Central American Species of Stevia*. Ph. D. dissertation, University of Texas at Austin, p. 609

Guerra-Ramírez, D., Cerda-García-Rojas, C. M., Puentes, A. M. and Joseph-Nathan, P. (1998) Longipinene diesters from *Stevia lucida*. *Phytochemistry* **48**, 151–154.

Heinrich, M. (1996) Ethnobotany of Mexican Compositae: an analysis of historical and modern sources. In *Compositae: Biology & Utilization*, Proceedings of the International Compositae Conference, Kew, 1994, P. D. S. Caligari and D. J. N. Hind (Eds), Royal Botanic Gardens, Kew, UK, **2**, 475–503.

Hernández, L. R., Catalán, C. A. N., Cerda-García-Rojas, C. M. and Joseph-Nathan, P. (1994) Sesquiterpene lactones from *Stevia breviaristata*. *Phytochemistry* **37**, 1331–1335.

Hernández, L. R., Catalán, C. A. N., Cerda-García-Rojas, C. M. and Joseph-Nathan, P. (1995) Guaianolides from *Stevia gilliesii*. *Natural Products Letters* **6**, 215–221.

Hernández, L. R., Catalán, C. A. N., Cerda-García-Rojas, C. M. and Joseph-Nathan, P. (1996a) Sesquiterpene lactones from *Stevia vaga*. *Phytochemistry* **42**, 1369–1373.

Hernández, L. R., De Riscala, E. C., Catalán, C. A. N., Díaz, J. G. and Herz, W. (1996b) Sesquiterpene lactones and other constituents of *Stevia maimarensis* and *Synedrellopsis grisebachii*. *Phytochemistry* **42**, 681–684.

Hodnick, W. F., Duval, D. L. and Pardini, R. S. (1994) Inhibition of mitochondrial respiration and cyanide-stimulated generation of reactive oxygen species by selected flavonoids. *Biochemical Pharmacology* **47**, 573–580.

Hirono, S., Chou, W.-H., Kasai, R., Tanaka, O. and Tada, T. (1990) Sweet and bitter diterpene-glucosides from leaves of *Rubus suavissimus*. *Chemical and Pharmaceutical Bulletin* **38**, 1743–1744.

Ishikawa, H., Kitahata, S., Ohtani, K. and Tanaka, O. (1991) Transfructosylation of rebaudioside A (a sweet glycoside of *Stevia* leaves) with *Microbacterium* β-fructofuranosidase. *Chemical and Pharmaceutical Bulletin* **39**, 2043–2045.

Joseph-Nathan, P. and Cerda-García-Rojas, C. M. (1994) Molecular rearrangements in longipinane derivatives. *Pure and Applied Chemistry* **66**, 2361–2364.

Joseph-Nathan, P., Cerda, C. M., Del Río, R. E., Román, L. U. and Hernández, J. D. (1986) Conformation and absolute configuration of naturally occurring longipinene derivatives. *Journal of Natural Products* **49**, 1053–1060.

Joseph-Nathan, P., Cerda, C. M., Román, L. U. and Hernández, J. D. (1989) Preparation and characterization of naturally occurring longipinene esters. *Journal of Natural Products* **52**, 481–496.

Joseph-Nathan, P., Cerda-García-Rojas, C. M., Castrejón, S., Román, L. U. and Hernández, J. D. (1991) High-performance liquid chromatography and nuclear magnetic resonance analysis of longipinene derivatives from *Stevia subpubescens* var. *intermedia*. *Phytochemical Analysis* **2**, 77–79.

Joseph-Nathan, P., Meléndez-Rodríguez, M., Cerda-García-Rojas, C. M. and Catalán, C. A. N. (1996) Photochemical rearrangements of highly functionalized longipinene derivatives. *Tetrahedron Letters* **37**, 8093–8096.

Kaneda, N., Kohda, H., Yamasaki, K., Tanaka, O. and Nishi, K. (1978) Paniculosides-I-V, diterpene-glucosides from *Stevia ovata* Lag. *Chemical and Pharmaceutical Bulletin* **26**, 2266–2267.

Kinghorn, A. D. and Kim, J. (1993) Potently sweet compounds from plants: techniques of isolation and identification. In *Bioactive Natural Products. Detection, Isolation and Structural Determination*, S. M. Colegate and R. J. Molyneux (Eds), CRC Press, Boca Raton, Florida, pp. 173–193.

Kinghorn, A. D. and Soejarto, D. D. (1985) Current status of stevioside as a sweetening agent for human use. In *Economic and Medicinal Plant Research*. H. Wagner, H. Hikino and N. R. Farnsworth (Eds), Academic Press, London, **1**, 1–52.

Kinghorn, A. D., Soejarto, D. D., Nanayakkara, N. P. D., Compadre, C. M., Makapugay, H. C., Hovanec-Brown, J. M. *et al.* (1984) A phytochemical screening procedure for sweet *ent*-kaurene glycosides in the genus *Stevia*. *Journal of Natural Products* **47**, 439–444.

Kohda, H., Tanaka, O. and Nishi, K. (1976a) Diterpene-glycosides of *Stevia paniculata* Lag.: structures of aglycones. *Chemical and Pharmaceutical Bulletin* **24**, 1040–1044.

Kohda, H., Yamazaki, K. and Tanaka, O. (1976b) Methylripariochromene A from *Stevia serrata*. *Phytochemistry* **15**, 846–847.

Marles, R. J., Pazos-Sanou, L., Compadre, C. M., Pezzuto, J. M., Bloszyk, E. and Arnason, J. T. (1995). Sesquiterpene lactones revisited. Recent developments in the assessment of biological activities and structure relationships. In *Phytochemistry of Medicinal Plants*. J. T. Arnason, R. Mata and J. T. Romeo (Eds), Plenum Press, New York, pp. 333–356.

Martínez, M., Muñoz-Zamora, A. and Joseph-Nathan, P. (1988) Conformational analysis of achillin and leukodin. *Journal of Natural Products* **51**, 221–228.

Martínez-Vázquez, M., Gallegos, R. E. and Joseph-Nathan, P. (1990) Eudesmanolides from *Stevia* aff. *tomentosum*. *Phytochemistry* **29**, 1689–1690.

Mata, R., Rodríguez, V., Pereda-Miranda, R., Bye, R. and Linares, E. (1991) A dammarane from *Stevia salicifolia*. *Phytochemistry* **30**, 3822–3823.

Mata, R., Rodríguez, V., Pereda-Miranda, R., Kaneda, N. and Kinghorn, A. D. (1992) Stevisalioside A, a novel bitter-tasting *ent*-atisene glycoside from the roots of *Stevia salicifolia*. *Journal of Natural Products* **55**, 660–666.

Montanaro, S., Bardón, A. and Catalán, C. A. N. (1996) Antibacterial activity of various sesquiterpene lactones. *Fitoterapia* **67**, 185–187.

Montes, A. L. (1969) Gas chromatography of essential oils of plants native to Central and Northern Argentina. *Anales de la Sociedad Científica Argentina* **187**, 21–48.

Murakami, T., Tanaka, N., Komazawa, Y., Saiki, Y. and Chen, C.-M. (1983) Chemiche und Chemotaxonomische Untersuchungen von Filices. XLI. Weitere Inhaltsstoffe von *Pteris purpureorachis* Copel. *Chemical and Pharmaceutical Bulletin* **31**, 1502–1504.

Oberti, J. C., Sosa, V. E., Herz, W., Siva Prasad, J. and Goedken, V. L. (1983) Crystal structure and stereochemistry of achalensolide, a new guaianolide from *Stevia achalensis*. *Journal of Organic Chemistry* **48**, 4038–4043.

Oberti, J. C., Gil, R. R., Sosa, V. E. and Herz, W. (1986) A guaianolide from *Stevia breviaristata*. *Phytochemistry* **25**, 1479–1480.

Okada, N., Shirata, K., Niwano, M., Koshino, H. and Uramoto, M. (1994) Immunosuppressive activity of a monoterpene from *Eucommia ulmoides*. *Phytochemistry* **37**, 281–282.

Ortega, A., Martínez, R. and García, C. L. (1980) Los diterpenos de *Stevia salicifolia* Cav. estructura del stevinsol y salicifoliol. *Revista Latinoamericana de Química* **11**, 45–48.

Ortega, A., Morales, F. J. and Salmón, M. (1985) Kaurenic acid derivatives from *Stevia eupatoria*. *Phytochemistry* **24**, 1850–1852.

Perez, G. R. M., Perez, G. S. and Perez, G. C. (1995) Chemical composition of the essential oil of *Stevia porphyrea* flowers. *Tecnología de Alimentos* **30**, 5–7 [*Chemical Abstracts* **124**, 211464w].

Picman, A. K. and Schneider, E. F. (1993) Inhibition of fungal growth by selected sesquiterpene lactones. *Biochemical Systematics and Ecology*, **21**, 307–314.

Quijano, L., Calderón, J. S., Gómez, F., Vega, J. L. and Ríos, T. (1982) Diterpenes from *Stevia monardaefolia*. *Phytochemistry* **21**, 1369–1371.

Rajbhandari, A. and Roberts, M. F. (1983) The flavonoids of *Stevia rebaudiana*. *Journal of Natural Products* **46**, 194–195.

Rajbhandari, A. and Roberts, M. F. (1984) Flavonoids of *Stevia nepetifolia*. *Journal of Natural Products* **47**, 559–560.

Rajbhandari, A. and Roberts, M. F. (1985a) The flavonoids of *Stevia microchaeta, Stevia monardifolia*, and *Stevia origanoides*. *Journal of Natural Products* **48**, 502–503.

Rajbhandari, A. and Roberts, M. F. (1985b) The flavonoids of *Stevia cuzcoensis, Stevia galeopsidifolia, Stevia serrata*, and *Stevia soratensis*. *Journal of Natural Products* **48**, 858–859.

Ríos, T., Romo de Vivar, A. and Romo, J. (1967) Stevin, a new pseudoguaianolide isolated from *Stevia rhombifolia* H. B. K. *Tetrahedron* **23**, 4265–4269.

Robinson, H. and King, R. M. (1977) Eupatorieae – systematic review. In *The Biology and Chemistry of the Compositae*. V. H. Heywood, J. B. Harborne and B. L. Turner (Eds), Academic Press, London, **1**, 437–485.

Robles, M., Aregullin, M., West, J. and Rodriguez, E. (1995) Recent studies on the zoopharmacognosy, pharmacology and neurotoxicology of sesquiterpene lactones. *Planta Medica* **61**, 199–203.

Román, L. U., Del Río, R. E., Hernández, J. D., Joseph-Nathan, P., Zabel, V. and Watson, W. H. (1981) Structure, chemistry and stereochemistry of rastevione, a sesquiterpenoid from the genus *Stevia*. *Tetrahedron* **37**, 2769–2778.

Román, L. U., Del Río, R. E., Hernández, J. D., Cerda, C. M., Cervantes, D., Castañeda, R. *et al.* (1985) Structural and stereochemical studies of naturally occurring longipinene derivatives. *Journal of Organic Chemistry* **50**, 3965–3972.

Román, L. U., Hernández, J. D., Castañeda, R., Cerda, C. M. and Joseph-Nathan, P. (1989) Isolation and preparation of two longipinene derivatives from *Stevia subpubescens*. *Phytochemistry* **28**, 265–268.

Román, L. U., Mora, Y. and Hernández, J. D. (1990) *Stevia serrata*, a source of chamazulene. *Fitoterapia* **61**, 84.

Román, L. U., Hernández, J. D., del Río, R. E., Bucio, M. A., Cerda-García-Rojas, C. M. and Joseph-Nathan, P. (1991) Wagner–Meerwein rearrangements of longipinane derivatives. *Journal of Organic Chemistry* **56**, 1938–1940.

Román, L. U., Hernández, J. D., Cerda-García-Rojas, C. M., Domínguez-López, R. M. and Joseph-Nathan, P. (1992) Molecular rearrangements in the longipinene series. *Journal of Natural Products* **55**, 577–588.

Román, L. U., Loeza-Coria, M., Hernández, J. D., Cerda-García-Rojas, C. M., Sánchez-Arreola, E. and Joseph-Nathan, P. (1993) Preparation of a new longipinane derivative from *Stevia serrata*. *Journal of Natural Products* **56**, 1148–1152.

Román, L. U., Morán, G., Hernández, J. D., Cerda-García-Rojas, C. M. and Joseph-Nathan, P. (1995a) Longipinane derivatives from *Stevia viscida*. *Phytochemistry* **38**, 1437–1439.

Román, L. U., Torres, J. M., Reyes, R., Hernández, J. D., Cerda-García-Rojas, C. M. and Joseph-Nathan, P. (1995b) *ent*-Kaurane glycoside from *Stevia subpubescens*. *Phytochemistry* **39**, 1133–1137.

Román, L. U., Zepeda, G. L., Morales, N. R., Hernández, J. D., Cerda-García-Rojas, C. M. and Joseph-Nathan, P. (1995c) Molecular rearrangement of rastevione mesylate into arteagane derivatives. *Journal of Natural Products* **58**, 1808–1816.

Román, L. U., Zepeda, G. L., Morales, N. R., Flores, S., Hernández, J. D., Cerda-García-Rojas, C. M. and Joseph-Nathan, P. (1996) Mechanistic studies of the longipinane to arteagane rearrangement. *Journal of Natural Products* **59**, 391–395.

Salmón, M., Díaz, E. and Ortega, A. (1973) Christinine, a new epoxyguaianolide from *Stevia serrata* Cav. *Journal of Organic Chemistry* **38**, 1759–1761.

Salmón, M., Díaz, E. and Ortega, A. (1977) Epoxilactonas de *Stevia serrata* Cav. *Revista Latinoamericana de Química* **8**, 172–175.

Salmón, M., Ortega, A. and Díaz, E. (1975) Structure and stereochemistry of a new germacrane sesquiterpene lactone. *Revista Latinoamericana de Química* **6**, 45–48.

Salmón, M., Ortega, A., García de la Mora, G. and Angeles, E. (1983) A diterpenic acid from *Stevia lucida*. *Phytochemistry* **22**, 1512–1513.

Sánchez-Arreola, E., Cerda-García-Rojas, C. M., Joseph-Nathan, P., Román, L. U. and Hernández, J. D. (1995) Longipinene derivatives from *Stevia serrata*. *Phytochemistry* **39**, 853–857.

Schmeda-Hirschmann, G., Zdero, C. and Bohlmann, F. (1986) Melampolides and germacranolides from *Stevia amambayensis*. *Phytochemistry* **25**, 1755–1756.

Seligmann, P. (1996) Flavonoids of the Compositae as evolutionary parameters in the tribes which synthesize them: a critical approach. In *Compositae: Systematics*. Proceedings of the International Compositae Conference, Kew, 1994, D. J. N. Hind and H. J. Beentje (Eds), Royal Botanic Gardens, Kew, UK, **1**, 159–167.

Sholichin, M., Yamasaki, K., Miyama, R., Yahara, S. and Tanaka, O. (1980) Labdane-type diterpenes from *Stevia rebaudiana*. *Phytochemistry* **19**, 326–327.

Sigstad, E. E., Catalán, C. A. N., Gutiérrez, A. B., Díaz, J. G., Goedken, V. L. and Herz, W. (1991) Guaianolides and germacranolides from *Stevia grisebachiana*. *Phytochemistry* **30**, 1933–1940.

Singh, J., Arosa, A. K., Kurana, A. and Kad, G. L. (1995) A very short and convenient synthesis of 2-methyl-6-(2′-oxo-4′-methyl-cyclohex-3′-en-1′-yl)-2-heptene and (±)-12-hydroxy-2-oxobisabol-3-ene. *Proceedings of the Indian Academy of Sciences (Chemical Sciences)* **107**, 203–205.

Soejarto, D. D., Kinghorn, A. D. and Farnsworth, N. R. (1982) Potential sweetening agents of plant origin. III. Organoleptic evaluation of *Stevia* leaf herbarium samples for sweetness. *Journal of Natural Products* **45**, 590–599.

Sosa, V. E., Oberti, J. C., Prasad, J. S. and Herz, W. (1984) Flavonoids and eupahakonenin B from *Stevia satureiaefolia*. *Phytochemistry* **23**, 1515–1516.

Sosa, V. E., Gil, R., Oberti, J. C., Kulanthaivel, P. and Herz, W. (1985) Sesquiterpene lactones and flavones from *Stevia procumbens*. *Journal of Natural Products* **48**, 340–341.

Sosa, V. E., Oberti, J. C., Gil, R. R., Rúveda, E. A., Goedken, V. L., Gutiérrez, A. B. *et al.* (1989) 10-Epideoxy-cumambrin B and other constituents of *Stevia yaconensis* var. *subeglandulosa*. *Phytochemistry* **28**, 1925–1929.

Tanaka, T., Tanaka, O., Lin, Z.-W. and Zhou, J. (1985) Sweet and bitter principles of the Chinese plant drug, bai-yun-shen: revision of the assignment of the source plant and isolation of two new diterpene glycosides. *Chemical and Pharmaceutical Bulletin* **33**, 4275–4280.

Wang, Y., Hamburger, M., Gueho, M. and Hostettmann, K. (1989) Antimicrobial flavonoids from *Psiadia trinervia* and their methylated and acetylated derivatives. *Phytochemistry* **28**, 2323–2327.

Willaman, J. J. and Li, H. L. (1970) Alkaloid-bearing plants and their contained alkaloids, 1957–1968. *Lloydia* **33**, 1–286.

Wollenweber, E. and Valant-Vetschera, K. M. (1996) New results with exudate flavonoids in Compositae. In *Compositae: Systematics*. Proceedings of the International Compositae Conference, Kew, 1994, D. J. N. Hind and H. J. Beentje (Eds), Royal Botanic Gardens, Kew, UK, **1**, 169–185.

Wollenweber, E., Mann, K., Hochwart, S. and Yatskievych, G. (1989) Exudate flavonoids in miscellaneous Asteraceae. *Phytochemical Bulletin* **21**, 19–23.

Yamasaki, K., Kohda, H., Kobayashi, T., Kasai, R. and Tanaka, O. (1976) Structures of *Stevia* diterpene-glucosides: application of ^{13}C NMR. *Tetrahedron Letters* **13**, 1005–1008.

Yamasaki, K., Kohda, H., Kobayashi, T., Kaneda, N., Kasai, R., Tanaka, O. and Nishi, K. (1977) Application of ^{13}C nuclear magnetic resonance spectroscopy to chemistry of glycosides: structures of paniculo-sides-I, -II, -III, -IV, and -V, diterpene glucosides of *Stevia paniculata* Lag. *Chemical and Pharmaceutical Bulletin* **25**, 2895–2899.

Zdero, C. and Bohlmann, F. (1986) New terpenoids from Mexican *Stevia* species. *Progress in Essential Oil Research*, Proceedings of the 16th International Symposium on Essential Oils, 1985, E.-J. Brunke (Ed.), De Gruyter, Berlin, pp. 249–261. [*Chemical Abstracts* **106**, 84853y].

Zdero, C., Bohlmann, F. and Schmeda-Hirschmann, G. (1987) Beyerene derivatives and other terpenoids from *Stevia aristata*. *Phytochemistry* **26**, 463–466.

Zdero, C., Bohlmann, F., King, R. M. and Robinson, H. (1988) The first 12,8β-germacrolide and other constituents from Bolivian *Stevia* species. *Phytochemistry* **27**, 2835–2842.

Zdero, C., Bohlmann, F. and Niemeyer, H. M. (1991) Guaianolides and other constituents from *Stevia* species. *Phytochemistry* **30**, 693–695.

Zygadlo, J. A., Ariza Espinar, L., Velasco Negueruela, A. and Perez Alonso, M. J. (1997) Volatile constituents of *Stevia achalensis* Hieronymus. *Flavour and Fragance Journal* **12**, 297–299.

6 Synthetic investigations on steviol, stevioside, and rebaudioside A, and their applications as starting materials

Darrick S. H. L. Kim

INTRODUCTION

Stevioside (**2**) (Figure 6.1) is the major sweet-tasting *ent*-kaurene glycoside of *S. rebaudiana*, and has been reported to be 250–300 times sweeter than sucrose (Crammer and Ikan 1986). Synthetic studies on steviol (**1**) (Figure 6.1), the aglycone of the sweet-tasting glycoside stevioside, have been carried out in order to try and improve on the sweetness properties of stevioside. Although stevioside (**2**) is potently sweet, it has an unpleasant aftertaste which limits its use. Structure-taste (sweetness) relationship studies in terms of the enzyme-induced transglycosylation of steviol glycosides have been reviewed in detail by Tanaka (1997), and this work is covered in Chapter 7. Since the discovery of its gibberellin-like activity (Ruddat *et al.* 1963) and the unique C/D ring structure in steviol (**1**), it has become a synthetic target for a number of chemists. The structure of steviol (**1**), including its absolute stereochemistry, was determined by degradation studies coupled with ORD measurements (Djerassi *et al.* 1961; Mosettig *et al.* 1963). In this chapter the approaches toward the chemical syntheses of steviol, stevioside, rebaudioside A, and their derivatives will be discussed. In addition, synthetic applications using these compounds as starting materials will be surveyed.

SYNTHESES OF STEVIOL

Steviol (**1**) has a bicyclo[3.2.1]octane unit as the C/D rings and a hydroxyl group at the bridge-head C-13 position. Among the reported total syntheses of steviol, Mori's (Mori and Matsui

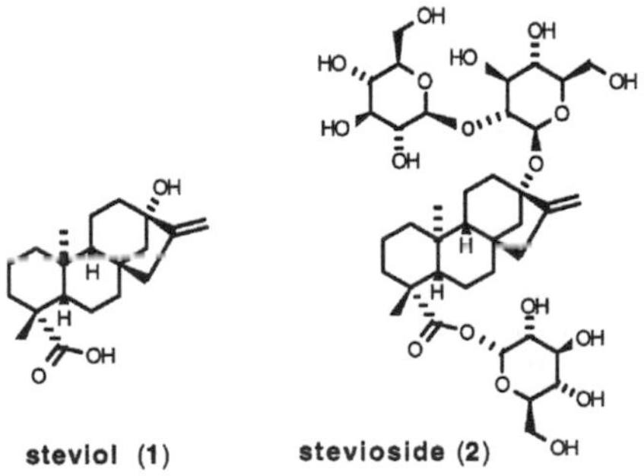

Figure 6.1 Structures of steviol and stevioside.

1965; 1966; 1970; Mori *et al*. 1970a; 1970b; Nakahara *et al*. 1971; Nakahara 1982) and Ziegler's (Ziegler and Kloek 1971; 1977) approaches appear to be the most comprehensive.

Mori's approach toward the total synthesis of steviol

The total synthesis of steviol (**1**) by Mori (Mori and Matsui 1965; 1966; 1970; Mori *et al*. 1970a; 1970b; Nakahara *et al*. 1971; Nakahara 1982) is described in Scheme 1. The synthesis involves the annelation of 6-methoxy-1-methyl-2-tetralone (**4**) with methyl acroylacetate (**3**) to afford **5**. Compound **5** was methylated and converted to thioketal **6**. Desulfurization of **6**, followed by Birch reduction afforded an α,β-unsaturated ketone (**7**). Compound **7** was selectively reduced with lithium tri-*t*-butoxyaluminum hydride and converted to vinyl ether **8**. The vinyl ether **8** was subjected to a Claisen rearrangement to give aldehyde **9**. Conversion of the aldehyde **9** to the corresponding acetal, followed by hydroboration-oxidation and Jones oxidation afforded a mixture of two isomeric ketones (**10** and **11**). Wittig reaction of **11** with ethylidenetriphenyl-phosphorane gave olefin **12**. Hydroboration-oxidation of **12**, followed by the Jones oxidation afforded the 13-acetyl acetal **13**. The axial acetyl group in **13** was isomerized to the equatorial position to yield **14**. Treatment of **14** with dilute aqueous HCl/acetone afforded the ketone **15**

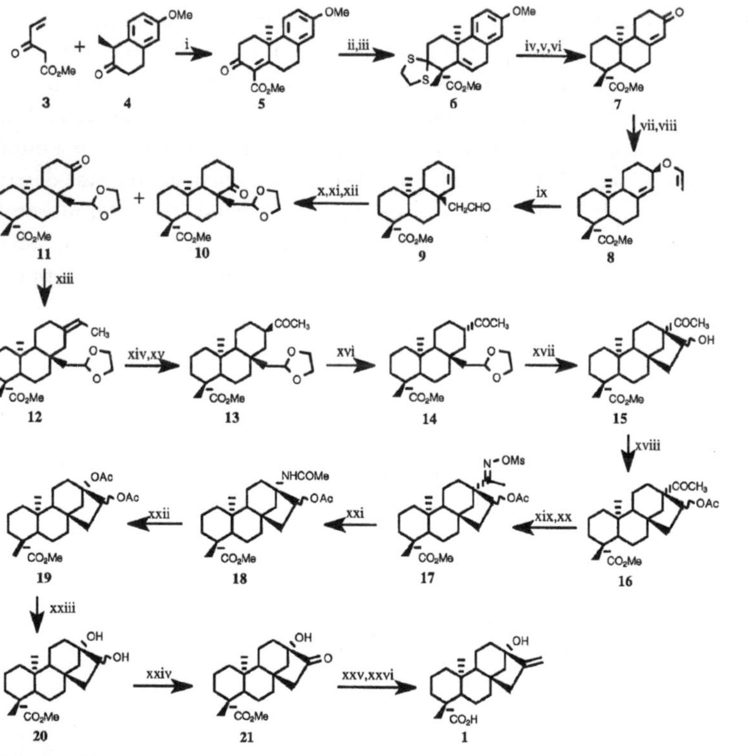

(i) Triton B, MeOH; (ii) *t*-BuOK, *t*-BuOH, MeI, PhH; (iii) HSCH$_2$CH$_2$SH, BF$_3$; (iv) Raney Ni; (v) H$_2$, Pd-C; (vi) Birch reduction; (vii) (*t*-BuO)$_3$AlHLi, THF; (viii) Hg(OAc)$_2$, vinyl ethyl ether, reflux; (ix) sealed tube, 200 °C; (x) ethylene glycol, *p*-TsOH, PhH, reflux; (xi) BH$_3$-THF, OOH⁻; (xii) 8 N Jones reagent; (xiii) Ph$_3$P=CHCH$_3$; (xiv) BH$_3$-THF, OOH⁻; (xv) 8 N Jones reagent; (xvi) NaOMe, MeOH; (xvii) 3 N HCl, acetone then NaOMe, MeOH; (xviii) Ac$_2$O, pyridine; (xix) HONH$_3$Cl, NaOAc, MeOH; (xx) MsCl, pyridine; (xxi) aqueous dioxane, reflux; (xxii) NaNO$_2$, AcOH-Ac$_2$O,-2°C, 1 hour; (xxiii) NaOH, MeOH; (xxiv) 8 N Jones reagent; (xxv) Ph$_3$P=CH$_2$; (xxvi) *t*-BuOK, DMSO, 100 °C.

Scheme 1 Mori's total synthesis of steviol.

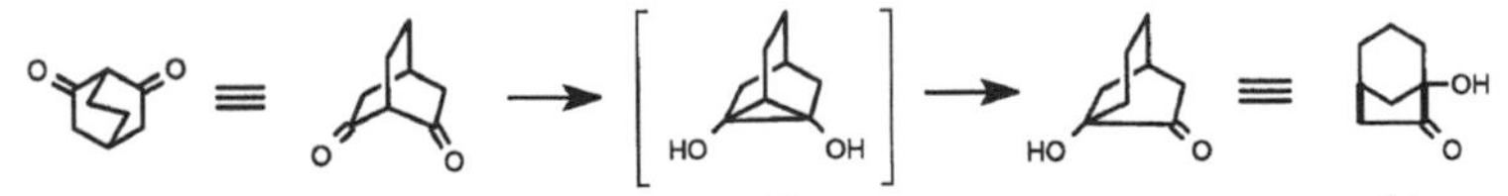

(6.1)

(i) 3 N HCl, acetone, reflux; (ii) 8 N Jones reagent; (iii) Zn-Hg, dilute HCl, toluene, reflux; (iv) $Ph_3P=CH_2$; (v) *t*-BuOK, DMSO, 100 °C.

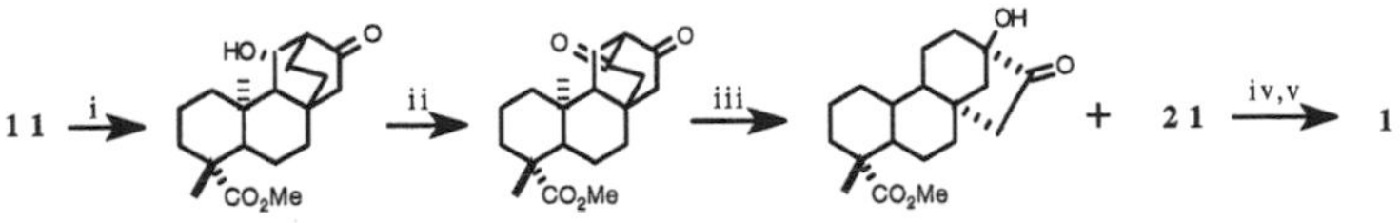

Scheme 2 Mori's total synthesis of steviol (alternative approach).

which was acetylated to **16**. The ketone **16** was converted to an oxime and mesylated to **17** and then subjected to a Beckmann rearrangement to afford the acetamide **18**. *N*-Nitrosation of the acetamide **18** with sodium nitrite in acetic anhydride-acetic acid afforded diacetate **19** which was hydrolyzed to the diol **20**. Jones oxidation of the diol **20** yielded a ketone **21** which was condensed with methylenetriphenylphosphorane and hydrolyzed to afford the desired (±)-steviol (**1**).

Based on an observed acid-induced rearrangement of the bicyclic dicarbonyl compound **22** to the C/D ring structure **24** of steviol and the gibberellins (Mori *et al*. 1970a; 1970b; Nakahara 1982; Ziegler and Kloek 1971; 1977) (Equation 6.1), steviol was prepared by an alternative route (Scheme 2). Acid-induced cyclization of **11** to **25** and Jones oxidation of **25** gave the diketone intermediate **26**. Treatment of **26** with zinc–mercury in the presence of HCl induced a rearrangement to afford a mixture of products (**21** and **27**) with the C/D ring structure in the opposite orientation. The desired ketone **21** was converted to steviol (**1**) following the steps described in Scheme 1.

Ziegler's approach toward the total synthesis of steviol

The total synthesis of steviol by Ziegler and Kloek (1971; 1977) involved the stereocontrolled photo-addition of allene **29** to cyclopent-1-ene-1-carboxaldehyde **28** as the key synthetic step to establish C/D ring structure **32** (Equation 6.2 and Scheme 3). Alkylation by potassium enolate of Hagemann's ester **36** with the mesylate of the *m*-methoxyphenylethyl alcohol **37** and subsequent decarboxylation afforded **38**. The enone **38** was subjected to a phosphorus pentaoxide assisted cyclization to afford **39**. A Wittig reaction of **39** and subsequent hydrolysis gave aldehyde **40**. Stereocontrolled α-methylation of the aldehyde **40**, followed by a treatment with Jones reagent afforded the tricyclic acid **41**. Birch reduction of the acid **41** to enone **42**, followed by thioketalization and desulfurization gave **43**. Ozonolysis of **43** and decomposition of the ozonolysis product with dimethylsulfide afforded the ketoacetal **44**. Acid-induced hydrolysis of **44** to the corresponding aldehyde, followed by pyrrolidine-involved aldol condensation afforded enamine **45**. Acid-induced hydrolysis of **45** yielded the key intermediate **46**. As in the model study (Equation 6.2), photoaddition of allene with **44** afforded the desired intermediate **47** which was reduced to an alcohol **48**. Mesylation of **48**, followed by a treatment with 2,6-lutidine in refluxing acetone and hydrolysis, afforded (±)-steviol (**1**).

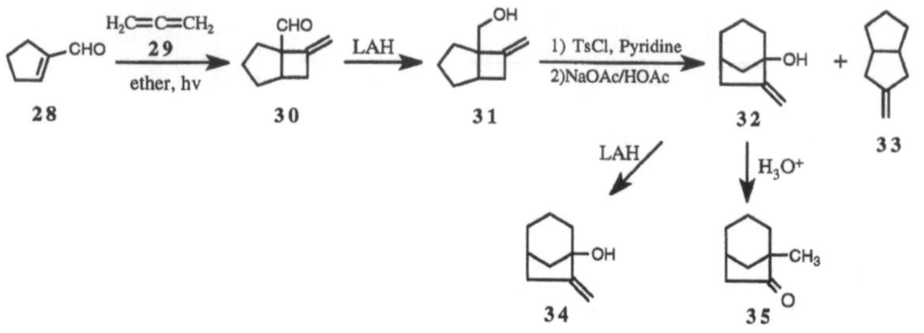

(6.2)

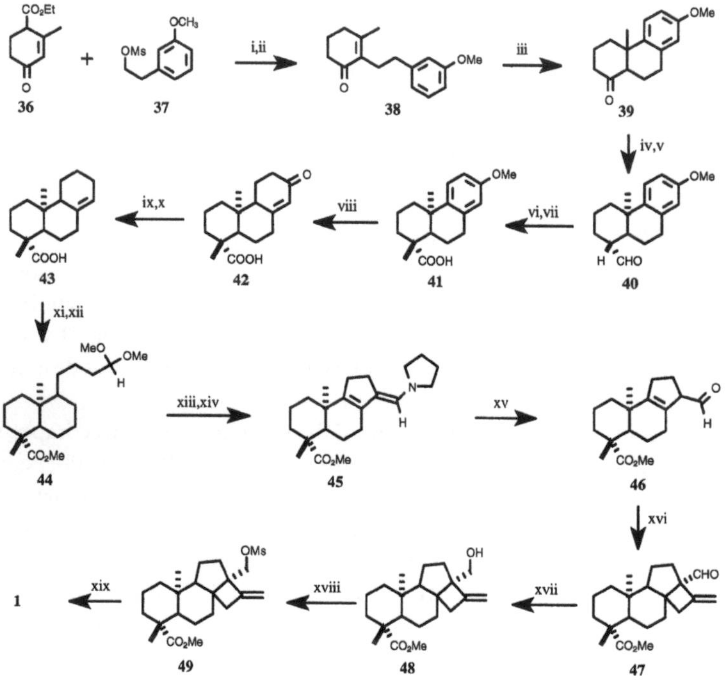

(i) *t*-BuOK, *t*-BuOH, toluene, 80 °C; (ii) H$_2$SO$_4$, EtOH, reflux; (iii) H$_3$PO$_4$, 110 °C; (iv) Cl^{-+}PPh$_3$CH$_2$OMe, DMSO; (v) HCl, THF; (vi) *t*-BuOK, *t*-BuOH, MeI; (vii) 8 N Jones reagent; (viii) Birch reduction; (ix) ethanedithiol, BF$_3$·Et$_2$O, AcOH; (x) Li°, NH$_3$, THF, –33 °C; (xi) diazomethane, Et$_2$O; (xii) ozone, MeOH, S(CH$_3$)$_2$; (xiii) HCl, THF, H$_2$O; (xiv) pyrrolidine, Ph-H, reflux, -H$_2$O; (xv) aqueous AcOH, NaOAc; (xvi) allene, –78 °C, *hv*; (xvii) EtOH, –78 °C, NaBH$_4$; (xviii) MsCl, pyridine; (xix) 2,6-lutidine, acetone, H$_2$O, reflux.

Scheme 3 Ziegler's total synthesis of steviol.

Cook's approach toward the partial synthesis of steviol

The synthesis of steviol by Cook and Knox (1970; 1976a; 1976b) involves *ent*-kaur-16-en-19-oic acid (**50**) as the starting material (Scheme 4). Ozonolysis of methyl ester derivative **51** and subsequent oxidation gave keto-ester **52**. Baeyer-Villiger oxidation of **52** gave γ-lactone **53** which was sequentially hydrolyzed, methylated, and oxidized to the corresponding keto-ester **54**. The treatment of **54** with sodium–liquid ammonia gave the acyloin-like cyclization product, diol-acid **55**, which was oxidized to **56**. Silyl ether protection of **56** and subsequent Wittig reaction with methylenetriphenylphosphorane, followed by a dilute acid work up gave steviol (**1**).

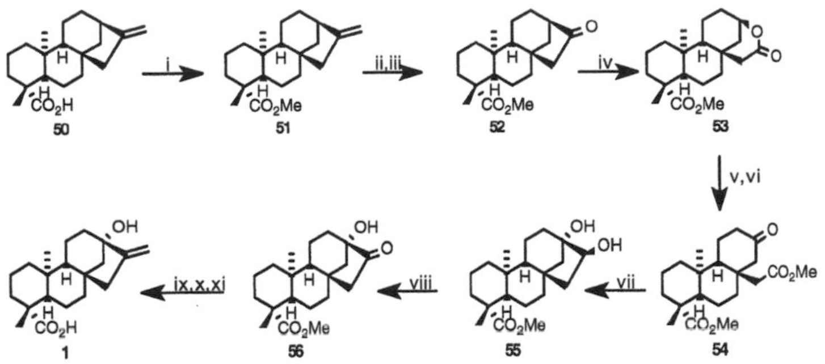

(i) CH₂N₂, ether; (ii) ozonolysis; (iii) Jones oxidation; (iv) PhCO₃H, CHCl₃, *p*-TsOH; (v) KOH, MeOH;
(vi) CrO₃-pyridine; (vii) Na°, NH₃, THF; (viii) Jones oxidation; (ix) TMSCl, TMS₂NH, pyridine;
(x) Ph₃PCH₃I, *t*-BuOK, *t*-BuOH, THF; (xi) H₃⁺O.

Scheme 4   Cook's partial synthesis of steviol.

Synthesis of Methyl (±)-7-oxopodocarp-8-en-16-oate

Methyl (±)-7-oxopodocarp-8-en-16-oate (**7**), a key intermediate in the synthesis of steviol, has been produced synthetically (Mori and Matsui 1966) (Scheme 5). Ethyl-1-methylcyclohexan-2-one-1-carboxylate (**57**) was converted into the ester **58** by treatment with sodium ethoxide followed by methylation with iodomethane. Addition of β-phenylethyl magnesium bromide

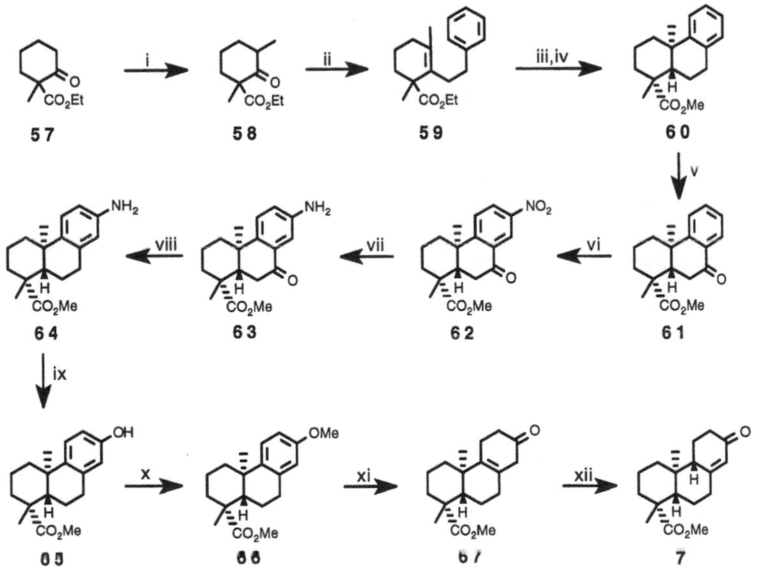

(i) NaOEt/EtOH and then MeI/toluene; (ii) C₆H₅CH₂CH₂MgBr, -H₂O; (iii) AcOH, H₂SO₄, reflux;
(iv) diazomethane, diethyl ether; (v) Jones oxidation; (vi) H₂SO₄, HNO₃; (vii) HCl, EtOH, Pd-C,H₂;
(viii) AcOH, HClO₄, Pd-C,H₂; (ix) 10% HCl, NaNO₂, urea; (x) NaOH, MeOH, dimethyl sulfate;
(xi) Na°, *t*-BuOH, NH₃ and then Li°, THF, EtOH; (xii) HCl, H₂O, reflux.

Scheme 5   Synthesis of methyl (±)-7-oxopodocarp-8-en-16-oate by Mori and Matsui.

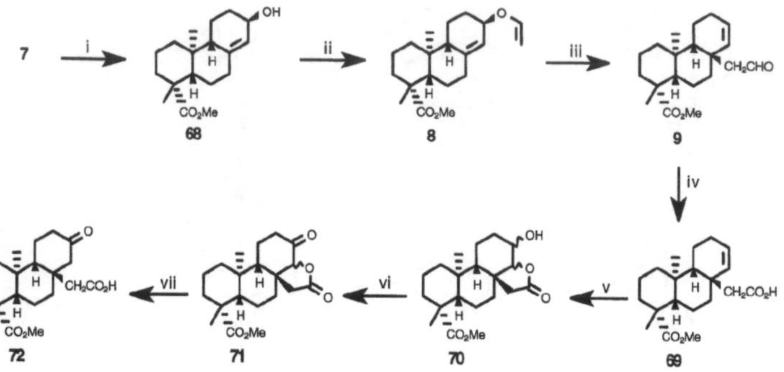

(i) (*t*-BuO)₃AlHLi; (ii) Hg(OAc)₂, ethyl vinyl ether; (iii) decalin, 200 °C; (iv) AgO; (v) HCO₃H, -OH and then acid-induced lactonization; (vi) Jones oxidation; (vii) Zn, AcOH.

Scheme 6 Synthesis of a (±)-seco acid by Mori and Matsui.

to the keto ester **58** followed by dehydration of the resulting alcohol afforded an unsaturated ester **59**. Treatment of **59** with sulfuric acid in acetic acid afforded (±)-desoxypodocarpic acid which was methylated to **60** by treatment with ethereal diazomethane. Chromic acid oxidation of **60** gave methyl (±)-9-oxodesoxypodocarpate (**61**) which was nitrated to **62**, and then hydrogenated to an amino ketone **63**, and further hydrogenated to an amino ester **64**. Compound **64** was diazotized and hydrolyzed to **65**. Methylation of **65** with dimethyl sulfate gave methyl (±)-7-methoxydesoxypodocarpate (**66**), and Birch reduction of **66** gave (±)-7-oxopodocarp-13-en-16-oic acid (**67**). Treatment of **67** with ethanolic hydrochloric acid gave **7**.

Synthesis of (±)-8α-carboxymethyl podocarpan-13-one-4β-carboxylate

A seco acid, (±)-8α-carboxymethyl podocarpan-13-one-4β-carboxylate (**72**), which retained all the asymmetric centers of steviol (**1**), except at C-13, has been obtained as one of the ozonolysis products of steviol methyl ester. A stereoselective total synthesis of this (±)-seco acid has been described (Mori and Matsui 1965) (Scheme 6).

Methyl (±)-podocarp-8(14)-en-13-one-4β-carboxylate (**7**) was reduced with lithium tri-*t*-butoxyaluminum hydride to the corresponding hydroxy ester **68** which was converted to a crystalline vinyl ester **8** by a mercuric acetate-catalyzed vinyl transetherification reaction in ethyl vinyl ether. The ether **8** was heated at 200 °C in decalin to give aldehyde **9** which was oxidized with silver oxide to an acid (**69**). Performic acid oxidation, alkaline hydrolysis, and acid-catalyzed lactonization of **69** afforded a hydroxy lactone **70** which was oxidized with Jones reagent to the corresponding keto lactone **71**. Treatment of keto lactone **71** with zinc dust in acetic acid gave (±)-8α-carboxymethyl podocarpan-13-one-4β-carboxylate (**72**), which was identical with that of the seco acid obtained from methyl steviol degradation.

CONVERSION OF STEVIOL TO STEVIOSIDE

Steviol (**1**) has been glycosylated to afford stevioside (**2**) (Ogawa *et al*. 1978; 1980) (Scheme 7). First, steviol was converted to steviolbioside from steviol methyl ester (**73**). Glycosidation of

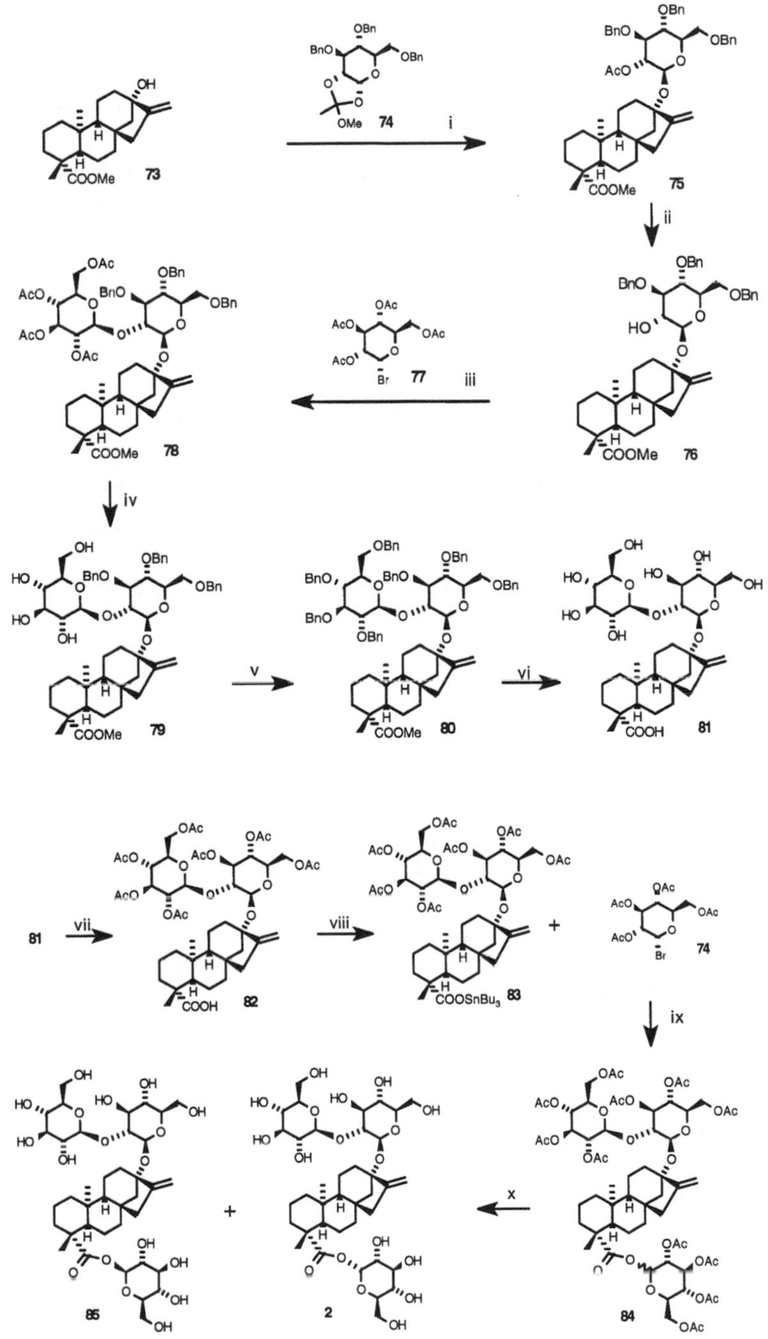

(i) MgBr₂, 95°C, vacuum; (ii) NaOMe, MeOH; (iii) CF₃SO₃Ag, 2,4,6-collidine; (iv) NaOMe, MeOH; (v) NaH, DMF, benzyl bromide; (vi) NaOMe, MeOH; (vii) Ac₂O, pyridine; (viii) (Bu₃Sn)₂O; (ix) Bn₄NBr, Cl(CH₂)₂Cl; (x) NaOMe, MeOH.

Scheme 7 Conversion of steviol to stevioside.

73 with 3,4,6-tri-*O*-benzyl-1,2-*O*-(1-methoxyethylidene)-α-D-glucopyranoside (**74**) in the presence of mercuric bromide afforded **75**. Saponification of **75** into **76** and subsequent glycosidation using 2,3,4,6-tetra-*O*-acetyl-α-D-glucopyranosyl bromide (**77**) and silver triflate-2,4,6-collidine afforded **78**, a derivative of steviol methyl ester (**73**). Saponification of **78** into **79** and subsequent benzylation gave 13-*O*-[2-*O*-(2,3,4,6-tetra-*O*-benzyl-β-D-glucopyranosyl)-3,4,6-tri-*O*-benzyl-β-D-glucopyranosyl] steviol methyl ester (**80**). Treatment of **80** with Na° in liquid ammonia afforded steviolbioside (**81**). Tributyl stannylation of heptaacetyl steviolbioside (**82**), obtained by acetylation of steviolbioside (**81**), gave the tributyltin carboxylate **83**. Subsequent reaction of **83** with **84** in refluxing toluene afforded an α and β mixture of the glucosyl ester. Deacetylation of this glucosyl ester mixture with methanolic sodium methoxide afforded the α-D-glucosyl ester (**85**) and the β-D-glucosyl ester, stevioside (**2**).

Synthesis of sweet-tasting stevioside analogues by glycosidation

In an effort to discover sweet stevioside analogues, Esaki *et al.* (1984) and Kamiya *et al.* (1979) glycosidated steviolbioside (**81**) with various mono- and disaccharides (**86–94**) in the presence of silver carbonate-Celite in 1,2-dichloroethane (Scheme 8). Their syntheses involved coupling reactions between hepta-*O*-acetyl steviolbioside with selected *O*-acetyl protected mono- or disaccharide halides which, after deacetylation, afforded the desired products (**95–103**).

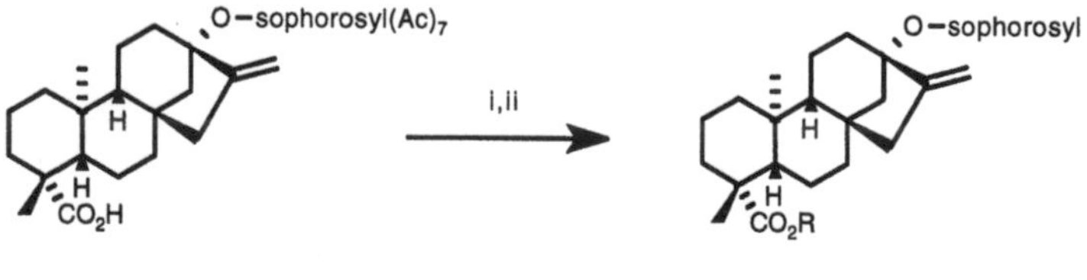

X = 2,3,4-Tri-*O*-acetyl-α-D-xylopyranosyl bromide (**86**)
2,3,4-Tri-*O*-acetyl-β-L-arabinopyranosyl bromide (**87**)
2,3,4,6-Tetra-*O*-acetyl-α-D-mannopyranosyl bromide (**88**)
2,3,4,6-Tetra-*O*-acetyl-α-L-glucopyranosyl bromide (**89**)
2,3,4-Tri-*O*-acetyl-α-L-rhamnopyranosyl chloride (**90**)
2,3,4-Tri-*O*-acetyl-α-L-quinovopyranosyl bromide (**91**)
Hexa-*O*-acetyl-2-*O*-α-L-rhamnopyranosyl-β-D-glucopyranosyl bromide (**92**)
Hexa-*O*-acetyl-2-*O*-α-L-rhamnopyranosyl-β-D-galactopyranosyl bromide (**93**)
Hexa-*O*-acetyl-2-*O*-α-L-quinovopyranosyl-β-D-glucopyranosyl bromide (**94**)

R = β-D-xylopyranosyl (**95**)
α-L-arabinopyranosyl (**96**)
α-D-mannopyranosyl (**97**)
β-L-glucopyranosyl (**98**)
α-L-rhamnopyranosyl (**99**)
β-L-quinovopyranosyl (**100**)
2-*O*-α-L-rhamnopyranosyl-β-D-glucopyranosyl (**101**)
2-*O*-α-L-rhamnopyranosyl-β-D-galactopyranosyl (**102**)
2-*O*-α-L-quinovopyranosyl-β-D-glucopyranosyl (**103**)

(i) **x**, silver carbonate-Celite, 1,2-dichloroethane; (ii) deacetylation.

Scheme 8 Synthesis of sweet-tasting stevioside analogues by glycosidation.

Conversion of stevioside to rebaudioside A

Rebaudioside A (**108**) is the sweetest of the *ent*-kaurene glycosides isolated from *S. rebaudiana* to date, being approximately 350–450 times sweeter than sucrose (Crammer and Ikan 1986). Stevioside (**2**) has been synthetically converted to rebaudioside A (**108**) (Kaneda *et al.* 1977) (Scheme 9). The synthetic pathway followed a selective removal of a glucose unit from stevioside (**2**) at the C-13 position by an enzyme, and systematic reintroduction of two glucose units of different linkage to the remaining glucose unit at the C-13 position.

The conversion of stevioside (**2**) to rebaudioside A (**108**) involved selective hydrolysis of the terminal glucosyl linkage of the β-sophorosyl moiety of **2**. This was achieved using a crude preparation of amylase prepared from *Aspergillus oryzae*, by incubating at 37 °C in McIlvain buffer (pH 4.0) for 80 hours that yielded a sweet desgluco compound (**104**) in quantitative yield. Saponification of **104** with 5% NaOH in MeOH afforded steviolmonoside (**105**), which was converted into the 4′,6′-benzylidene derivative **106** by treatment with benzaldehyde in 98% HCOOH at room temperature. Glucosylation of **106** with excess **107** in chlorobenzene, debenzylation with 30% AcOH at 80 °C, and deacetylation with BaO in MeOH at 0 °C afforded **108**.

Synthesis of sweet-tasting stevioside analogues with a sodiosulfopropyl group

Due to the potential acute toxicity of steviol (**1**) (Vignais *et al.* 1966) and the suggested possible breakdown of stevioside (**2**) to steviol *in vivo* and absorption of steviol through the gastrointestinal tract (Wingard *et al.* 1980), a series of sodiosulfopropyl group-attached stevioside analogues, which retain the sweet taste of stevioside but are biologically stable, were prepared (DuBois *et al.* 1981; 1984) (Scheme 10). The 19-*O*-β-D-glucosyl moiety of stevioside (**2**) was replaced by a sodiosulfopropyl moiety to give **114**, in which the sweet taste was not only retained but also improved considerably. Two analogues, 2′,3′,19-tri-*O*-(sodiosulfopropyl)-steviolmonoside (**110**) and 2″,3″,3′,19-tetra-*O*-(sodiosulfopropyl)-steviolbioside (**114**), of the biologically

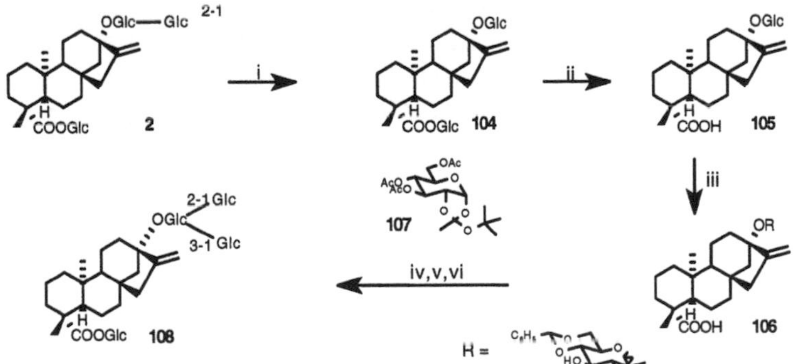

(i) amylase, McIlvain's buffer; (ii) 5% NaOH/MeOH; (iii) benzaldehyde, 98% HCO₂H; (iv) **107**, C₆H₅Cl; (v) 30% AcOH; (vi) 0.5 N BaO, MeOH.

Scheme 9 Conversion of stevioside to rebaudioside A by Kaneda.

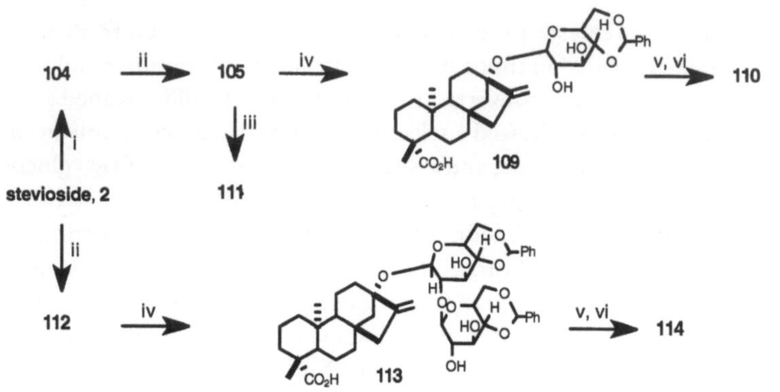

104 $\xrightarrow{\text{ii}}$ 105 $\xrightarrow{\text{iv}}$ [structure] $\xrightarrow{\text{v, vi}}$ 110

stevioside, 2   111

112 $\xrightarrow{\text{iv}}$ [structure] $\xrightarrow{\text{v, vi}}$ 114

104 Rubusoside (R$_1$ = R$_2$ = β-D-glucosyl)
105 Steviolmonoside (R$_1$ = H, R$_2$ = β-D-glucosyl)
110 R$_1$ = (CH$_2$)$_3$SO$_3$Na; R$_2$ = 2,3-di-*O*-(sodiosulfopropyl)-β-D-glucosyl
111 R$_1$ = (CH$_2$)$_3$SO$_3$Na; R$_2$ = β-D-glucosyl
112 Steviolbioside (R$_1$ = H, R$_2$ = β-D-sophorosyl)
114 R$_1$ = (CH$_2$)$_3$SO$_3$Na; R$_2$ = 2,3,3-tri-*O*-(sodiosulfopropyl)-β-D-sophorosyl

(i) Samzyme R; (ii) NaOH; (iii) K$_2$CO$_3$, 1,3-propanesultone, DMF; (iv) PhCH(OCH$_3$)$_2$, HClO$_4$, DMF;
(v) KCH$_2$SOCH$_3$, 1,3-propanesultone, Me$_2$SO; (vi) H$_3$O$^+$.

Scheme 10 Synthesis of steviol analogues with a sodiosulfopropyl group by DuBois.

labile diterpenoid triglycoside stevioside were prepared. Under conditions simulating those of the human GI tract, the analogues **110** and **114** were found to be completely stable while stevioside (**2**) was converted completely to the aglycone, steviol (**1**).

SYNTHETIC APPLICATIONS UTILIZING STEVIOSIDE AND STEVIOL

ent-Kaurene is the core carbon skeleton of a number of natural products that include steviol, stevioside, grandiflorenic acid, the gibberellins, hibaene, and erythroxydiol A. Due to the presence of this unique C/D ring structure, these compounds have been synthetic challenge to a number of chemists. Since stevioside (**2**) is readily available, steviol (**1**) has been applied as the key starting material in the synthesis of a number of natural products with an *ent*-kaurene skeleton.

Grandiflorenic acid synthesis

Starting from stevioside (**2**), the syntheses of grandiflorenic acid (**123**) and its analogues **124** and **125** have been carried out (Cheng and Zhou 1993a) (Scheme 11). Grandiflorenic acid (**123**), isolated from a Mexican medicinal plant, zoapatle (*Montanoa tomentosa*), has been reported to possess abortifacient activity (Lozoya *et al.* 1983). Methyl steviol (**73**), obtained by methylation of steviol (**1**), was acetylated and the double bond migrated on treatment with I$_2$ to give **116**. SeO$_2$ oxidation of compound **116** afforded **117** which was hydrolyzed and oxidized by Swern oxidation to give **119**. Acetylation of **119** and subsequent hydrogenation afforded **120**. Bromination of **120** with pyrrolidone hydrotribromide and subsequent reduction with LiBH$_4$ gave **122** as the key intermediate of grandiflorenic acid synthesis in nine steps (8.4% yield from steviol).

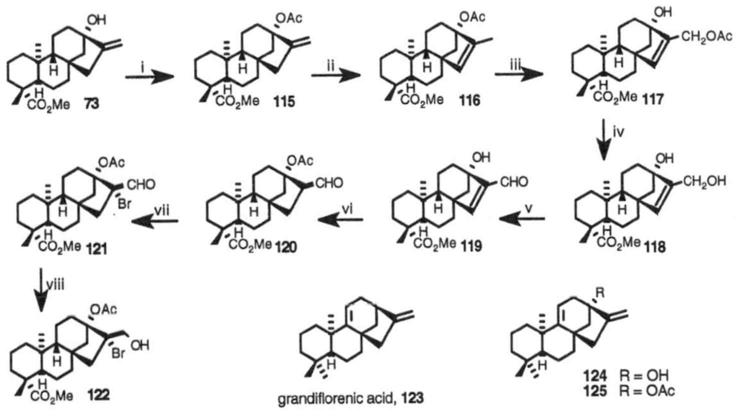

(i) Ac$_2$O, pyridine, DMAP, rt; (ii) I$_2$, benzene, reflux; (iii) SeO$_2$ (cat.), *t*-BuOOH, CH$_2$Cl$_2$-AcOH (1:1); (iv) K$_2$CO$_3$, MeOH-THF-H$_2$O, rt; (v) Swern oxidation and then acetylation; (vi) 10% Pd/CaCO$_3$, EtOH, H$_2$; (vii) pyrrolidone hydrotribromide, CH$_2$Cl$_2$; (viii) LiBH$_4$, Et$_2$O, rt.

Scheme 11 Grandiflorenic acid synthesis by Cheng and Zhou.

Synthesis of *ent*-kaur-16-en-19-oic acid

During the course of synthesizing grandiflorenic acid (**123**) (Lozoya *et al.* 1983), *ent*-kaur-16-en-19-oic acid (**50**) and its double bond isomer have been synthesized from stevioside (**2**) through two skeletal rearrangements in nine steps in the total yield of 9% and 17%, respectively (Cheng and Zhou 1993b) (Scheme 12). The initial step of the synthesis involved the conversion of stevioside (**2**) to **126**. Compound **126** was rearranged to **127** by treatment with BF$_3$-OEt$_2$ which was mesylated to give **128**. Demesylation of **128** was attempted with

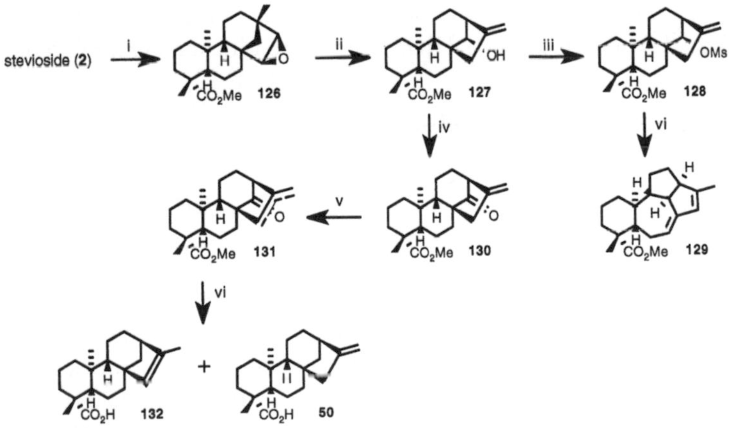

(i) BF$_3$-OEt$_2$, ether, rt; (ii) MsCl, TEA, CH$_2$Cl$_2$, DMAP; (iii) NaI, Zn, DME, reflux; (iv) (COCl)$_2$, DMSO, CH$_2$Cl$_2$, –78 °C; (v) LiI, collidine, Ph$_3$P, reflux; (vi) ethylene glycol, Na°, NH$_2$NH$_2$, reflux.

Scheme 12 Synthesis of *ent*-kaur-16-en-19-oic acid by Cheng and Zhou.

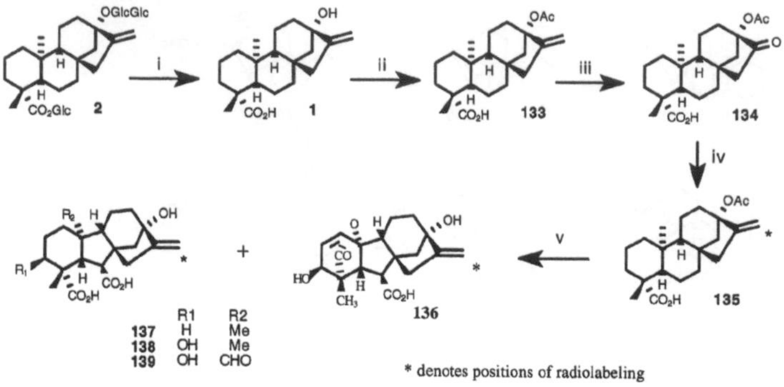

(i) pectinase; (ii) Ac$_2$O, pyridine; (iii) OsO$_4$-NaIO$_4$; (iv) (methyl-d_3)Ph$_3$PI, *t*-BuOK, *t*-BuOH, THF; (v) *Gibberella fujikuroi.*

Scheme 13 Synthesis of radiolabeled gibberellic acid from steviol by Gianfagna.

NaI and zinc to afford a rearrangement product, **129**. Swern oxidation of **127**, followed by treatment with LiI, collidine, and Ph$_3$P gave **131** in quantitative yield. Wolff-Kishner reduction of **131** afforded a mixture of **132** and *ent*-kaur-16-en-19-oic acid (**50**).

Synthesis of radiolabeled gibberellic acids from steviol

Radiolabeled gibberellic acids from steviol have been prepared by feeding chemically synthesized radiolabeled steviol acetate (**133**) to the fungus *Gibberella fujikuroi* (Gianfagna *et al.* 1983) (Scheme 13). Steviol (**1**) was obtained by enzymatic hydrolysis of stevioside (**2**) and the product was acetylated with acetic anhydride-pyridine under reflux. The resulting acetate **133** was oxidized with OsO$_4$-NaIO$_4$ to *ent*-13-acetoxy-16-oxo-17-norkauran-19-oic acid, **134**. Labeled steviol acetate (**135**) was produced by the Wittig reaction from **134** using [^{2}H]-(methyl) triphenyl phosphonium halide. Labelled steviol acetate (**135**) was hydrolyzed in base to steviol for a fungal feeding experiment to *G. fujikuroi* to afford four radiolabeled gibberellic acids [^{2}H]GA$_1$ (**136**), [^{2}H]GA$_{53}$ (**137**), [^{2}H]GA$_{18}$ (**138**), and [^{2}H]GA$_{23}$ (**139**), after incubating for four days.

Studies on the total synthesis of gibberellic acids

In a study on the total synthesis of gibberellic acids, the formation of the bicyclo [3.2.1] C/D ring system present in stevioside (**2**) has been investigated (Corey *et al.* 1970) (Scheme 14). The crystalline tricarbocyclic ketal **141** was synthesized from the corresponding ketone **140** which in turn was obtained by a Robinson annelation sequence. Reaction of 2-(*N*-pyrrolidyl)indene with methyl vinyl ketone in THF afforded a Michael (keto-enamine) adduct which was subjected to cyclization in dioxane–glacial acetic acid–water–sodium acetate trihydrate (10:1:1:1) to form the tricyclic ketone **140**. The crude semisolid **140** was treated without purification with ethylene glycol and *p*-toluenesulfonic acid in benzene at reflux to afford the ketal **141**. Reaction of ketal **141** with *t*-butyl nitrite and *t*-BuOK in dry *t*-BuOH resulted in the formation of the oxime **142**. Treatment of the reaction mixture with 2 N NaOH prior to acidification with 2 N HCl afforded the keto ketal **143**. Compound **143** and oxime **142** were readily separated on silica gel and the oxime could be converted to **143** by sequential treatment with *t*-BuNO$_2$-*t*-BuOH solution, aqueous base, and aqueous acid. Hydrogenation of the ketone **143** using Pd/C and ethyl

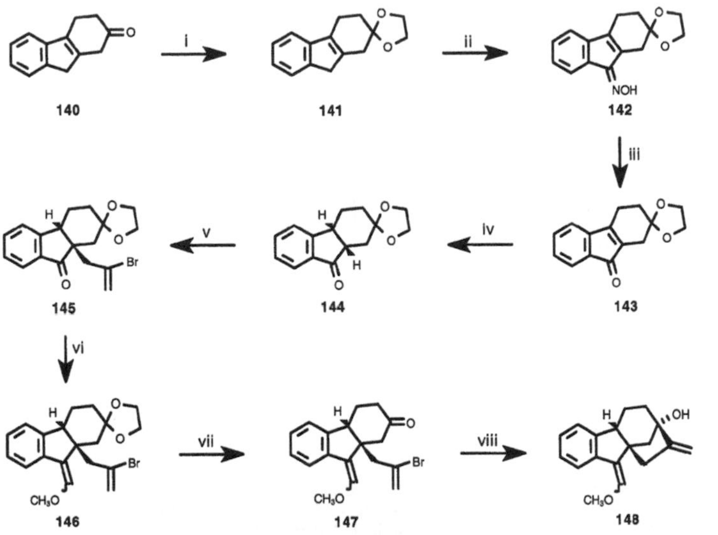

(i) ethylene glycol, TsOH, C$_6$H$_6$, reflux; (ii) *t*-BuONO, *t*-BuOK, *t*-BuOH; (iii) HO⁻; (iv) Pd/C, H$_2$; (v) 2,3-dibromopropene, *t*-BuOK, *t*-BuOH; (vi) CH$_3$OCHP(Ph)$_3$, THF; (vii) AcOH, H$_2$O; (viii) (*n*-Bu)$_2$CuLi, ether.

Scheme 14 Synthesis of 7-hydroxymethylene-8-methylenegibba-1,3,4a(10a)-triene by Corey.

acetate led to the formation of the saturated keto ketal **144**. Alkylation of **144** by 2,3-dibromopropene and *t*-BuOK in *t*-BuOH produced the tricarbocyclic bromide **145** stereospecifically. Reaction of **145** with methoxymethylenetriphenylphosphorane in THF afforded the enol ether **146**. Exposure of **146** to 80% acetic acid–20% water resulted in the formation of the bromo ketone **147**. Reaction of **147** with six equivalents of di-*n*-butyl copper-lithium in ether gave the desired cyclization product, 7-hydroxymethylene-8-methylenegibba-1,3,4a(10a)-triene (**148**).

Total synthesis of (±)-hibaene

Ireland and coworkers converted ketoacetal **149** to the hydroxy olefin **156**, a model for the C/D ring system of steviol (**1**) (Bell *et al.* 1966b). Acid-catalyzed rearrangement of **156** provided entry to the hibaene-stachene system, and the synthesis of racemic hibaene **158** was achieved (Scheme 15).

Introduction of the C-13 acetyl grouping was accomplished through hydroboration-oxidation of the C-13 ethylene derivative **150** which in turn was obtained from the C-13 keto acetal **149** with ethylidenetriphenylphosphorane. An aldol-type cyclization proceeded well after the C-13 acetyl group had been epimerized to the more stable β (equatorial) position with base. The conversion of the C-13 acetyl functionality to a hydroxyl group was carried out by utilizing the Beckmann rearrangement of the derived oxime and then replacement of the C-13 nitrogen by acetate via rearrangement of the N-nitroso derivative. After hydrolysis and oxidation of the derived diol, the hydroxy olefin **156** was readily obtained from the acyloin **155** by condensation with methylenetriphenylphosphorane. This process was accomplished in a 46% overall yield from the C-13 keto acetal **149** to the hydroxy olefin **156**, and serves as a quite satisfactory procedure for the generation of this bridgehead hydroxylated system.

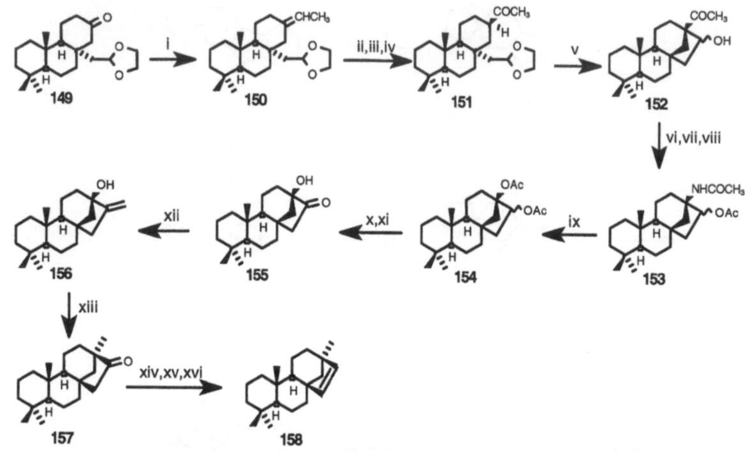

(i) CH$_3$CH$_2$P(Ph)$_3$Br, *t*-BuOK, *t*-BuOH, THF; (ii) NaBH$_4$, BF$_3$-Et$_2$O, THF and then 10% KOH, 30% H$_2$O$_2$, reflux;
(iii) CrO$_3$-acetone; (iv) NaOMe, MeOH; (v) 3 N HCl, acetone, reflux; (vi) HONH$_3$Cl, NaOAc, MeOH, reflux;
(vii) TsCl, pyridine; (viii) dioxane, reflux; (ix) NaOAc. HOAc, NO$_2$; (x) NaOH, MeOH, reflux; (xi) CrO$_3$-acetone;
(xii) CH$_3$P(Ph$_3$)Br, *t*-BuOK, *t*-BuOH, THF; (xiii) 10 N HCl, MeOH, reflux; (xiv) NaBH$_4$, EtOH; (xv) TsCl, pyridine;
(xvi) collidine, reflux.

Scheme 15 Total synthesis of (±)-hibaene by Bell.

The conversion of the hydroxy olefin **156** to hibaene **158** via the ketone **157** follows familiar pathways. This method has been used to converted the C-13 keto alcohol **149** to a model for the C/D ring system of steviol (**1**) and gibberellic acid and hibaene system.

Hydroxylation of the bicyclo[3.2.1]octane system

It was found that the bicyclo[3.2.1]octane system of kaurene can be hydroxylated by treatment with molecular oxygen under light and subsequent LAH reduction (Bell *et al.* 1966b) (Scheme 16).

The oxidation of olefin **159** was achieved by application of a photosensitized oxygenation procedure. The oxygenation of olefins invariably appeared to take place with allylic-type rearrangement of the double bond in highly stereoselective manner. In order to obtain the desired secondary alcohol **161**, an endocyclic olefin intermediate **160**, prepared from iodine

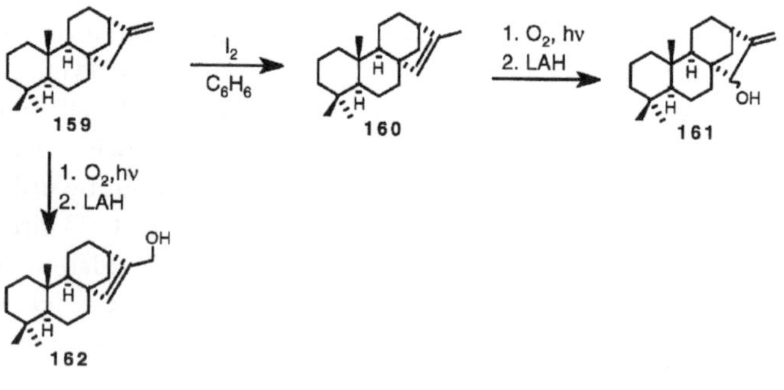

Scheme 16 Hydroxylation of the bicyclo[3.2.1]octane system by Bell.

assisted double bond isomerization of **159**, was employed. Photosensitized oxygenation of the endocyclic olefin **160**, followed by an allylic-type rearrangement of the double bond afforded only the desired secondary alcohol **161**.

In general, kaurene (**159**) was isomerized to isokaurene **160** by iodine in benzene to an equilibrium mixture of the two olefins **159** and **160**, rich in isokaurene **160**. Photosensitized oxygenation of the mixture of olefins **159** and **160** resulted in a mixture of products in which the secondary alcohol was readily separated from the primary alcohol.

Total synthesis of (±)-kaurene and (±)-atisirene

The synthesis of (±)-kaurene (**159**) from the olefinic aldehyde **163** via hydroboration of the double bond and aldol-type cyclization of the C-14 ketoacetal **165** to form a bridged system has been reported (Bell *et al*. 1966a) (Scheme 17). The route makes available as well the substituted bicyclo[2.2.2]octane derivatives through similar cyclization of the isomeric C-13 ketoacetal **149** from the hydroboration and leads to (±)-atisirene (**169**).

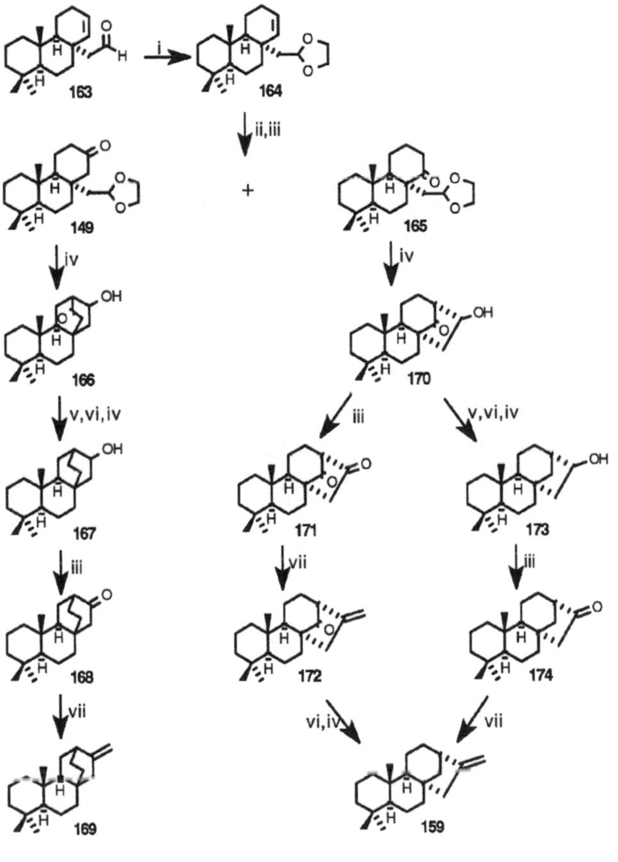

(i) ethylene glycol, TsOH, C_6H_6, reflux; (ii) BH_3-THF, H_2O_2,OH⁻; (iii) CrO_3-acetone; (iv) H_3O^+; (v) DHP, H⁺; (vi) N_2H_4, NaOR; (vii) $(C_6H_5)P^+CH_2^-$.

Scheme 17 Total synthesis of (±)-kaurene and (±)-atisirene by Bell.

The hydroboration and subsequent Jones oxidation of **164** afforded two products, **149** and **165**, which are a C-13 and a C-14 ketoacetal, respectively. The product of acetal cleavage with aqueous acid–acetone solution of both ketoacetals **149** and **165** gave the corresponding ketols **166** and **170** formed by aldol-type cyclization. Oxidation of the ketol **170** to a single diketone **171** was obtained after the asymmetry of the hydroxyl-bearing carbon was destroyed. In the presence of methylenetriphenylphosphorane, the olefinic ketone **172** was produced in high yield. Wolff-Kishner reduction of **172** under forcing conditions afforded (±)-kaurene **159**. In order to prevent reverse aldol-type cleavage of the ketol **170**, the alcohol was first protected as the tetrahydropyranyl ether and then forcing Wolff-Kishner reduction afforded **173**. Oxidation with Jones reagent led to a single ketone **174**. Condensation of the ketone **174** with methylene-triphenylphosphorane again generated (±)-kaurene (**159**).

Ketol **166** was converted to the monoketone **168** by way of the alcohol **167** which was obtained by modified Wolff-Kishner reduction of the derived tetrahydropyranyl ether. Oxidation of the alcohol **167** led to a single compound. Condensation of ketone **168** with methylenetriphenylphosphorane generated atisirene (**169**).

Synthesis of erythroxydiol A (hydroxymonogynol)

(−)-Erythroxydiol A (hydroxymonogynol) (**183**), isolated from *Erythroxylon monogynum*, has been prepared from (−)-steviol (Mori and Matsui 1970; Mori *et al*. 1972) (Scheme 18). (−)-Steviol methyl ester (**73**) was oxidized with *m*-chloroperbenzoic acid in benzene-dioxane to give an epoxide **175**, which was treated with a trace of HCl in aqueous acetone to afford a beyerane ketol, **176**. Jones oxidation of the ketol **176** gave a keto acid **177**, which was esterified with diazomethane to the corresponding methyl ester **178**. This methyl ester was treated with NaBH₄ in ethanol to give the hydroxy-ester **179**, which was mesylated (MsCl/pyridine) and chlorinated (collidine, reflux) to the chloro compound **181**. Compound **181** was heated with lithium bromide and lithium carbonate in DMF to give the ester **182**, which was reduced with LiAlH₄ to afford (±)-erythroxydiol A (**183**).

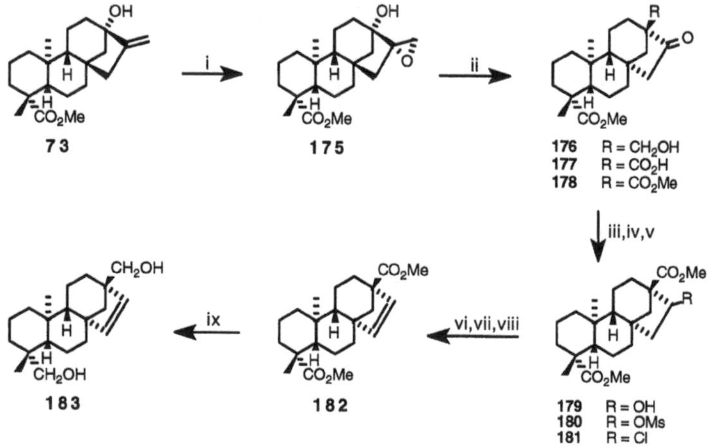

(i) *m*-CPBA, benzene-dioxane; (ii) HCl, acetone; (iii) CrO₃-acetone; (iv) diazomethane; (v) NaBH₄, ethanol; (vi) MsCl, pyridine; (vii) collidine, reflux; (viii) LiBr, LiCO₃, DMF; (ix) LiAlH₄.

Scheme 18 Synthesis of erythroxydiol A (hydroxymonogynol).

A structure–activity relationship study by Cook and Knox

For a gibberellin-like structure–activity relationship study, acyloin-like cyclization involving the syntheses of steviol (**1**) and A-ring modified steviol analogues has been performed by Cook and Knox (1970; 1976a; 1976b) (Schemes 19 and 20).

Synthesis of one of the steviol analogues commenced with OsO_4-$NaIO_4$ oxidation of a diol-ene (**184**) which gave the nor-keto-diol **185** (see Scheme 19). Baeyer-Villiger oxidation of **185** with perbenzoic acid and p-TsOH afforded the γ-lactone **186**, for which the 1,3-diol system was protected as an ethylene group **187**. Alkaline hydrolysis and acid work up of **187** and subsequent methylation (diazomethane) gave hydroxy-ester **188**. Hydroxy-ester **188** was oxidized with chromium-pyridine to keto-ester **189** and was treated sodium naphthalenide to give the acyloin-like cyclization product ketol **190**. Wittig reaction of **190** with methylenetriphenyl-phosphorane gave hydroxy-ene **191**.

Synthesis of another gibberellin analogue involved the γ-lactone **186** as the starting material (see Scheme 20). Compound **186** was treated with triphenylmethyl chloride to give monotrityl ether **192**, and treatment with phosphorus oxychloride gave the alkene **193** which was

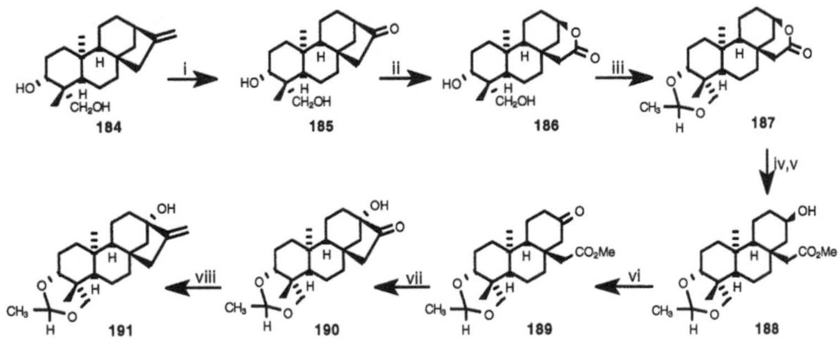

(i) OsO_4, $NaIO_4$, dioxane; (ii) $PhCO_3H$, p-TsOH, $CHCl_3$; (iii) paraldehyde, HCl, ether; (iv) KOH and then H_3^+O; (v) CH_2N_2, ether; (vi) CrO_3-pyridine; (vii) Na°, naphthalene, THF and then CH_3CO_2H; (viii) Ph_3PCH_3I, t-BuOK, t-BuOH, THF.

Scheme 19 Synthesis of gibberellin analogues by Cook and Knox (1).

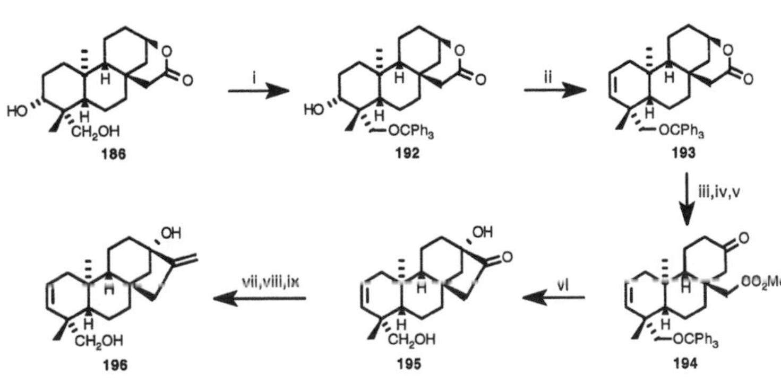

(i) Ph_3CCl, pyridine; (ii) $POCl_3$, pyridine; (iii) NaOH; (iv) CH_2N_2, ether; (v) CrO_3-pyridine; (vi) Na°, naphthalene, THF; (vii) TMSCl, $(TMS)_2NH$, pyridine; (viii) Ph_3PCH_3I, t-BuOK, t-BuOH, THF; (ix) EtOH.

Scheme 20 Synthesis of gibberellin analogues by Cook and Knox (2).

subsequently saponificated, methylated, and oxidized to the corresponding keto-ester **194**. The acyloin-like cyclization of **194** using sodium naphthalenide gave **195**. Treatment of ketol **195** with trimethylsilyl chloride and a subsequent Wittig reaction with methylenetriphenyl-phosphorane followed by heating in ethanol afforded diene-diol **196**.

SUMMARY AND CONCLUSIONS

Stevioside, the major-sweet tasting *ent*-kaurene glycoside of *S. rebaudiana*, which is reported to be 150–300 times sweeter than sucrose, has been a target of commercial interest in the sweet-ener industry. In this chapter the approaches taken toward the chemical syntheses of steviol, stevioside, rebaudioside A, and several of their derivatives have been compiled in order to provide a general reference source for scientists who are interested in the chemistry of *ent*-kaur-ene based sweeteners. In order to try and improve on the sweetness properties of stevioside, several synthetic studies have been carried out. In addition, the bicyclo[3.2.1]octane ring unit, which exists as the common C/D ring structure of steviol and the gibberellins has been a synthetic target for a number of chemists. Although a substantial amount of work has been put into the chemistry of *ent*-kaurene based sweeteners, it would be too expensive to produce this class of sweeteners through chemical synthesis for commercial profit.

REFERENCES

Bell, R. A., Ireland, R. E. and Partyka, R. A. (1966a) Experiments directed toward the total synthesis of terpenes. VIII. The total synthesis of (±)-kaurene and (±)-atisirene. *Journal of Organic Chemistry 31*, 2530–2536.

Bell, R. A., Ireland, R. E. and Mander, L. N. (1966b) Experiments directed toward the total synthesis of terpenes. IX. The total synthesis of (±)-hibaene and the oxygenation of some tetracyclic diterpenes. *Journal of Organic Chemistry* **31**, 2536–2542.

Cheng, Y. X. and Zhou, W. S. (1993a) Study on the synthesis of tetracyclic diterpenoids 7. *Chinese Chemical Letters* **4**, 291–294.

Cheng, Y. X. and Zhou, W. S. (1993b) Study on the tetracyclic diterpenoids 5. Synthesis of *ent*-kaur-16-en-19-oic acid. *Acta Chimica Sinica* **51**, 819–924.

Cook, I. F. and Knox, J. R. (1970) A synthesis of steviol. *Tetrahedron Letters*, 4091–4093.

Cook, I. F. and Knox, J. R. (1976a) The synthesis of 13-hydroxylated *ent*-kaur-16-ene derivatives using an acyloin-like cyclization of keto esters. *Tetrahedron* **32**, 363–367.

Cook, I. F. and Knox, J. R. (1976b) Bridged-ring products from the acyloin-like cyclization of diterpenoid keto-esters. *Tetrahedron* **32**, 369–375.

Corey, E. J., Narisada, M., Hiraoka, T. and Ellison, R. A. (1970) Studies on the total synthesis of gibberellic acids. A simple route to the tetracarbocyclic network. *Journal of the American Chemical Society* **92**, 396–397.

Crammer, B. and Ikan, R. (1986) Sweet glycosides from the *Stevia* plant. *Chemistry in Britain* **22**, 915–916, 918.

Djerassi, C., Quitt, P., Mosettig, E., Combie, R. C., Rutledge, P. S. and Briggs, L. H. (1961) Optical rota-tory dispersion studies. LVIII. The complete absolute configurations of steviol, kaurene and the diter-pene alkaloids of the garryfoline and atisine groups. *Journal of the American Chemical Society* **83**, 3720–3722.

DuBois, G. E., Dietrich, P. S., Lee, J. F., McGarraugh, G. V. and Stephenson, R. A. (1981) Diterpenoid sweeteners. Synthesis and sensory evaluation of stevioside analogues nondegradable to steviol. *Journal of Medicinal Chemistry* **24**, 1269–1271.

DuBois, G. E., Bunes, L. A., Dietrich, P. S. and Stephenson, R. A. (1984) Diterpenoid sweeteners. Syn-thesis and sensory evaluation of biologically stable analogues of stevioside. *Journal of Agricultural and Food Chemistry* **32**, 1321–1325.

Esaki, S., Tanaka, R. and Kamiya, S. (1984) Synthesis and taste of certain steviol glycosides. *Agricultural and Biological Chemistry* **48**, 1831–1834.

Gianfagna, T., Zeevaart, J. A. D. and Lusk, W. J. (1983) Synthesis of [^{2}H]-gibberellins from steviol using the fungus *Gibberella fujikuroi. Phytochemistry* **22**, 427–430.

Kamiya, S., Konishi, F. and Esaki, S. (1979) Synthesis and taste of some analogs of stevioside. *Agricultural and Biological Chemistry* **43**, 1863–1867.

Kaneda, N., Kasai, R., Yamasaki, K. and Tanaka, O. (1977) Chemical studies on sweet diterpene-glycosides of *Stevia rebaudiana*: Conversion of stevioside into rebaudioside-A. *Chemical and Pharmaceutical Bulletin* **25**, 2466–2467.

Lozoya, X., Enriquez, R. G., Bejar, E., Estrada, A. V., Giron, H., Ponce-Monter, H. and Gallegos, A. J. (1983) The zoapatle V. The effect of kauradienoic acid upon uterine contractility. *Contraception* **27**, 267–279.

Mori, K. and Matsui, M. (1965) Total synthesis of methyl (±)-8α-carboxymethylpodocarpan-13-one-4β-carboxylate, a degradation product of steviol. *Tetrahedron Letters* 2347–2350.

Mori, K. and Matsui, M. (1966) Diterpenoid total synthesis-II. An alternative route to methyl (±)-7-oxo-podocarp-8-en-16-oate. *Tetrahedron* **22**, 879–884.

Mori, K. and Matsui, M. (1970) Synthesis of erythroxydiol A (hydroxymonogynol). *Tetrahedron Letters* 3287–8288.

Mori, K., Matsui, M. and Sumiki, Y. (1970a) A new method for the construction of a bicyclo[3.2.1]octane ring system with a bridgehead hydroxyl group. A partial synthesis of (–)-epiallogibberellic acid. *Tetrahedron Letters* 429–432.

Mori, K., Nakahara, Y. and Matsui, M. (1970b) Total synthesis of (±)-steviol. *Tetrahedron Letters* 2411–2414.

Mori, K., Nakahara, Y. and Matsui, M. (1972) Diterpenoid total synthesis-XIX. (+)-Steviol and erythroxydiol A: rearrangements in bicyclooctane compounds. *Tetrahedron* **28**, 3217–3226.

Mosettig, E., Beglinger, U., Dolder, F., Licht, H., Quitt, P. and Waters, J. A. (1963) The absolute configuration of steviol and isosteviol. *Journal of the American Chemical Society* **85**, 2305–2309.

Nakahara, Y., Mori, K. and Matsui, M. (1971) Diterpenoid total synthesis. Part XVI. Alternative synthetic routes to (±)-steviol and (±)-kaur-16-en-19-oic acid. *Agricultural and Biological Chemistry* **35**, 918–928.

Nakahara, Y. (1982) Synthetic studies of physiologically active natural products with characteristic ring structures. *Nippon Nogeikagaku Kaishi* **56**, 943–955.

Ogawa, T., Nozaki, M. and Matsui, M. (1978) A stereocontrolled approach to the synthesis of glycosyl esters. Partial synthesis of stevioside from steviobioside. *Carbohydrate Research* **60**, C7–C10.

Ogawa, T., Nozaki, M. and Matsui, M. (1980) Total synthesis of stevioside. *Tetrahedron* **36**, 2641–2648.

Ruddat, M., Lang, A. and Mosettig, E. (1963) Gibberellin activity of steviol, a plant terpenoid. *Naturwissenschaften* **50**, 23.

Tanaka, O. (1997) Improvement of taste of natural sweeteners. *Pure and Applied Chemistry* **69**, 675–683.

Vignais, P. V., Duee, E. D., Vignais, P. M. and Huet, J. (1966) Effects of atractyligenin and its structural analogues on oxidative phosphorylation and on the translocation of adenine nucleotides in mitochondria. *Biochimica et Biophysicia Acta* **118**, 465–483.

Wingard, R. E., Jr, Brown, J. P., Enderlin, F. E., Dale, J. A., Hale, R. L. and Seitz, C. T. (1980) Intestinal degradation and absorption of the glycosidic sweeteners stevioside and rebaudioside A. *Experientia* **36**, 519–520.

Ziegler, F. E. and Kloek, J. A. (1971) 1-Hydroxy-7-methylene bicyclo[3.2.1]octane: A gibbane-steviol C/D ring model. *Tetrahedron Letters* 2201–2203.

Ziegler, F. E. and Kloek, J. A. (1977) The stereocontrolled photoaddition of allene to cyclopent-1-ene-1-carboxaldehydes. A total synthesis of (±)-steviol methyl ester and isosteviol methyl ester. *Tetrahedron* **33**, 373–380.

7 Methods to improve the taste of the sweet principles of *Stevia rebaudiana*

Kazuhiro Ohtani and Kazuo Yamasaki

STEVIOL GLYCOSIDES

Stevioside (**1**) is a natural sweetener isolated from the herb, *S. rebaudiana* (Bertoni) Bertoni (Compositae) which is indigenous to Paraguay and has long been used to sweeten local beverages. It is a diterpene glucoside having the *ent*-kaurene diterpene aglycone, steviol, and is sweeter than sugar by about 150 times. The final structure elucidation of stevioside (**1**) was performed by Mosettig *et al.* (1963). More than ten years later, several congeners of stevioside were isolated from the same plant by two Japanese groups, such as rebaudiosides A (**2**) (Kohda *et al.* 1976), C (**3**) (Sakamoto *et al.* 1977a), D (**4**) and E (**5**) (Sakamoto *et al.* 1977b), and dulcoside A (**6**) (Kobayashi *et al.* 1977). All of these glycosides have the same aglycone, steviol (13-hydroxy-*ent*-kaur-16-en-19-oic acid), but have different sugar moieties. All compounds are sweet, but the magnitude and quality of the taste differ from each other (Figure 7.1). Among these, rebaudioside A (**2**) has the greatest degree of sweetness, and the quality of its taste is pleasant. In most individual plants of *S. rebaudiana*, the yield of stevioside (**1**) is the highest (2–10%), rebaudioside A (**2**) follows next (*ca.* 1%), and the other glycosides are minor components.

Rebaudioside A

Several attempts were made to increase the yield of rebaudioside A (**2**) by selective breeding of *S. rebaudiana*, and some individual plants exceeded the yield of stevioside (**1**) as mentioned elsewhere in this volume. Also, a semisynthetic approach was successful in converting stevioside to rebaudioside A (**2**). Since rebaudioside A (**2**) has only one additional glucosyl moiety at the C-3 position of the inner glucose at C-13 of stevioside, selective β-glucosylation of stevioside at this point would afford rebaudioside A. A strategy is needed as to how to block the unwanted hydroxyl group in the three glucosyl moieties. As shown in Figure 7.2, the outer glucose at C-13 was selectively hydrolyzed by the crude enzyme, 'Takadiastase Y', in almost a quantitative fashion to afford the intermediate, steviol 13,19-di-*O*-β-glucoside (**7** = rubusoside, later isolated from *Rubus suavissimus*). Next, the ester glucosyl moiety at C-19 was removed by alkaline hydrolysis to afford steviolmonoside (**49**). After blockage of the C-4 and C-6 positions of the glucosyl moiety by a benzylidene derivative, other glucose units were introduced forcibly at C-2 and C-3 of the C-13 glucose unit and at the C-19 free carboxyl moiety by reaction with 3,4,6-tri-*O*-acetyl-α-D-glucopyranose 1,2-(*tert*-butyl orthoacetate). Rebaudioside A (**2**) was obtained in good yield after removal of the protective groups at C-4 and C-6 of the glucose moiety with BaO in MeOH (Kaneda *et al.* 1977).

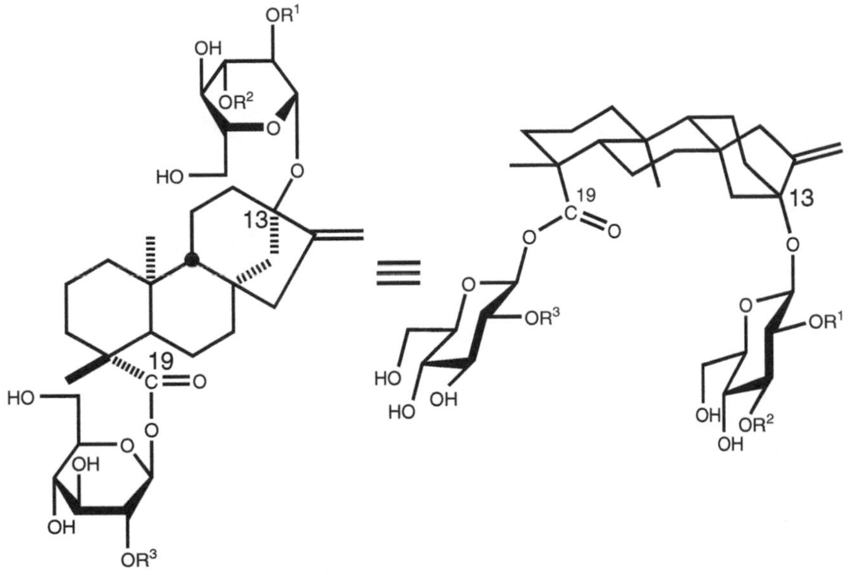

	R^1	R^2	R^3	RS^a	QT^b
stevioside (1)	β–D–Glc	H	H	143	0
rebaudioside A (2)	β–D–Glc	β–D–Glc	H	242	+2
rebaudioside C (3)	α–L–Rha	β–D–Glc	H	nd	−1
rebaudioside D (4)	β–D–Glc	β–D–Glc	β–D–Glc	221	+3
rebaudioside E (5)	β–D–Glc	H	β–D–Glc	174	+1
dulcoside A (6)	α–L–Rha	H	H	nd	−2

^{a}RS: relative sweetness to sucrose, nd: not determined

^{b}QT: quality of taste, +: better, −: worse

RS and QT can be comparable only in single figure.

Figure 7.1 Sweet steviol glycosides from leaves of *Stevia rebaudiana*.

Rubusoside and related glycosides from *Rubus suavissimus*

In a continuation of their search for natural sweeteners, Tanaka *et al.* (1981) isolated rubuso-side (**7** = steviol 13,19-di-*O*-β-D-glucoside) from the leaves of a type of the Chinese rasp-berry, R. *suavissimus* S. Lee (Rosaceae). Although *Rubus* is taxonomically distinct from *Stevia* (Compositae), the structure of **7** was identical with the intermediate obtained from the synthesis of rebaudioside A (**2**) from stevioside (**1**) (see Figure 7.2). Rubusoside (**7**) tastes sweet, but the magnitude of sweetness is not high [relative sweetness to sucrose (RS) = 114], and the hedonic quality [quality of taste (QT)] is not good. From the same species, Hirono *et al.* (1990) and Ohtani *et al.* (1992) isolated a number of other related diter-pene glycosides, namely, a series of suaviosides (**8–41**). In Figure 7.3 (A–C), the structures and tastes of these compounds are shown.

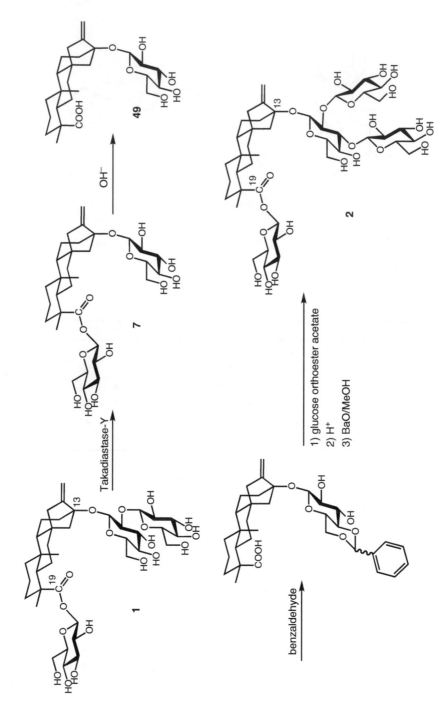

Figure 7.2 Synthesis of rebaudioside A (**2**) from stevioside (**1**).

(A)

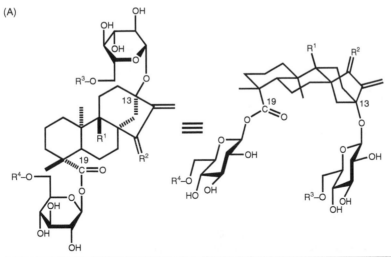

	QT[a]	R[1]	R[2]	R[3]	R[4]
rubusoside (**7**)	+++	H	H$_2$	H	H
suavioside B (**8**)	++	OH	H$_2$	H	H
suavioside C1 (**9**)	B	H	H$_2$	p–coumaroyl	H
suavioside C2 (**10**)	−	H	H$_2$	H	p–coumaroyl
suavioside D1 (**11**)	−	H	H$_2$	caffeoyl	H
suavioside D2 (**12**)	−	H	H$_2$	H	caffeoyl
15β–hydroxyrubusoside (**13**)	++	H	β–OH	H	H
15-oxorubusoside (**14**)	++	H	O	H	H
suavioside Q1 (**15**)	+++	H	H$_2$	α–D–Glc	H
suavioside Q2 (**16**)	+++	H	H$_2$	H	α–D–Glc
suavioside R1 (**17**)	+++	H	H$_2$	β–D–Glc	H
suavioside R2 (**18**)	+++	H	H$_2$	H	β–D–Glc
suavioside S1 (**19**)	+++	H	H$_2$	α–D–Gal	H
suavioside S2 (**20**)	+++	H	H$_2$	H	α–D–Gal

[a]QT: quality of taste

(B)

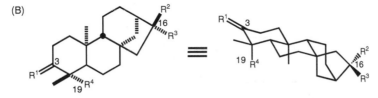

	QT[a]	R[1]	R[2]	R[3]	R[4]
paniculoside IV (**44**)	−	H$_2$	OH	CH$_2$OH	COO–β–D–Glc
suavioside A (**21**)	+	α–OH	OH	CH$_2$O–β–D–Glc	CH$_3$
suavisode E (**22**)	−	H$_2$	CH$_2$OH	OH	COO–β–D–Glc
suavioside K (**23**)	B	H$_2$	CH$_2$O–β–D–Glc	OH	COO–β–D–Glc
sugeroside (**24**)	−	O	OH	CH$_2$O–β–D–Glc	CH$_3$

Figure 7.3 Continued

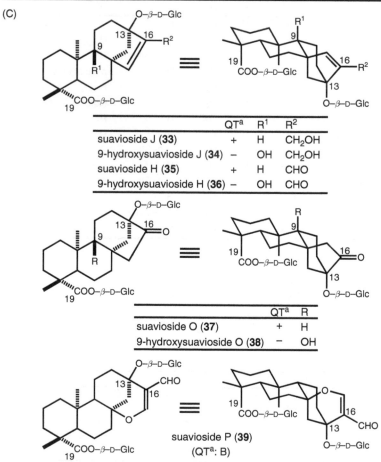

	QT[a]	R[1]	R[2]	R[3]	R[4]	R[5]
suavioside F (25)	B	β–D–Glc	H₂	OH	CH₃	H
suavioside G (26)	+	β–D–Glc	H₂	OH	CH₃	β–D–Glc
suavioside I (27)	+	H	H₂	OH	CH₂OH	β–D–Glc
suavioside L (28)	+	β–D–Glc	H₂	H	CH₂OH	β–D–Glc
15-oxosuavioside L (29)	B	β–D–Glc	O	H	CH₂OH	β–D–Glc
15-oxo-16-*epi*-suavioside L (30)	B	β–D–Glc	O	CH₂OH	H	β–D–Glc
16β-hydroxysuavioside L (31)	+	β–D–Glc	H₂	OH	CH₂OH	β–D–Glc
16α-hydroxysuavioside L (32)	+	β–D–Glc	H₂	CH₂OH	OH	β–D–Glc

(C)

	QT[a]	R[1]	R[2]
suavioside J (33)	+	H	CH₂OH
9-hydroxysuavioside J (34)	–	OH	CH₂OH
suavioside H (35)	+	H	CHO
9-hydroxysuavioside H (36)	–	OH	CHO

	QT[a]	R
suavioside O (37)	+	H
9-hydroxysuavioside O (38)	–	OH

suavioside P (39)
(QT[a]: B)

Figure 7.3 Continued

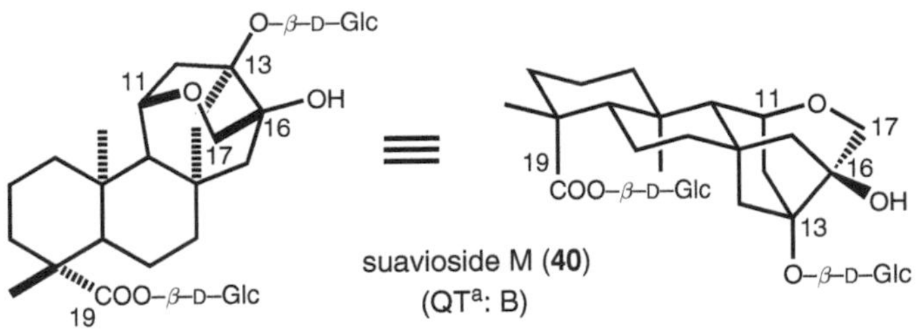

Figure 7.3 Diterpene glycosides from the leaves of *Rubus suavissimus*.

STRUCTURE–ACTIVITY RELATIONSHIPS

Stevia rebaudiana and *R. suavissimus* supplied good lead compounds to study for structure–sweetness relationships, since the former contains several glycosides (**1–6**) of the same aglycone, steviol, and the latter contains many glycosides (**7–40** and **44**), most of which have common sugar moieties (13- and 19-*bis*-β-D-glucoside, **5**) with slightly different aglycones. Several other structurally related glucosides, paniculosides I–V (**41–45**), isolated from *S. paniculata* by Yamasaki *et al.* (1977) and from *S. ovata* by Kaneda *et al.* (1978) were also considered. In addition to these natural glycosides, chemically modified compounds (**46–50**) were also evaluated for their sweetening activity (Figure 7.4).

Aglycone moiety

Starting from rubusoside (**7**) (RS = 114), the relative sweetnesses of the compounds with slightly modified skeletons were compared. The sugar units at C-13 and C-19 of the *ent*-kaurene skeleton seemed to be essential to express sweetness, since all of the monodesmosides, paniculosides I–V (**41–45**) as well as suavioside E (**22**), K (**23**) and sugeroside (**24**) were tasteless. The only exceptions were the slightly sweet suavioside A (**21**) and I (**27**) which lack a sugar unit at C-19 and C-13, respectively. This supported the deduction that both sugar units (or at least a hydrophilic functional group) at C-13 and C-19 are necessary for sweetness. Also, stevioside 19-β-D-O-glucoside (**48**), chemically derived from steviol, tasted weakly sweet (RS < 50) with concomitant bitterness, although the solubility in water was low. Hydrogenation of the double bond (C-15) caused a remarkable decrease in the sweetness of the mixture of the 16α- and β-methyl congeners obtained, and dihydrorubusoside (**46**), with RS < 50, also tasted bitter. Essentially the same trend was observed in the case of stevioside (one more additional glucose at C-13), with RS = 143, with the hydrogenated compound, dihydrostevioside (**47**), exhibiting less sweetness (RS < 50) (Kasai *et al.* 1981). Hydroxylation at C-9 of the steviol skeleton decreased the sweetness, which was observed in the case of suavioside B (**8**). Hydroxylation of the double bond (C-16) of rubusoside also decreased the sweetness. All of the mono- and dihydroxylated derivatives at the C-16 and/or C-17 positions of the rubusoside congeners (**26–28** and **31–32**) showed only slight sweetness, or bitterness, and the 19-desgluco-16-OH derivative **25** tasted bitter. Introduction of a keto group at C-15 (**29** and **30**) also changed the sweet taste to a bitter taste, while introduction of an endocyclic double bond at C-15(16) (**33–36**) resulted in the reduction of sweetness. In the case of the 9,17-dihydroxylated compound **34**, it became tasteless. Suaviosides P (**39**) and M (**40**), which have modified steviol skeletons, exhibited bitter tastes.

QT	R¹	R²
42	–	H	H

(table for A)

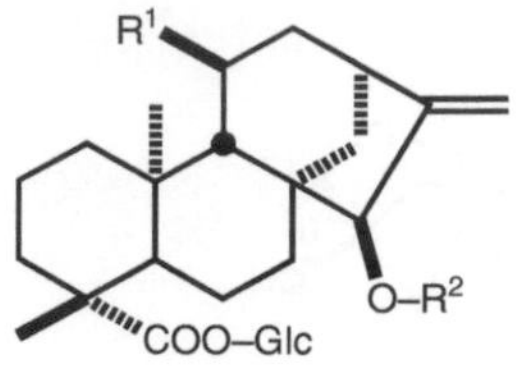

	QT	R¹	R²
42	–	H	H
43	–	OH	H
45	–	H	β-D-Glc

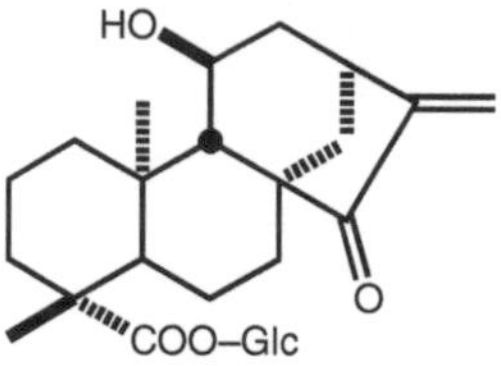

44 (QT:–) 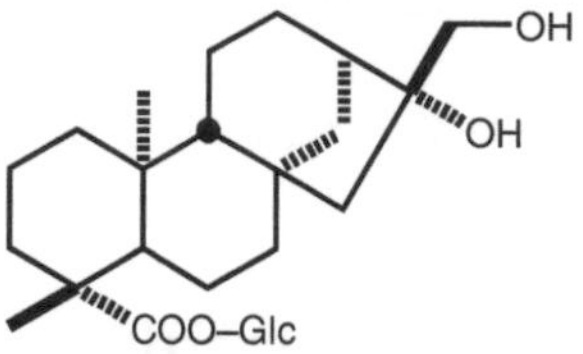   **41** (QT:–)

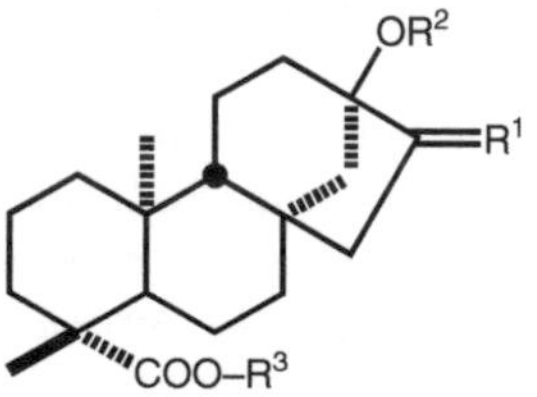

	QT	R¹	R²	R³
46	+	CH₃, H	β-D-Glc	β-D-Glc
47	+	CH₃, H	β-D-Glc(2–1)-ββ-D-Glc	β-D-Glc
48	+	CH₂	H	β-D-Glc
49	–	CH₂	β-D-Glc	H
50	–	CH₂	β-D-Glc(2–1)-β-D-Glc	H

Figure 7.4 Diterpene glucosides from *Stevia paniculata* and *S. ovata* (A) and artificial glucosides derived from steviol glucosides (B).

The generalized conclusion obtained so far is that no other diterpenoid skeleton exhibited better sweetness than that of steviol, and essentially every part of the steviol molecule seems to be necessary for the expression of the sweet taste when suitably glycosylated.

Sugar moieties

Concerning the effect of the sugar moieties of steviol glycosides, several natural compounds from *S. rebaudiana* and *R. suavissimus* were examined. As was partly mentioned in the previous discussion, two saccharide moieties at C-13 and C-19 ester (bisdesmoside) seem to be important. These two sugar moieties, which are separated by the aglycone, are spatially very close to each other. From the results shown in Figure 7.1, when there are one or two C-19 ester sugars, the magnitude of sweetness and the quality of taste of the resultant compound increases according to the number of glucosyl units attached to C-13 to some extent.

TRANSGLUCOSYLATION

The above findings prompted us to try and improve the sweetness of stevioside (**1**) and its congeners by selectively adding sugar units to the C-13 sugar residues. Since stevioside is used as a food additive in Japan, non-synthetic biological (enzymatic) treatment was preferable. In the food industry, several glucosidases have been used to modify the structures of polysaccharides. Some of these enzymes not only hydrolyze but also transfer the sugar moieties to other molecules. For example, cyclomaltodextrin glucanotransferase (CGTase) reacts with starch to produce cyclodextrins (cyclic oligomers of α-1,4 glucose), as well as to transfer α-glucosyl units from starch to the OH-4 of a glucosyl moiety (trans-α-1,4-glucosylation) (The Amylase Research Society of Japan 1988) (Figure 7.5).

CGTase and starch

Stevioside (**1**) was treated with the CGTase with soluble starch as donor, yielding a complex mixture of products which were mono-, di-, tri- and polyglucosylated at the 19-*O*-glucosyl unit and/or the terminal glucosyl unit of the 13-*O*-sophorosyl moiety (Kasai *et al*. 1981) (Figure 7.6). Separation and structure identification of all the mono- (**S1a** and **S1b**), di- (**S2a, S2b,** and **S2c**) and tri- (**S3a, S3b, S3c,** and **S3d**) glucosylated products were achieved by Fukunaga *et al*. (1989), who also evaluated the intensity of the sweetness and the quality of the taste. As expected, significant improvement in the quality of taste and intensity of sweetness was observed for most of the glucosylated products, especially for **S1a** and **S2a**, which are mono- and di-glucosylated at the terminal glucosyl moiety of the 13-*O*-sophorosyl group, respectively. In contrast, glucosylation at the 19-*O*-glucosyl group (**S1b, S2c,** and **S3d**) resulted in a decrease of intensity and/or quality of sweetness. This fundamental finding was further developed to selectively obtain better sweeteners using more specific transglucosylation systems.

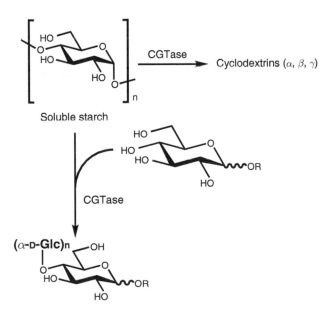

Figure 7.5 Function of cyclodextrin glucanotransferase (CGTase).

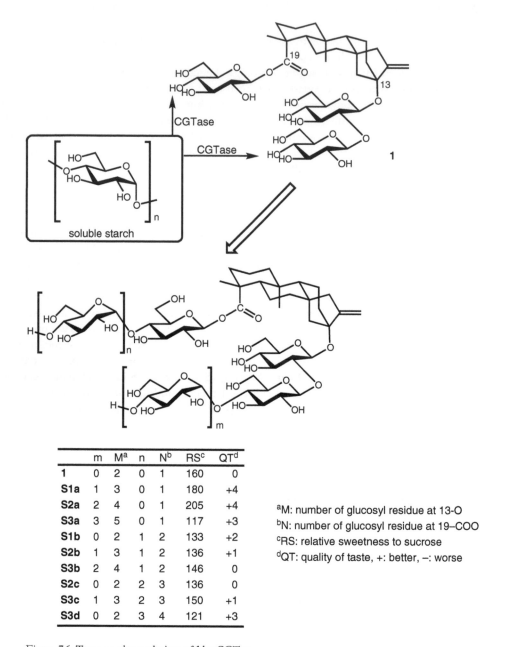

	m	M^a	n	N^b	RS^c	QT^d
1	0	2	0	1	160	0
S1a	1	3	0	1	180	+4
S2a	2	4	0	1	205	+4
S3a	3	5	0	1	117	+3
S1b	0	2	1	2	133	+2
S2b	1	3	1	2	136	+1
S3b	2	4	1	2	146	0
S2c	0	2	2	3	136	0
S3c	1	3	2	3	150	+1
S3d	0	2	3	4	121	+3

[a]M: number of glucosyl residue at 13-O
[b]N: number of glucosyl residue at 19–COO
[c]RS: relative sweetness to sucrose
[d]QT: quality of taste, +: better, –: worse

Figure 7.6 Trans-α-glucosylation of **1** by CGTase.

Pullulanase and pullulan

Through several screening procedures for selecting a better enzyme-donor system, one of the glucosidases, pullulanase from *Klebsiella* sp., proved to be effective. The products **S1a**, **S2a**, and **S2c** were also obtained from stevioside by pullulan and crude pullulanase. Although the yields were rather low, the selectivity in terms of the yield of the desirable **S1a** and **S2a** was higher

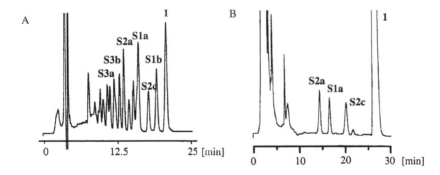

Figure 7.7 HPLC chromatogram of transglucosylated products with CGTase (A) and pullulanase (B). HPLC conditions: (A) column, YMC-pack ODS-AM302 (4 mm i.d. × 15 cm); eluent, 50% MeOH, flow rate, 0.8 ml/min; column temperature, 60 °C; detection, UV 210 nm; (B) column, TSK-gel ODS-120T (4 mm i.d. × 15 cm): eluent 50% MeOH, flow rate, 1.0 ml/min; column temperature 80 °C, detection, UV 210 nm.

than in the case of CGTase. Figure 7.7 illustrates the HPLC chromatograms of the products from the above two enzymatic reactions (Lobov *et al.* 1991).

Rubusoside as the starting material

Rubusoside (**7**), a congener of stevioside (**1**) with one less sugar unit, was also treated with the CGTase-starch system, to undergo trans-α-1,4-glucosylation. From the reaction mixture, mono- (**R1a** and **R1b**), di- (**R2a**, **R2b**, and **R2c**), tri- (**R3a**, **R3b**, **R3c**, and **R3d**) and three of five tetraglucosylated products (**R4a**, **R4b**, and **R4e**) were separated (Darise *et al.* 1984) and identified (Figure 7.8) (Ohtani *et al.* 1991a). Two of the tetraglucosylated products (**R4c** and **R4d**) were difficult to purify and were obtained as a mixture. Strong enhancement of the intensity of sweetness was observed for the products **R2a**, **R3a**, **R3b**, and **R4b** which were di- or triglucosylated at the 13-*O*-glucosyl moiety. On the other hand, tetraglucosylation at the 13-*O*-glucosyl moiety as well as glucosylation at the 19-*O*-glucosyl moiety led to a decrease in the intensity of sweetness. These results were similar to the case of stevioside (**1**) as the starting material, and strongly suggested that for enhancement of the intensity of sweetness of steviol glycosides, the elongation of the 13-*O*-glucosyl moiety up to a total of four glucosyl units, accompanied by the suppression of glucosylation at the 19-*O*-glucosyl moiety, are desirable.

Shortening of the long α-1,4-glucosyl chain

When undesirable products are obtained having 13-*O*-sugar units that are too long, it is necessary to shorten the sugar chain. β-Amylase hydrolyzes a α-1,4-glucosyl chain from the non-reducing end to release maltose (4α-glucosyl glucose). By treatment with this enzyme, three or more additional α-1,4-glucosyl chains of the 13-*O*-glucosyl moiety of transglucosylated steviol derivatives are converted into a mono- or di-α-1,4-glucosyl chain. In Japan, transglucosylated stevioside in which the α-glucosyl chain is shortened by this treatment (Kasai *et al.* 1981) is currently available commercially.

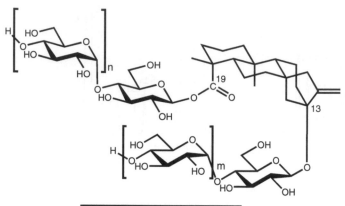

	m	M^a	n	N^b	RS^c
7	0	1	0	1	114
R1a	1	2	0	1	132
R2a	2	3	0	1	278
R3a	3	4	0	1	214
R4a	4	5	0	1	115
R1b	0	1	1	2	102
R2c	1	2	1	2	95
R3b	2	3	1	2	182
R4b	3	4	1	2	202
R2b	0	1	2	3	99
R3c	1	2	2	3	110
R4c	2	3	2	3	184
R4d	1	2	3	4	–
R3d	0	1	3	4	58
R4e	0	1	4	5	49

[a]M: number of glucosyl residue at 13-O
[b]N: number of glucosyl residue at 19–COO
[c]RS: relative sweetness to sucrose

Figure 7.8 Trans-α-glucosylation of **7** by CGTase.

Selective syntheses of improved sweeteners

Several attempts were carried out to produce more desirable sweeteners as selectively as possible starting from stevioside (**1**), rubusoside (**7**), and even from a mixture of several stevioside congeners for economic reasons. The tactical priority was the effective transfer of one or two glucosyl units to the 13-O-glucosyl (or sophorosyl) unit of the starting steviol glucoside.

Trans-α-1,4-glucosylation of steviolmonoside and steviolbioside

The first strategy for an effective transglucosylation was to remove the 19-glucose from the starting material. By alkaline treatment, stevioside (**1**) and rubusoside (**7**) afforded steviolbio-

Table 7.1 Solubulizing effect of cyclodextrins (CDs) on **49** and **50** in 50 mM acetate buffer (pH 5.4)

Sample	CD	Concentration of CD (mg/ml)	Solubility of 49 or 50 (mg/ml)
49	none		0.69
(10 mg/ml)	α-CD	15.1	1.26
	β-CD	17.7	2.48
	γ-CD	20.2	7.01
50	none		0.21
(10 mg/ml)	α-CD	20.3	0.28
	β-CD	23.6	0.90
	γ-CD	27.0	>10.0

side (**50**) and steviolmonoside (**49**), respectively. Transglucosylation of both compounds by a CGTase-starch system was successful, but proceeded very slowly due to the low solubility in a buffer solution. The solubility of both compounds was increased remarkably in the presence of γ-cyclodextrin (γ-CD) (Table 7.1). Trans-α-1,4-glucosylation of steviolbioside (**50**) or steviolmonoside (**49**) by the CGTase-starch system was carried out with the aid of γ-CD to afford the mixture of transglucosylated products in better yields (Figure 7.9). After acetylation, each product was subjected to chemical β-glucosylation of the 19-COOH unit followed by

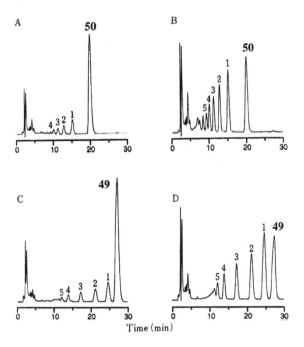

Figure 7.9 HPLC chromatograms of transglucosylated products of **50** and **49** with or without γ-cyclodextrin (γ-CD). (A) **50** without γ-CD; (B) **50** with γ-CD, (C) **49** without γ-CD; (D) **49** with γ-CD. Peak numbers 1–5 in each chromatogram indicate the number of transferred glucosyl units. Conditions: column, YMC-pack ODS-AM302 (4 mm i.d. × 15 cm); eluent, MeOH-0.05% TFA (60:40), flow rate, 0.8 ml/min; column temperature, 60 °C; detection, UV 210 nm.

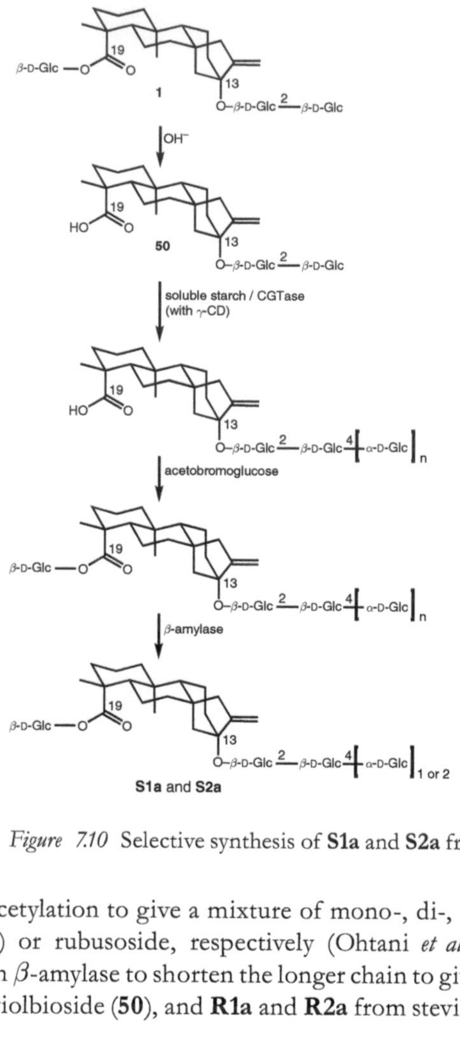

β-D-Glc —O   19   O

1   13

O–β-D-Glc —2— β-D-Glc

↓ OH⁻

HO   19   O

50   13

O–β-D-Glc —2— β-D-Glc

↓ soluble starch / CGTase
  (with γ-CD)

HO   19   O

13

O–β-D-Glc —2— β-D-Glc —4— α-D-Glc]ₙ

↓ acetobromoglucose

β-D-Glc —O   19   O

13

O–β-D-Glc —2— β-D-Glc —4— α-D-Glc]ₙ

↓ β-amylase

β-D-Glc —O   19   O

13

O–β-D-Glc —2— β-D-Glc —4— α-D-Glc]₁ or 2

S1a and S2a

Figure 7.10 Selective synthesis of **S1a** and **S2a** from **1**.

deacetylation to give a mixture of mono-, di-, tri-, and poly-glucosylated stevioside (Figure 7.10) or rubusoside, respectively (Ohtani *et al.* 1991b). The reaction mixture was treated with β-amylase to shorten the longer chain to give the excellent sweeteners, **S1a** and **S2a** from steviolbioside (**50**), and **R1a** and **R2a** from steviolmonoside (**49**).

Trans-α-1,4-glucosylation of 19-O-β-galactosyl esters

Structure – sweetness correlations were investigated for the synthesized steviol glycosides with some of the glucosyl units replaced by other sugar units (Kamiya *et al.* 1979; Esaki *et al.* 1984) (Figure 7.11). Kitahata *et al.* (1978) revealed, in their study on acceptor specificity of transglucosylation by a CGTase-starch system, that transglucosylation hardly occurs with galactose, mannose and ribose. The 19-O-glucosyl groups of stevioside (**1**) and rubusoside (**7**) were chemically replaced by a β-D-galactosyl group (Figure 7.12) through steviolbioside (**50**) and steviolmonoside (**49**), respectively. Then, both the galactosyl esters (**SGal** and **RGal**) were subjected to trans-α-1,4-glucosylation by the CGTase-starch system to give mono-, di-, tri-, and tetra-α-glucosylated products: **SGal-1, -2, -3,** and **-4** from **SGal**, and **RGal-1, -2, -3,** and **-4** from **RGal** (Mizutani *et al.* 1989). Results of the evaluation of the sweetness of the products are summarized in Figure 7.12. Replacement of the 19-O-glucosyl group by a β-galactosyl group (**SGal** and **RGal**) led to a deterioration of the taste qualities. Elongation of the 13-O-glycosyl moiety

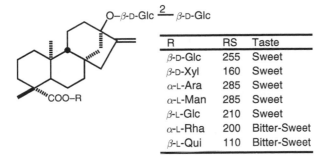

R	RS	Taste
β-D-Glc	255	Sweet
β-D-Xyl	160	Sweet
α-L-Ara	285	Sweet
α-L-Man	285	Sweet
β-L-Glc	210	Sweet
α-L-Rha	200	Bitter-Sweet
β-L-Qui	110	Bitter-Sweet

Figure 7.11 Sweetness of synthetic steviol glycosides.

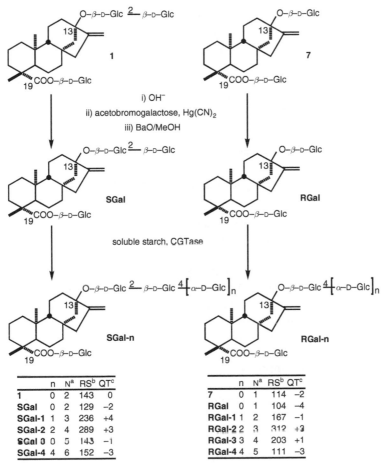

i) OH⁻
ii) acetobromogalactose, Hg(CN)₂
iii) BaO/MeOH

soluble starch, CGTase

	n	N^a	RS^b	QT^c
1	0	2	143	0
SGal	0	2	129	−2
SGal-1	1	3	236	+4
SGal-2	2	4	289	+3
SGal-3	3	5	143	−1
SGal-4	4	6	152	−3

	n	N^a	RS^b	QT^c
7	0	1	114	−2
RGal	0	1	104	−4
RGal-1	1	2	167	−1
RGal-2	2	3	312	+3
RGal-3	3	4	203	+1
RGal-4	4	5	111	−3

[a] N: number of glucosyl residue at 13-O
[b] RS: relative sweetness to sucrose
[c] QT: quality of taste, +: better, −: worse

Figure 7.12 Transglucosylation products of **SGal** and **RGal** with CGTase.

of **SGal** and **RGal** up to total of four glucosyl units improved the sweetness remarkably: **SGal-1** and -2 from **SGal**, and **RGal-1**, -2, and -3 from **RGal**, while greater transglucosylation resulted in a change of the taste for the worse. The improved sweeteners, **SGal-1** and -2 as well as **RGal-1** and -2 were obtained by treatment of the reaction mixtures with α-amylase to shorten the sugar chain.

Protection of the 19-O-glucosyl group against trans-glucosylation

Another strategy for a selective transglucosylation of the 13-O-sugar is to protect the appropriate position (C-4) of the 19-O-glucosyl group with a different sugar, such as galactose which resists transglucosylation. For this purpose, transgalactosylation of rubusoside (**7**) by several α- and β-galactosidases have been investigated (Kitahata *et al.* 1989a,b). It was found that treatment of rubusoside by *Bacillus circulans* α-galactosidase and lactose for 60 minutes afforded a desirable product (**RGal-C1**) which was formulated as 13-O-β-D-glucosyl-19-O-[β-D-galacto-syl-$(1 \rightarrow 4)$-β-D-glucosyl]-steviol (Figure 7.13) together with small amounts of undesirable by-products, such as those which were β-galactosylated at C-4 and C-6 positions of the 13-O-glucosyl moiety and at the C-6 position of the 19-O-glucosyl moiety. In the course of the reaction, the yield of **RGal-C1** reached a maximum in 60 minutes, and then decreased followed by an increase of the by-products. **RGal-C1** was subjected to trans-α-1,4-glucosyla-tion to afford selectively glucosylated products at the 13-O-glucosyl moiety. Degalactosylation with β-galactosidase followed by treatment with β-amylase gave the improved sweeteners, **R1a** and **R2a**, exclusively (Figure 7.14) (Ohtani *et al.* 1991b).

Transglycosylation by other enzyme systems

Trans-α-1,6- and 1,3-glucosylation by biozyme L and glucosyltransferase from Streptococcus mutans

Aiming for a more effective transglucosylation system, several commercially available gluco-syltransferases were screened. Treatment of stevioside (**1**) with maltose and 'Biozyme L' (a crude β-amylase preparation produced by *Aspergillus* spp.) afforded three products: (i) **1m** (6-O-α-glucosylated at the C-19-O-glucose); (ii) **2m** (6-O-α-glucosylated at C-13-O-glucose); and (iii) **3m** (3-O-α-glucosylated at the C-13-O-glucose) which were formulated as shown in Figure 7.15 (Lobov *et al.* 1991). The relative intensities of sweetness of **1m** and **2m** were less than that of stevioside, while a remarkable improvement of the quality of taste was observed for **1m**. This is the first finding where the quality of taste of stevioside has been improved with an additional glucose attached at the C-19-O-glucosyl group. (In the food industry the quality of taste is sometimes more important than the magnitude of sweetness.) In contrast, **3m** tasted bitter even if it was glucosylated at the C-13 sophorosyl unit. This unexpected finding suggests that the site of glucosylation subtly affects the structure-activity relationship of the sweet taste.

Trans-α-1,6-glucosylation by dextrin dextranase

On treatment with starch hydrolysate (by isoamylase) and dextrin dextranase produced by *Acetobacter capsulatus* ATCC 11894, stevioside (**1**) yielded three products in remarkably high yields. Of these products, one was identical with the product **2m**. The other two, tentatively called **SG1** and **SG2**, have branched sugar units at C-13 of steviol and are illustrated in Figure

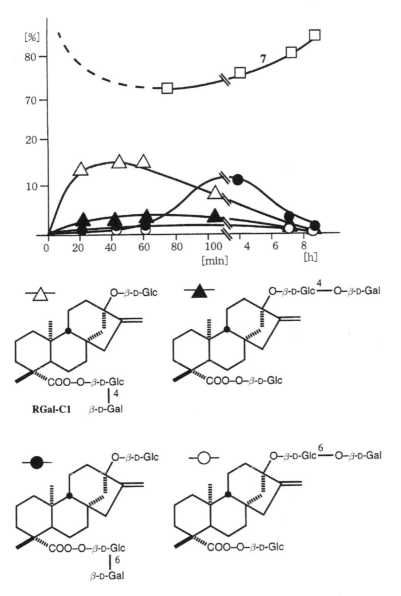

Figure 7.13 Transgalactosylation products of **7** with lactose and β-galactosidase from *Bacillus circulans*.

7.16 (Yamamoto *et al.* 1994). The relative intensities of sweetness of all these branched α-1-6-glucosylated compounds were significantly lower than that of stevioside.

Trans-β-1,3-glucosylation

Stevioside (**1**) was treated with curdlan and the enzyme from *Streptomyces* sp. W19-1 to give three *trans* β-1,3-glucosylated products: (i) **SβG1a**: 13-*O*-β-sophorosyl-19-*O*-β-laminaribiosyl-steviol;

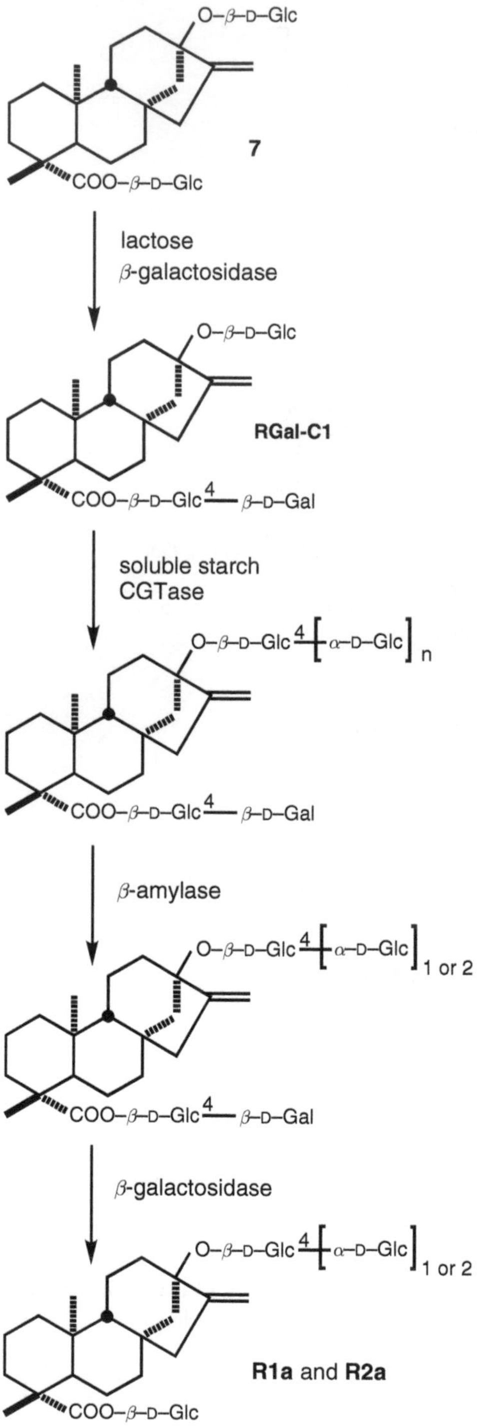

Figure 7.14 Selective transglucosylation of **7** with β-galactosidase.

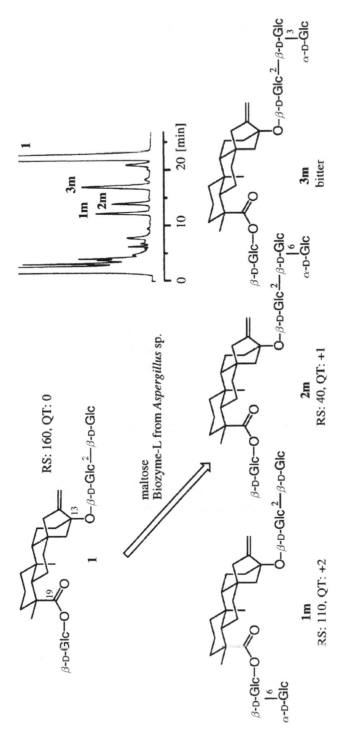

Figure 7.15 Transglucosylation of **1** with maltose and Biozyme L.

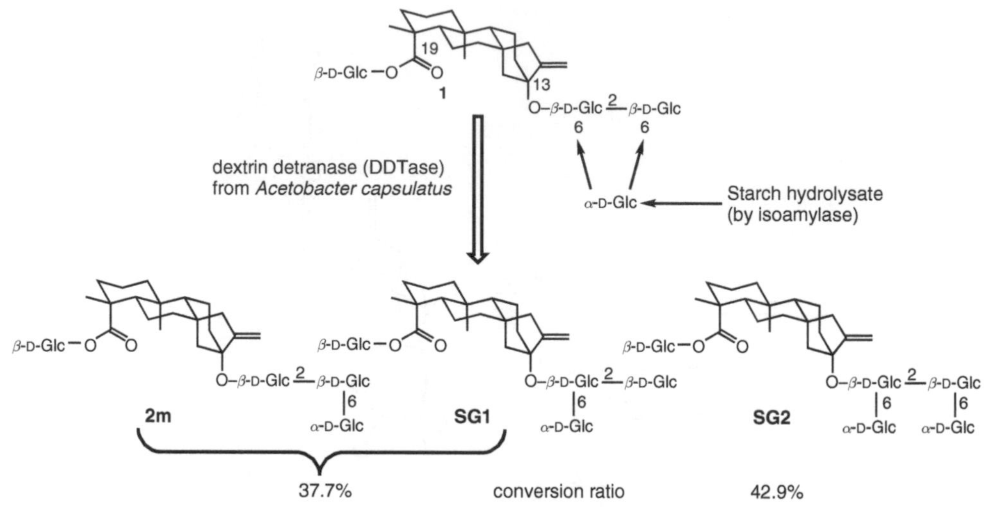

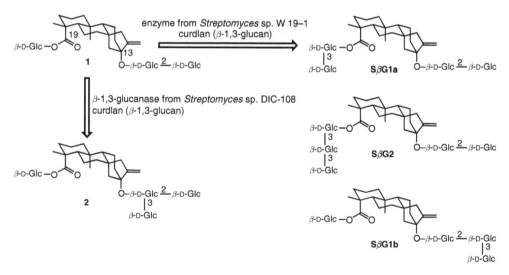

Figure 7.16 Trans-α-glucosylation of **1** with dextrin dextranase.

(ii) **SβG1b**: 13-*O*-[β-glucosyl-(1 → 3)-β-glucosyl-(1 → 2)-β-glucosyl]-19-*O*-β-glucosyl-steviol; and (iii) **SβG2**: 13-*O*-β-sophorosyl-19-*O*-β-laminaritriosyl-steviol (Kusama *et al.* 1986). Improvement of the quality of taste was noted for **SβG1a** and **SβG2** as well as a mixture of the products, while the relative intensities of sweetness of these products were lower than stevioside. Evaluation of sweetness of **SβG1b** has not been conducted due to its low yield (Figure 7.17). An attractive transglucosylation method was reported on the formation of rebaudioside A (**2**) from stevioside (**1**). Treatment of stevioside (**1**) by 1,3-glucan and an enzyme from *Streptomyces* sp. DIC-108 yielded rebaudioside A (**2**), the better sweetener (Anonymous, 1994).

Figure 7.17 Trans-β-1,3-glucosylation of **1** with curdlan and β-1,3-glucanase.

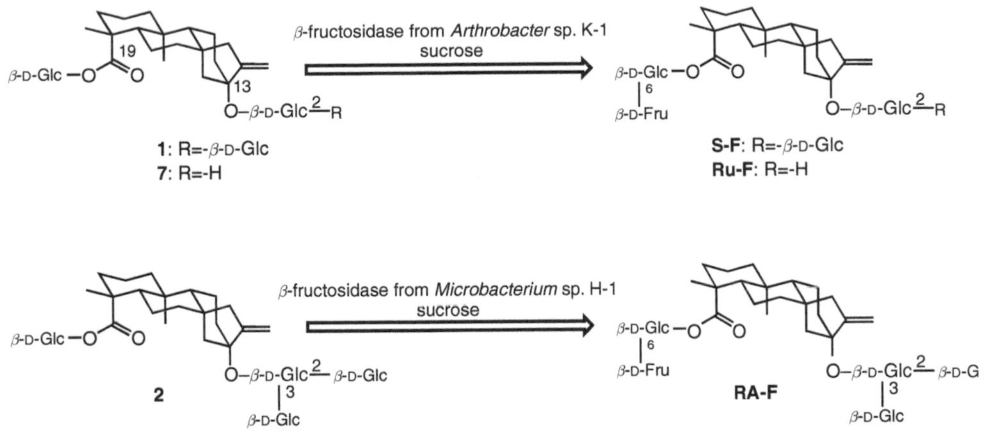

Figure 7.18 Transfructosylation of **1, 2** and **7** with sucrose and transfructosidase.

Trans-β-1,6-glucosylation by cultivation with Actinomycetes strain K-128

An Actinomycetes strain K-128 isolated from soil was cultured in a medium containing stevioside and curdlan to give a trans-β-1,6-glucosylated product, 13-[β-glucosyl-(1-6)-β-glucosyl-(1 → 2)-β-glucosyl]-19-O-β-glucosyl-steviol (Kusakabe *et al*. 1992). The evaluation of the taste was not described.

Trans-β-2,6-fructofuranosylation

Incubation of stevioside (**1**) and rubusoside (**7**) with sucrose and β-fructofuranosidase from *Arthrobacter* sp. K-1 afforded in high yields their respective products which were trans-β-2,6-fructofuranosylated at the 19-O-glucosyl moiety (Figure 7.18). These were tentatively designated as **S-F**, from stevioside, and **Ru-F**, from rubusoside (Ishikawa *et al*. 1990). The relative intensity of sweetness was not enhanced, however, a significant improvement of quality of taste was observed for both compounds. Rebaudioside A (**2**) was subjected to trans-β-2,6-fructofranosylation by sucrose and β-fructofuranosidase from *Microbacterium* sp. H-1 to give a product, **RA-F** which was β-2,6-fructosylated at the 19-O-glucosyl group (Ishikawa *et al*. 1991). In this case, the relative intensity of sweetness and quality of taste were not significantly improved. It is noteworthy that the fructofuranosyl linkage is rather unstable, being hydrolyzed on standing in foods. Treatment of **S-F** with the CGTase-starch system yielded *trans*-α-1,4-glucosylated products (**S-F-α- Glc**) at the 13-O-sophorosyl moiety (Hirai 1995). Significant improvement both in relative intensity of sweetness and quality of taste was noted for the di-α-glucosylated product.

SUMMARY AND CONCLUSIONS

The sweetness of several steviol glycosides differs substantially reflecting relatively small variations in their structures. Up to now, no other sweet *ent*-kaurene diterpene glycosides having an aglycone different from steviol have better taste qualities. On the other hand, some semisynthetic steviol glycosides having different sugar moieties exceed the quality of sweetness of the natural products from *S. rebaudiana*. Generally, the quality and magnitude of sweetness reaches a maximum with three to four monosaccharide units at the C-13 position of steviol and one to two monosaccharide units at the C-19 position. Since chemical reactions are ineffective because of a lack of selectivity and poor yields, several attempts to improve the quality and potency of sweetness of steviol glycosides using enzymatic transglycosylation have been made. Starting from stevioside (**1**), rubusoside (**7**), or mixtures of natural steviol glycosides, several enzymatic reactions with suitable sugar donors selectively afforded products with extended C-13 sugar units. However, in most cases, it was necessary to reduce the length of the sugar chains with glycosidases. These methods have been used in the Japanese food industry. For further studies, better selection of the combination of appropriate transglycosidases and effective donors is needed although many trials have been done already with several α-transglucosidases. Fructose may be one of the key sugar units which can improve the sweetness of steviol glycosides in a more effective and economical manner.

ACKNOWLEDGEMENTS

Most of the research described in this chapter was conducted under the guidance of Dr Osamu Tanaka, the project leader and Emeritus Professor of Hiroshima University, and formerly President of Suzugamine Women's College, to whom our thanks are due. We also thank many contributors whose names appear in the list of reference below.

REFERENCES

Anonymous (1994) Production of rebaudioside A. Japan Tokkyo Koho (B2) H6-73468, 12pp.

Darise, M., Mizutani, K., Kasai, R., Tanaka, O., Kitahata, S., Okada, S. *et al.* (1984) Enzymic transglucosylation of rubusoside and the structure–sweetness relationship of steviol-bisglucosides. *Agricultural and Biological Chemistry* **48**, 2483–2488.

Esaki, S., Tanaka, R. and Kamiya, S. (1984) Synthesis and taste of certain steviol glycosides. *Agricultural and Biological Chemistry* **48**, 1831–1834.

Fukunaga, Y., Miyata, T., Nakayasu, N., Mizutani, K., Kasai, R. and Tanaka, O. (1989) Enzymic transglucosylation products of stevioside: Separation and sweetness-evaluation. *Agricultural and Biological Chemistry* **53**, 1603–1607.

Hirai, K. (1995) Characteristic and biological effect of Santencha™, herbal tea from leaves of *Rubus suavissimus*. *Japanese Food Science* **1**, 57–62.

Hirono, S., Chou, W., Kasai, R., Tanaka, O. and Tada, T. (1990) Sweet and bitter diterpene-glucosides from leaves of *Rubus suavissimus*. *Chemical and Pharmaceutical Bulletin* **38**, 1743–1744.

Ishikawa, H., Kitahata, S., Ohtani, K., Ikuhara, C. and Tanaka, O. (1990) Production of stevioside and rubusoside derivatives by transglucosylation of β-fructosidase. *Agricultural and Biological Chemistry* **54**, 3137–3143.

Ishikawa, H., Kitahata, S., Ohtani, K. and Tanaka, O. (1991) Transfructosylation of rebaudioside A (a sweet glycoside of Stevia leaves) with *Microbacterium* β-fructosidase. *Chemical and Pharmaceutical Bulletin* **39**, 2043–2045.

Kamiya, S., Konishi, F. and Esaki, S. (1979) Synthesis and taste of some analogs of stevioside. *Agricultural and Biological Chemistry* **43**, 1863–1867.

Kaneda, N., Kasai, R., Yamasaki, K. and Tanaka, O. (1977) Chemical studies on sweet diterpene-glycosides of *Stevia rebaudiana*: conversion of stevioside into rebaudioside A. *Chemical and Pharmaceutical Bulletin* **25**, 2466–2467.

Kaneda, N., Kohda, H.,Yamasaki, K.,Tanaka, O. and Nishi, K. (1978) Paniculosides-I–V, diterpene-glu-cosides from *Stevia ovata*. *Chemical and Pharmaceutical Bulletin* **26**, 2266–2267.

Kasai, R., Kaneda, N.,Tanaka, O.,Yamasaki, K., Sakamoto, I., Morimoto, K. *et al.* (1981) Sweet diterpene-glycosides of leaves of *Stevia rebaudiana* Bertoni. Synthesis and structure – sweetness relationship of rebaudiosides-A, -D, -E and their related glycosides. *Nippon Kagakukaishi* **1981**, 726–735.

Kitahata, S., Okada, S. and Fukui, T. (1978) Acceptor specificity of the transglucosylation catalyzed by cyclodextrin glycosyltransferase. *Agricultural and Biological Chemistry* **42**, 2369–2374.

Kitahata, S., Ishikawa, H., Miyata, T. and Tanaka, O. (1989a) Production of rubusoside derivatives by transgalactosylation of various β-galactosidases. *Agricultural and Biological Chemistry* **53**, 2923–2928.

Kitahata, S., Ishikawa, H., Miyata, T. and Tanaka, O. (1989b) Production of rubusoside derivatives by transgalactosylation of various α-galactosidases. *Agricultural and Biological Chemistry* **53**, 2929–2934.

Kobayashi, M., Horikawa, S., Degrandi, I., Ueno, J. and Mitsuhashi, H. (1977) Dulcosides A and B, new diterpene glycosides from *Stevia rebaudiana*. *Phytochemistry* **16**, 1405–1408.

Kohda, H., Kasai, R.,Yamasaki, K., Murakami, K. and Tanaka, O. (1976) New sweet diterpene glucosides from *Stevia rebaudiana*. *Phytochemistry* **15**, 981–983.

Kusakabe, I.,Watanabe, S., Morita, R.,Terahara, M. and Murakami, K. (1992) Formation of transfer pro-duct from stevioside by the cultures of actinomycete. *Bioscience, Biotechnology, and Biochemistry* **56**, 233–237.

Kusama, S., Kusakabe, I., Nakamura,Y., Eda, S. and Murakami, K. (1986) Transglucosylation into stevio-side by the enzyme system from *Streptomyces* sp. *Agricultural and Biological Chemistry* **50**, 2445–2451.

Lobov, S.V., Kasai, R., Ohtani, K.,Tanaka, O. and Yamasaki, K. (1991) Enzymic production of sweet stev-ioside derivatives:Transglucosylation by glucosidases. *Agricultural and Biological Chemistry* **55**, 2959–2965.

Mizutani, K., Miyata,T., Kasai, R.,Tanaka, O., Ogawa, S. and Doi, S. (1989) Study on improvement of sweetness of steviol bisglucosides: Selective enzymic transglucosylation of the 13-*O*-glucosyl moiety. *Agricultural and Biological Chemistry* **53**, 395–398.

Mosettig, E., Beglinger, U., Doldcr, F., Lichiti, H., Quitt, P. and Waters, J. A. (1963) The absolute config-uration of steviol and isosteviol. *Journal of the American Chemical Society* **85**, 2305–2307.

Ohtani, K., Aikawa,Y., Ishikawa, H., Kasai, R., Kitahata, S., Mizutani, K. *et al.* (1991a) Further study on the 1,4-α-transglucosulation of rubusoside, a sweet steviol-bisglucoside from *Rubus suavissimus*. *Agricul-tural and Biological Chemistry* **55**, 449–453.

Ohtani, K., Aikawa,Y., Fujisawa,Y., Kasai, R.,Tanaka, O. and Yamasaki, K. (1991b) Solubilization of ste-violbioside and steviolmonoside with γ-cyclodextrin and its application to selective synthesis of better sweet glycosides from stevioside and rubusoside. *Chemical and Pharmaceutical Bulletin* **39**, 3172–3174.

Ohtani, K., Aikawa,Y., Kasai, R., Chou,W.,Yamasaki, K. and Tanaka, O. (1992) Minor diterpene glyco-sides from sweet leaves of *Rubus suavissimus*. *Phytochemistry* **31**, 1553–1559.

Sakamoto, I.,Yamasaki, K. and Tanaka, O. (1977a) Application of ^{13}C NMR spectroscopy to chemistry of natural glycosides: Rebaudioside-C, a new sweet diterpene glycoside of *Stevia rebaudiana*. *Chemical and Pharmaceutical Bulletin* **25**, 844–846.

Sakamoto, I.,Yamasaki, K. and Tanaka, O. (1977b) Application of ^{13}C NMR spectroscopy to chemistry of plant glycosides: rebaudiosides-D and -E, new diterpene-glucosides of *Stevia rebaudiana* Bertoni. *Chem-ical and Pharmaceutical Bulletin* **25**, 3437–3439.

Tanaka,T., Kohda, H.,Tanaka, O., Chen, F., Chou,W. and Leu, J. (1981) Rubusoside (β-D-glucosyl ester of 13-*O*-β-D-glucosyl-steviol), a sweet principle of *Rubus chingii* Hu (Rosaceae). *Agricultural and Biological Chemistry* **45**, 2165–2166. (Note that, subsequently, the species name of this plant was amended as shown in the text of the present chapter.)

The Amylase Research Society of Japan (Ed.) (1988) *Handbook of Amylases and Related Enzymes*, Pergamon Press, Oxford, UK, p. 154.

Yamamoto, K.,Yoshikawa, K. and Okada, S. (1994) Effective production of glucosyl-stevioside by α-1,6 transglucosylation of dextrin dextranase. *Bioscience, Biotechnology, and Biochemistry* **58**, 1657–1661.

Yamasaki, K., Kohda, H., Kobayashi,T., Kaneda, N., Kasai, R.,Tanaka, O. *et al.* (1977) Application of ^{13}C nuclear magnetic resonance spectroscopy to chemistry of glycosides: Structures of paniculosides-I, -II, -III, -IV and -V, diterpene glucosides of *Stevia paniculata* Lag. *Chemical and Pharmaceutical Bulletin* **25**, 2895–2989.

8 Pharmacology and toxicology of stevioside, rebaudioside A, and steviol

Ryan J. Huxtable

INTRODUCTION

Of the three compounds to be discussed in this chapter, stevioside and rebaudioside A are major natural glycosides found in the leaves of *S. rebaudiana* (henceforth in this chapter expressed as *Stevia*), while the aglycone, steviol is a biosynthetic precursor in the leaves (Kohda *et al*. 1976; Shibata *et al*. 1995; Kim *et al*. 1996) and a putative mammalian metabolite of stevioside. These compounds are structurally related to *ent*-kaurenoic acid (Figure 8.1).

Stevia leaves contain naturally high levels of the glycosides, and selective breeding has increased these levels further. Typical concentrations range from 5 to 10% w/w of the dried leaf for stevioside, 2–4% for rebaudioside A, 1–2% for rebaudioside C, and 0.4–0.7% for dulcoside A (Kinghorn and Soejarto 1985). Newer, commercially developed strains may contain an excess of 14% diterpene glycosides (Phillips 1987).

Stevioside, in the form of the pure compound or of *Stevia* leaf extracts, has been widely used as a food additive, particularly in Brazil, Korea and Japan. It has been estimated, e.g. that somewhere between 85 and 170 metric tons of stevioside were consumed in Japan in 1987. This is equivalent to approximately 1,700 tons of leaf (Kinghorn and Soejarto 1991). The absence of reports of adverse reactions from these countries is *prima facie* evidence of lack of gross toxicity. Safety concerns, therefore, must address issues of chronic or accumulative toxicity, subtle toxic actions occurring at low frequency, and the possible existence of particularly sensitive groups (e.g. pregnant women, babies, the elderly, the sick, those taking particular drugs, or those exhibiting particular genetic or pharmacogenetic traits).

Considering the extent to which *Stevia* and stevioside are used, studies relevant to toxicity and pharmacology are jejune. In considering the available literature, it is necessary to interpret the findings in the light of actual or anticipated human consumption. For stevioside, *per capita* consumption among users has been estimated to be 300 mg/day (0.37 mM) (Sakaguchi and Kan 1982). Much of the literature is in the form of incomplete reports or abstracts, or has been published in hard to obtain journals.

A number of earlier reviews on *Stevia* and its glycosides have appeared (Kinghorn and Soejarto 1985; Bakal and O'Brien Nabors 1986; Pezzuto 1986; Crammer and Ikan 1987; Phillips 1987; Kinghorn and Soejarto 1991; Hanson and De Oliveira 1993).

STUDIES ON EXTRACTS

Many of the earlier toxicological studies of *Stevia* glycosides involved either water or ethanolic extracts of *Stevia* leaves containing stevioside and rebaudioside A at various concentrations

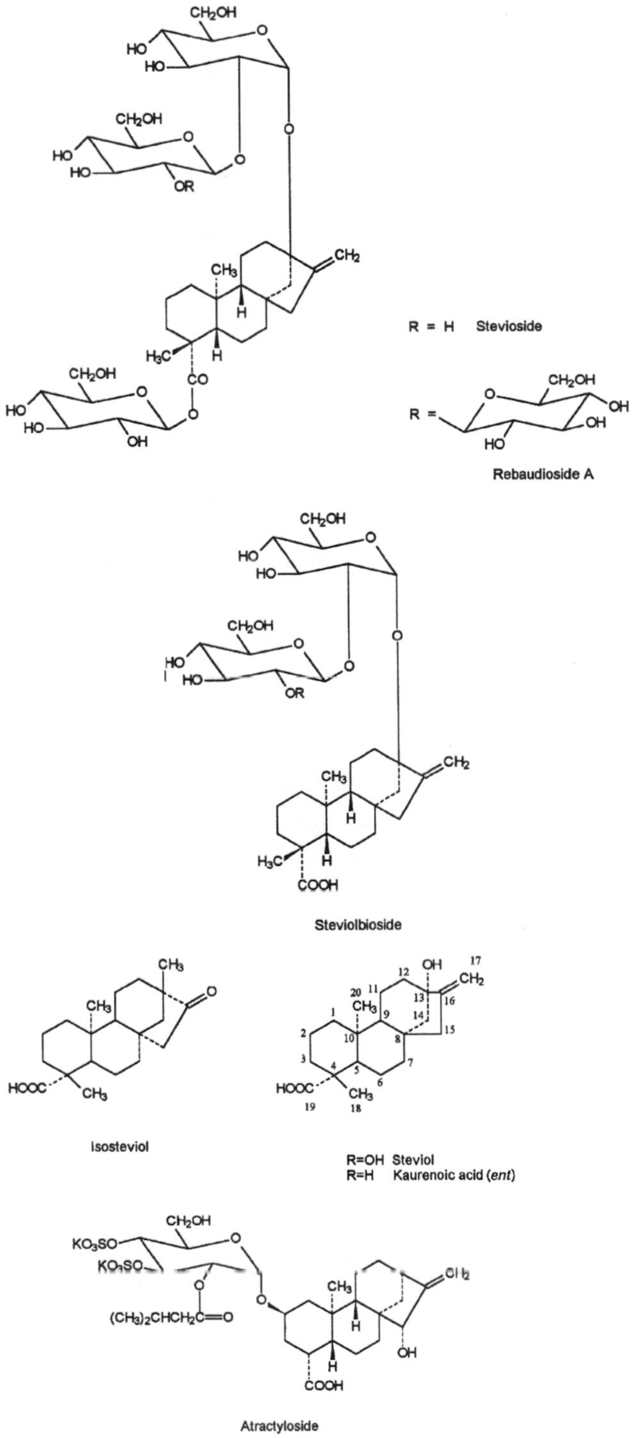

Figure 8.1 Structures of some *Stevia* glycosides and related glycosides and aglycones.

(Planas and Kuc 1968; Oviedo *et al.* 1970; Akashi 1975; Akashi and Yokoyama 1975; Lee *et al.* 1979; Takaki *et al.* 1984; Yamada *et al.* 1985; Medon and Ziegler 1986; Oliveira-Filho *et al.* 1989; Sincholle and Marcorelles 1989; Levy *et al.* 1994; Melis 1995). These studies are typically discussed in terms of the percentage of stevioside in the extract, but the potential contribution of other components in the leaf must always be kept in mind. A number of differences in the pharmacology of *Stevia* extracts and stevioside have been noted. Thus, the extract inhibits glutamate dehydrogenase; stevioside does not (Levy *et al.* 1994). The hypoglycemic effects reported for *Stevia* extracts are not seen with stevioside (Oviedo *et al.* 1970; Curi *et al.* 1986). The potent inhibition of 2,4-dinitrophenol-stimulated ATPase of rat liver mitochondria by crude leaf extract cannot be accounted for on the basis of stevioside content (Kelmer Bracht, Kemmelmeier *et al.* 1985). The presence in extracts of glycosides having an unesterified free carboxyl group [rebaudioside B and steviolbioside (see Figure 8.1)] may contribute to the toxicity of extracts relative to purified stevioside.

Lethality

Aqueous ethanolic extracts consisting 50% of stevioside had an acute LD_{50} (i.p.) in rats of 3400 mg/kg (Lee *et al.* 1979). In mice aqueous extracts comprising 50% stevioside had an acute LD_{50} of 17 g/kg (Akashi and Yokoyama 1975). Enrichment to 40–55% stevioside decreased lethality to >42 g/kg.

Subchronic and chronic studies

Stevia extracts containing 50% stevioside were fed to rats at doses up to 1 g/day for 56 days without detectable effects on biochemical or pathological endpoints, except for a slight depression in hepatic lactate dehydrogenase activity (Lee *et al.* 1979). A *Stevia* extract containing 75% stevioside and 16% rebaudioside A was fed to rats for up to 24 months at doses up to 550 mg/kg/day. The only effect noted was slight growth retardation (Yamada *et al.* 1985).

Pharmacological actions

Energy metabolism

Crude *Stevia* extracts inhibited mitochondrial glutamate dehydrogenase from beef and rat (Suttajit *et al.* 1993).

Carbohydrate metabolism

There have been repeated claims in hard-to-obtain sources that *Stevia* extracts improve glucose tolerance in alloxan-diabetic rabbits (von Schmeling *et al.* 1977), and both diabetic and non-diabetic humans (Miquel 1966; Alvarez *et al.* 1981). In one study, it was reported that powdered leaves as a 10% addition to a high carbohydrate diet decreased blood glucose and hepatic glycogen levels in rats after four weeks (Suzuki *et al.* 1977). One study reported in an abstract claims a 35% drop in blood glucose in human volunteers eight hours after consumption of an extract (Oviedo *et al.* 1970). On the other hand, 0.5–1.0 g/day of *Stevia* extract for 56 days was without effect on blood sugar (Lee *et al.* 1979).

Cardiovascular effects

Stevia extracts have been reported in humans to reduce heart rate and mean arterial blood pressure (Boeckh and Humboldt 1981). Aqueous extracts fed to rats for 40–60 days produced hypotension, the mean arterial pressure falling from 110 mm Hg to 90 mm Hg over the 40-day-treatment period (Melis 1995).

Renal effects

In rats, extracts of *Stevia* given by mouth for 40–60 days induced diuresis and natriuresis and increased renal plasma flow, indicating vasodilatation (Melis 1995; 1996). Glomerular filtration rate remained unchanged. The extract given was equivalent to 1.33 g dry leaves, twice a day. The stevioside content was judged to be too low to be responsible for the vasodilatation. Mean arterial pressure was decreased in both normal and Goldblatt hypertensive rats given extract equivalent to 2.67 g/dry leaves per day for 30 days (Melis 1996).

Reproduction

An early report that an aqueous extract of *Stevia* reduced fertility of female rats (Planas and Kuc 1968) has led to considerable controversy. The amount fed to the rats daily was eight times the dosage per kg purportedly used by Paraguayan Indians as an oral contraceptive. Although this report has been frequently cited, its findings have not been reproduced. No effects on fertility were seen in studies on rats or rabbits (Akashi and Yokoyama 1975). The contraceptive effect of *Stevia*, if any, does not appear to be due to the sweeteners in it. In one small study, a hot water extract was added to rat food to the level of 0.14% stevioside for 21 days (Akashi 1975). No reproductive effects were seen in either sex. No endocrine effects were seen in male rats (Sincholle and Marcorelles 1989), including a study in which 25–30-day-old rats were fed leaf extract for the next 60 days (Oliveira-Filho *et al.* 1989).

STEVIOSIDE

Stevioside ((4α) 13-[(2-O-β-D-glucopyranosyl-α-D-glucopyranosyl)oxy]kaur-16-en-18-oic acid β-D-glucopyranosyl ester; steviosin) is both a glucoside and a glucosyl ester analog of *ent*-kaurenoic acid (see Figure 8.1). The molecular composition is $C_{38}H_{60}O_{18}$ (molecular weight 804). The diglucosyl residue (2-O-β-D-glucopyranosyl-α-D-glucose) is also called sophorose.

Absorption, distribution and metabolism

In the rat, stevioside (125 mg/kg; p.o.) has a half-life of 24 hour, and is largely excreted in the feces in the form of steviol (Nakayama *et al.* 1986). Other metabolites include steviolbioside (see Figure 8.1). In this species, at least, metabolism appears to be mediated primarily by the gut microflora. Thus, [17-^{14}C]stevioside is converted to steviol by suspensions of rat intestinal microflora. Conversion is complete within two days (Wingard *et al.* 1980).

The distribution of a derivative, [^{131}I]iodostevioside (position of the label not reported), has been studied in rats following i.v. administration (Cardoso *et al.* 1996). Radioactivity rapidly accumulated first in the small intestine and then in the liver. Within two hours, 52% of the radioactivity administered appeared in the bile. The largest biliary component was

[^{131}I]iodosteviol (47% of total radioactivity), followed by [^{131}I]iodostevioside (37%) and an unidentified metabolite (15%).

Non-enzymatic conversion of stevioside to steviol does not occur (Pezzuto *et al.* 1985). Acid hydrolysis yields isosteviol (see Figure 8.1), while incubation for up to three months under conditions ranging pH 2–8 and 5 to 90 °C does not result in detectable formation of steviol. Stevioside appears to be poorly transported across the cell membrane (Yamamoto *et al.* 1985). No uptake was observed in suspensions of human red blood cells (Kelmer Bracht, Kemmelmeier *et al.* 1985). The volume of distribution of stevioside was identical to that of [^{14}C]sucrose. In the isolated, perfused rat liver, no metabolism of stevioside occurs over a two hour period at concentrations of 0.2 mM and 0.5 mM (Ishii-Iwamoto and Bracht 1995).

Lethality and whole animal toxicity

In whole animal studies, stevioside has shown low toxicity (Akashi and Yokoyama 1975; Fujita and Edahiro 1979; Lee *et al.* 1979; Yamada *et al.* 1985; Xili *et al.* 1992). From studies on crude *Stevia* extracts, it can be calculated that the acute LD_{50} of stevioside (i.p.) in rats is greater than 1700 mg/kg body weight (Lee *et al.* 1979). For oral administration in mice, an LD_{50} of 8.2 g/kg has been reported in one study (cited in Lee *et al.* 1979), and an LD_{50} of >15 g/kg has been calculated in another (for stevioside of 93–95% purity) (Akashi and Yokoyama 1975). In mice, no toxicity was detected following 2 g/kg of pure stevioside (Medon *et al.* 1982). No toxic effects were detected in mice two weeks following administration of doses up to 2 g/kg (by gastric intubation) (Medon *et al.* 1982). More recently, stevioside has been fed to rats at up to 5% of the diet for 104 weeks with no evidence of carcinogenicity (Toyoda *et al.* 1997).

Yamada *et al.* (1985) calculate a maximum no-effect level in rats of 550 mg/kg/day for a two-year toxicity and carcinogenicity study. Xili *et al.* (1992) fed stevioside (85% pure) to rats at a dose of 800 mg/kg/day for two years. The maximum no-effect level of stevioside was equivalent to 1.2% of the diet, or 794 mg/kg/day. By extrapolating this figure to humans and allowing a safety factor of 100, these authors calculate an acceptable daily intake (ADI) of 7.94 mg/kg/day. This is equivalent to 476 mg for a 60 kg adult or 160 mg for a 20 kg child. Assuming a mean daily intake of 50 g sucrose and a maximum replacement of 50% of this by stevioside, these authors calculate the 'likely maximum intake' of stevioside by humans to be 125 mg/day, or approximately 2 mg/kg (with the assumption that 1 mg stevioside replaces 200 mg sucrose). This is well below their calculated ADI, and so they conclude that stevioside is safe for human use. However, the assumptions in this paper seem dubious for a North American population. The ADI of 160 mg for a 20 kg child is equivalent to 32 g sucrose. A so-called 'thirst quencher' on the table as I write contains 14 g of sucrose/dextrose per 240 ml portion, while the various energy drinks in my collection contain up to 13 g sugar per 100 ml. With 50% replacement of sucrose by stevioside, many children in America, therefore, could potentially ingest well in excess of 2 mg/kg stevioside.

Pharmacological actions

Energy metabolism

Stevioside interferes with oxidative phosphorylation in isolated mitochondria (Vignais *et al.* 1966; Yamamoto *et al.* 1985; Kelmer Bracht, Alvarez *et al.* 1985; Kelmer Bracht, Kemmelmeier *et al.* 1985), with 50% inhibition seen at 1.2 mM (Vignais *et al.* 1966). This inhibition appears to be due to disruption of adenine dinucleotide translocation, a necessary process in shuttling high energy phosphate groups generated in mitochondria to their sites of consumption in the cell. In rat liver mitochondria, stevioside (5 mM) abolished coupled respiration (i.e. the increased O_2 uptake in the presence of ADP) (Kelmer Bracht, Alvarez *et al*. 1985). It also inhibited the stimulation of mitochondrial ATPase induced by the uncoupling agent, 2,4-dinitrophenol. An inhibition of 50% was achieved at 1.2 mM stevioside. At 1.5 mM, it is without effect on glutamate dehydrogenase activity of rat or bovine liver mitochondria (Levy *et al*. 1994). Stevioside is much less potent, however, than atractyloside, a compound of related structure, but having a free carboxyl group (see Figure 8.1). At 1 µM, this latter compound produces 50% inhibition of oxidative phosphorylation (Klingenberg and Heldt 1982).

Some of the mitochondrial actions of stevioside are summarized in Table 8.1. High levels of stevioside (1.5 g/kg body weight; s.c.) caused severe disruption of mitochondrial cristae in kidney tubules (Toskulkao *et al*. 1994a). The mitochondrial actions of stevioside are seen only on isolated organelles, and are not observed in intact cells. This suggests that stevioside does not permeate cell membranes, although this point does not appear to have been directly examined (Yamamoto *et al*. 1985) apart from the distribution studies discussed above (Yamamoto *et al*. 1985; Ishii-Iwamoto and Bracht 1995).

Effects on energy metabolism that do not involve mitochondria have been examined in cells lacking these organelles. Erythrocytes rely on glycolysis for ATP production. Stevioside has little effect on such cells (Kemmelmeier *et al*. cited in Kelmer Bracht, Kemmelmeier *et al*. 1985).

Carbohydrate metabolism

The effect of stevioside on energy metabolism in isolated mitochondria has spurred the search for related effects on intact cells. At a concentration of 3 mM, however, stevioside was

Table 8.1 Inhibitory constants (half-maximal effects: I_{50}) for *Stevia rebaudiana* glycosides and aglycones on mitochondrial activities

Parameter	I_{50} *values* (mM)			
	Isosteviol	*Steviol*	*Steviolbioside*	*Stevioside*
Coupled respiration (state III respiration)	0.07	0.05	0.25	2.5
Uncoupled respiration (with L-glutamate as substrate)	0.4	0.9	0.6	>2.0
DNP-stimulated ATPase	0.1	0.02	0.6	1.2
NADH oxidase activity	0.1	0.6	0.6	>5.0
L-Glutamate dehydrogenase	0.15	0.55	0.6	>5.0
ATP-dependent swelling	0.1	0.1	1.0	2.0
L-Glutamate-dependent swelling	0.06	0.13	0.3	>2.0
Succinate-dependent swelling	0.1	0.2	0.4	>2.0
TMPD[a]-ascorbate dependent swelling (in the presence of antimycin A)	0.05	0.1	0.8	>2.0
Succinate-dependent net proton efflux	0.1	–	0.4	>2.0
Succinate oxidase activity	0.15[b]	0.15[c]	–	–
Succinate dehydrogenase	0.15[b]	0.15[c]	–	–

Sources: Taken from Kelmer Bracht, Alvarez *et al*. 1985; Kelmer Bracht, Kemmelmeier *et al*. 1985.
Notes
a TMPD: N,N,N',N'-tetramethyl-p-phenylenediamine.
b Incomplete inhibition; maximally 35%.
c Incomplete inhibition; maximally 75%.

without effect on gluconeogenesis or O_2 uptake in renal cortical tubules (Yamamoto *et al.* 1985).

At the whole organ level, stevioside inhibits the monosaccharide transporter in the isolated, perfused rat liver (Ishii *et al.* 1987). The transporter carries glucose, fructose and galactose in both directions. Stevioside (0.8 mM) halves the transport rate of glucose into the liver. At 1.5 mM stevioside, there is 73% inhibition of 1 mM glucose uptake, combined with inhibition of fructose metabolism (Yamamoto *et al.* 1985). Stevioside also inhibits hepatic release of glucose. In livers undergoing glycogenolysis, the presence of stevioside leads to an increase in the intracellular:extracellular concentration gradient of glucose. Both the K_m and V_{max} of monosaccharide transport are altered, suggesting a mixed type of inhibition. In the rat kidney, also, stevioside given by *in vivo* infusion (8 mg/kg/h, or higher) decreased the renal tubular resorption of glucose (Melis 1992a).

In hamsters fed stevioside (2.5 g/kg/day) for 12 weeks, glucose absorption was inhibited (Toskulkao and Sutheerawattananon 1994). Doses of 1 g/kg/day were without effect. The effect was attributed to two actions: (i) a decrease in intestinal Na^+/K^+-ATPase activity; and (ii) a decreased absorptive area in the intestines. Body weight of the hamsters decreased over the course of the study. In contrast, stevioside (5 mM) has been reported to have no inhibitory effect on glucose absorption from the rat jejunum *in vitro* (Toskulkao *et al.* 1995b).

No change in blood sugar levels in rats was observed following administration of stevioside (7% of diet) for 56 days (Akashi and Yokoyama 1975). A similar result was obtained from a study in which stevioside (75–150 mg/kg/day) was fed for 30 days (Medon and Ziegler 1986). However, an increase in plasma glucose in rats was found when stevioside was given by intravenous infusion (Suanarunsawat and Chaiyabutr 1997). A loading dose of 100 mg/kg body weight followed by an infusion of 100 mg/kg body weight/hour for one hour resulted in a significant increase in plasma glucose level. The increase reached 47% at 200 mg/kg body weight loading and infusion. There was no change in plasma insulin levels. The authors attribute their findings to an increased glucose transport into cells.

In 24-hour-fasted rats, stevioside (0.2 mM) given orally with fructose (0.2 mM) as a gluconeogenic substrate led to increased glycogen deposition in the liver (Hubler *et al.* 1994). In fasted rats, stevioside in the drinking water (1 or 2 mM) also increased glycogen deposition in the liver in the absence of a gluconeogenic substrate (Hubler *et al.* 1994). It is difficult to reconcile these findings with inhibition of monosaccharide transport. Conversely, in the same species, 0.1% stevioside in a high carbohydrate diet decreased liver glycogen levels, but was without effect in a high fat diet (Suzuki *et al.* 1977). Clarification is needed for the effects of stevioside on carbohydrate metabolism in the intact animal.

Effects on blood pressure and renal function

Various effects have been claimed for stevioside on kidney function and blood pressure regulation (Melis 1992b). It lowers mean arterial blood pressure, decreases renal vascular resistance, produces diuresis, and increases fractional excretion of Na^+ and K^+ (Humboldt and Boeckh 1977; Boeckh and Humboldt 1981; Melis *et al.* 1985; Melis *et al.* 1986; Melis and Sainati 1991; Melis 1992a,b). The lack of effect on glomerular filtration rate implies that stevioside vasodilates both afferent and efferent arterioles (Melis 1992b).

One group has studied renal effects by using an *in vivo* priming dose of 8 or 16 mg/kg, followed by an infusion at the rate of 8 or 16 mg/kg/h (Melis and Sainati 1991). The higher dose led to a marked fall in blood pressure in anesthetized rats (from 110 to 72 mm Hg). All the above effects observed in rats were blocked by indomethacin, an inhibitor of cyclooxygenase, an enzyme involved in prostaglandin synthesis (Melis *et al.* 1985). This suggests that the actions of stevioside are mediated via a prostaglandin-dependent mechanism. The Ca^{2+}-channel antagonist, verapamil, increased the vasodilating effects of stevioside, while calcium chloride antagonized them (Melis and Sainati 1991). This has been interpreted as indicating that the effects of stevioside are due to Ca^{2+} channel blockade. However, one suspects the same effects of verapamil and calcium chloride would be seen regardless of the mechanism by which stevioside is acting.

In the Goldblatt hypertensive rat, stevioside given by the same protocol (16 mg/kg priming dose; 16 mg/kg/h infusion) also decreased mean arterial pressure and increased renal plasma flow (Melis 1992b). Glomerular filtration rate is also increased, implying a preferential vasodilation of afferent arterioles.

When administered *in vivo*, stevioside (1.5 g/kg, s.c.) in rats decreased the ability of isolated renal cortical tissue to accumulate *p*-aminohippuric acid nine hours later (Toskulkao *et al.* 1994b). Infusions (i.v.) of much lower levels, down to 8 mg/kg/h, also increased clearance of *p*-aminohippuric acid and glucose (Melis 1992a).

Stevioside itself had a clearance less than that of *p*-aminohippuric acid but greater than that of inulin. Inulin clearance is a measure of glomerular filtration rate. This finding, therefore, suggests that the glycoside is secreted by the renal tubular epithelium (Melis 1992a).

Nephrotoxic effects of stevioside have been observed in the rat following a single injection of 1.5 g/kg, s.c. (Toskulkao *et al.* 1994a). This is a dose equivalent to about 250 times the average daily intake of human consumers. Blood urea nitrogen began to rise three-hour later, an indication of inability to excrete nitrogen. Blood levels were highest at nine hour (52.4 ± 2.9 mg/% compared with 18.7 ± 0.7 in control animals), and were still elevated 48 hours post-injection. Other biochemical disturbances nine hours post-injection were elevated plasma creatinine, increased urinary glucose (from 4.9 to 47.6 mg/%), and increases in urinary levels of two enzymes: γ-glutamyltranspeptidase (from 0.12 to 0.99 IU/ml) and alkaline phosphatase. In association with these changes were electron microscopic indications of degeneration of kidney tubules with severe disruption of mitochondrial cristae. It is possible that these effects are a consequence of hydrolysis to steviol. Similar nephrotoxic effects have also been reported by others, both in the rat and hamster (Panichkul *et al.* 1988), although I have not had access to the full report.

These findings indicate that at the doses and in the species used, stevioside has vasoactive properties.

Reproductive and teratogenic effects

Stevioside (0.15–3% of diet) was fed to male rats for 60 days before mating and to female rats for 14 days prior to mating and for the first seven days of gestation (Mori *et al.* 1981). No effects were seen on either mating performance or fertility. The group receiving the highest exposure had a slight retardation in weight gain. Given by gavage once a day through days 6 through 16 of pregnancy, stevioside (0–1000 mg/kg/day) produced no fetal malformations or other toxic effects (Usami *et al.* 1995).

Chromosomal and mutagenic effects

Chromosomal abnormalities are only seen with stevioside at high concentrations. In Chinese hamster D-6 cells cultured with various concentrations of stevioside for 28 hours, aberrations were only seen at concentrations of 2% (24.8 mM) or higher (Nadamitsu *et al.* 1985). The commonest alteration was an increase in the number of gaps and interchanges. Sister chromatid exchanges also increased in a dose-dependent manner. In a Chinese hamster fibroblast cell line, stevioside (85% pure; 14.9 mM) did not induce chromosomal aberrations (Ishidate *et al.* 1984).

No chromosomal effects of stevioside were noted in cultured human lymphocytes (Suttajit *et al.* 1993). No mutagenic effects were noted in six *in vitro* (four bacterial and two mammalian) and one *in vivo* screens (Matsui *et al.* 1996).

When tested against two strains of *Salmonella typhimurium*, no mutagenic effects of stevioside were noted at a concentration of 25 mg/plate (Suttajit *et al.* 1993). At 50 mg/plate, mutagenicity was shown against one strain. No mutagenic action against *S. typhimurium* was noted in the presence of an activating system isolated from Aroclor-pretreated rats (Medon *et al.* 1982).

Conclusions

Stevioside has a number of pharmacological actions, including disruption of oxidative phosphorylation, inhibition of monosaccharide transport and impairment of kidney function. Some of these actions have been demonstrated on intact animals or isolated organs, others only on subcellular preparations. A number of authors have suggested therapeutic applications of certain of these actions, particularly of the hypotensive effect. In terms of use as a preferably inert food additive, possession of a potentially therapeutic property implies the potential for a toxic effect. For example, *Stevia* preparations have been proposed as being of benefit in weight control. In other circumstances, a decreased weight gain accompanying the use of a food additive might be considered a toxic response.

In an example of having their cake and wanting to eat it, the authors of one of the major whole animal studies of the safety of stevioside conclude that the material is safe and has no adverse effects (Xili *et al.* 1992). However, they go on to conclude that stevioside may have 'therapeutic value in the treatment of patients with diabetes-related obesity, hypertension or cardiac disease' (what is meant by the latter being undefined). If an agent can lower blood pressure or affect other aspects of cardiovascular functioning, it can clearly produce adverse or unwanted effects. Pharmacology and toxicology are the two faces of the same coin.

The absorption, distribution, metabolism and excretion of stevioside have been poorly studied, particularly in non-rodent species.

In view of the toxicity of the aglycone, steviol, more information is needed on the ability of mammals, or their intestinal microflora, to convert stevioside to steviol (Wingard *et al.* 1980). Are human microflora capable of this transformation? Do certain disease states or antibiotic regimens alter microflora and their ability to perform this transformation? In view of the toxicity associated with related compounds, such as atractyloside, having a free carboxyl group at the C-4 position, the mammalian ability to convert stevioside to steviolbioside (see Figure 8.1) also needs to be investigated. This glycoside has been postulated to be an intermediate in the mammalian conversion of stevioside to steviol (Pezzuto 1986).

In *in vitro* preparations, stevioside can interfere with oxidative phosphorylation and energy metabolism in mitochondria. However, the ability of this glycoside to penetrate cell membranes does not appear to have been sufficiently studied.

REBAUDIOSIDE A

Rebaudioside A ($C_{44}H_{70}O_{23}$; molecular weight 966) (see Figure 8.1) has a similar structure to stevioside. It differs only in having an extra glucopyranosyl residue attached to the sugar unit at C-13. Despite its relative abundance in *S. rebaudiana*, rebaudioside A has been little studied.

Absorption, distribution and metabolism

Rat intestinal microflora efficiently hydrolyze rebaudioside A to steviol (Wingard *et al.* 1980). Apart from this, reports in the mainstream literature on absorption, distribution and metabolism of rebaudioside A appear non-existent.

Lethality and whole animal toxicity

'Crude' rebaudioside A was given to mice by gastric intubation (2 g/kg) (Medon *et al.* 1982). No toxic effects or effects on organ weight were noted two weeks later.

Pharmacological actions

Chromosomal and mutagenic effects

No mutagenic effects were seen in a *Salmonella typhimurium* strain, either in the presence or absence of a metabolic activating system (Medon *et al.* 1982). Rebaudioside A was without effect on azoxymethane-induced colonic aberrant crypt foci in rats (Kawamori *et al.* 1995).

Conclusions

Almost nothing is known *per se* of the pharmacology and toxicology of rebaudioside A. It differs in structure from stevioside (see Figure 8.1) only in the presence of an additional glucosyl unit. If the cardiac glycosides are any guide, this type of structural variation may affect solubility and binding characteristics (and hence pharmacokinetic properties) without qualitative change in biochemical actions.

STEVIOL

Steviol is *ent*-13-hydroxykaur-16-en-19-oic acid ($C_{20}H_{30}O_3$; molecular weight 318) (see Figure 8.1). The compound is well absorbed by mouth (Pezzuto 1986).

Pharmacological actions

Energy metabolism

Steviol inhibits oxidative phosphorylation, with 40 µM producing 50% inhibition (Vignais *et al.* 1966). It is, therefore, more potent than the related aglycones atractyligenin and dihydrosteviol (see Figure 8.1), which produce 50% inhibition at 210 µM and 100 µM, respectively. Some of the mitochondrial actions of steviol are summarized on Table 8.1. The effects on oxidative phosphorylation are complex, involving at least three components: (i) inhibition of adenine nucleotide exchange; (ii) inhibition of NADH oxidase; and (iii) inhibition of L-glutamate dehydrogenase (Kelmer Bracht, Alvarez *et al.* 1985; Yamamoto *et al.* 1985). Adenine nucleotide exchange between intra- and extramitochondrial spaces is involved in the shuttling of high energy phosphate groups generated in the mitochondrion to their sites of consumption in the cytoplasm. Inhibition of this exchange, therefore, implies a profound disturbance of energy flow in the cell. Unlike atractyligenin, the inhibition produced by steviol is non-competitive (Vignais *et al.* 1966). Inhibition of nucleotide exchange is dependent on the presence of a free CO_2H group, with inhibitory action not being seen with the methylated analog, 13-hydroxystevane, or the glucosylated analog, stevioside (Vignais *et al.* 1966). The exocyclic methylene group is also involved, as reduction to dihydrosteviol decreases activity.

The complexity the action of steviol on mitochondrial function is indicated by the finding that 0.5 mM steviol inhibits DNP-stimulated ATPase by 92%, NADH oxidase by 45%, succinate oxidase by 42%, succinate dehydrogenase by 46% and glutamate dehydrogenase by 46% (Vignais *et al.* 1966; Kelmer Bracht, Alvarez *et al.* 1985). In the absence of an uncoupling agent, steviol at low concentrations (0.03 mM) stimulated mitochondrial ATPase activity (Kelmer Bracht, Alvarez *et al.* 1985).

In hamster intestine, steviol (1 mM) inhibited glucose uptake and altered the morphology of intestinal absorptive cells (Toskulkao *et al.* 1995a; Toskulkao *et al.* 1995b). At 2 mM, steviol also lowered mucosal mitochondrial NADH cytochrome C reductase and mucosal ATP concentration without affecting ATPase activity (Toskulkao *et al.* 1995b). The time course of fall in ATP correlates with the decrease in glucose transport. This effect on glucose transport, therefore, appears to be a consequence of inhibition of mitochondrial adenine nucleotide exchange. In fasted rats given a fructose load, steviol (0.2 mM) increased glycogen deposition in the liver (Hubler *et al.* 1994). In the absence of a sugar load, steviol was without effect on glycogen levels.

Effects on blood pressure and renal function

Steviol interferes with energy metabolism in rat renal tubules, blocking oxygen uptake and glucose production (Yamamoto *et al.* 1985). Gluconeogenesis is inhibited 50% at 0.3 mM steviol, and complete inhibition is produced at 1 mM. Oxygen uptake is inhibited 50% at 0.4 mM steviol in the presence of pyruvate (10 mM) as substrate. Steviol inhibited *p*-aminohippuric acid uptake in rat renal cortical slices (Toskulkao *et al.* 1994b).

Other metabolic effects

Steviol inhibits phorbol ester-stimulated ornithine decarboxylase activity in mouse skin (Okamoto *et al.* 1983). Steviol (200 nM) applied to the skin one hour before phorbol ester administration lowers the ornithine decarboxylase response by 63%. A stimulation in ornithine decarboxylase activity is an early event in trophic responses.

Chromosomal and mutagenic effects

There is general agreement that steviol is mutagenic in the forward mutation assay (an assay on a 'normal' or unmutated gene), although not in the reverse (Ames) assay (an assay on a gene containing a designed mutation) (Pezzuto *et al.* 1985; 1986; Matsui *et al.* 1989; Matsui, Matsui *et al.* 1996; Matsui, Sofuni *et al.* 1996). Steviol is mutagenic (Pezzuto *et al.* 1985) as a result of bioactivation (Pezzuto 1986; Pezzuto *et al.* 1986). In the presence of a metabolic activating system, it is mutagenic in both *Salmonella typhimurium* and in human and rat liver microsomes and rat liver S-9 fraction (Pezzuto 1986). In some strains of *S. typhimurium* and *E. coli*, however, steviol is not mutagenic, even in the presence of a metabolic activating system. 'Activated' steviol binds covalently to calf thymus DNA (Pezzuto 1986). In *S. typhimurium* TM677 strain, it induces mutations of the guanine phosphoribosyltransferase (*gpt*) gene (Matsui, Sofuni *et al.* 1996). Sequence studies showed that steviol induced mutations near a putative pausing site for DNA synthesis, leading to DNA duplication, deletion and untargeted mutagenesis by stimulation of misalignment and realignment of developing DNA strands.

The nature of the mutagenic metabolite is unknown. Further metabolism of steviol in rat liver is complex, with at least nine metabolites being detected (Figure 8.2). The presence of a C-13 hydroxyl and a C-16/C-17 double bond are required for activation. The major metabolite is 15α-hydroxysteviol, which is non-mutagenic both in the presence and absence of a metabolic activating system (Pezzuto 1986). Other metabolites include 7β-hydroxysteviol, 17-hydroxyisosteviol and *ent*-16-oxo-17-hydroxybeyeran-19-oic acid, the latter probably being formed from the 16,17-oxide (Pezzuto 1986; Compadre *et al.* 1988). It had been shown to be a product of acid rearrangement of steviol-16,17-oxide (Mori *et al.* 1972). The mutagenic substance has been proposed to be 15-oxosteviol, although this compound has not been detected as a metabolite of steviol (Pezzuto 1986; Compadre *et al.* 1988). 15-Oxosteviol has been reported as bactericidal and weakly mutagenic. It forms an adduct with cysteine via an allylic Michael addition. Others, however, have strongly disputed these findings, reanalyzing the original data to show that 15-oxosteviol is without mutagenic action (Procinska *et al.* 1991). On repetition of the earlier studies, no mutagenicity was found for 15-oxosteviol in *Salmonella* or *Escherichia*. These latter authors suggest that the apparent mutagenicity of steviol may be due to an impurity. Others disagree (Matsui, Matsui *et al.* 1996).

The nature of the mutagenic metabolite thus remains in doubt. It appears to be formed via a P450-mediated oxidation, as pretreatment of rats with Aroclor 1254 or phenobarbital induces metabolism (Pezzuto *et al.* 1985). Activation is NADPH-dependent (Pezzuto 1986). The compound appears to be a nucleophile, as mutagenicity is decreased by GSH, cysteamine and cysteine, although not by methionine (Pezzuto 1986). Neither isosteviol (see Figure 8.1) nor the 16α,17-epoxide of steviol (see Figure 8.2) are mutagenic (Pezzuto *et al.* 1985). The non-involve-

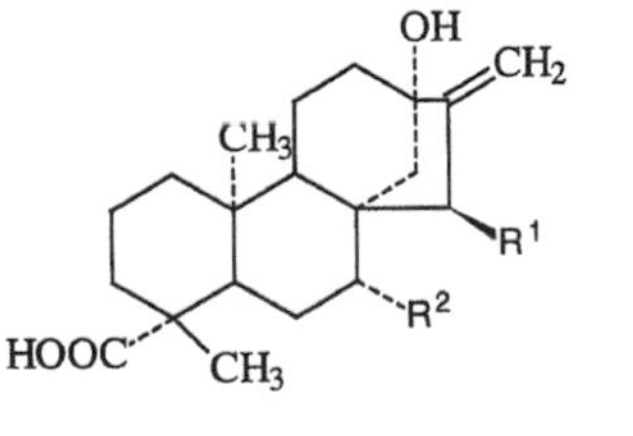

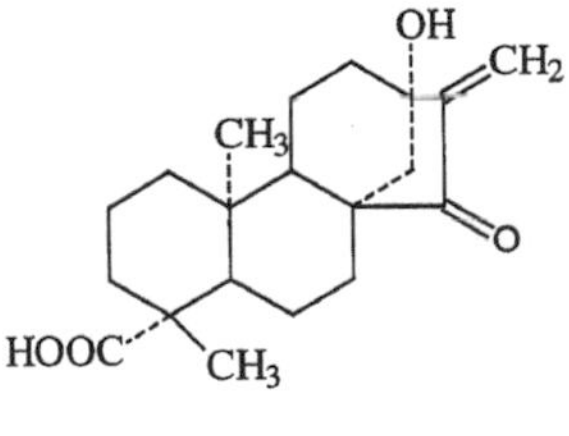

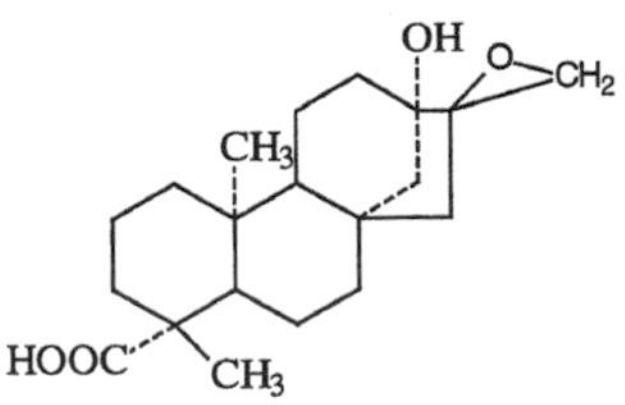

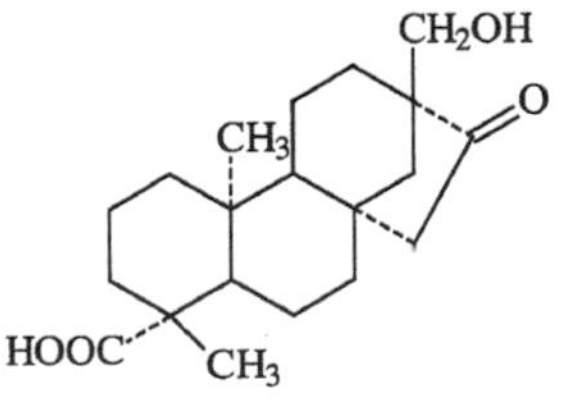

Figure 8.2 Structures of steviol metabolites and related compounds. I: $R^1 = OH$, $R^2 = H$ 15α-hydroxysteviol; $R^1 = H$, $R^2 = OH$ 7β-hydroxysteviol; II: 15-oxosteviol; III: steviol-16,17-oxide; IV: 17-hydroxyisosteviol.

ment of an epoxide is indicated by the finding that epoxide hydrolase does not inhibit mutagenicity (Pezzuto *et al*. 1986). The 13-hydroxy group is a requirement, as *ent*-kaurenoic acid (see
Figure 8.1) is not mutagenic (Pezzuto *et al*. 1986). Dihydrosteviol A and B, grandifloric acid and
steviol acetate are also non-mutagenic, indicating that unsaturation at C_{16}-C_{17} is also important
for expression of mutagenicity (Pezzuto 1986).

On metabolic activation, steviol is mutagenic in Chinese hamster lung fibroblasts *in vivo*, and
in a gene mutation assay in the same cells (Matsui, Sofuni *et al*. 1996). It was without chromosomal effects on cultured human lymphocytes (Suttajit *et al*. 1993).

Reproductive effects

At doses of 0.75 g/kg body weight/day, steviol is toxic to pregnant hamsters and their fetuses
when given on days six through ten of gestation (Wasuntarawat *et al*. 1998). Steviol produces
decreased maternal weight gain and high maternal mortality. The number of live births per
litter is decreased and the mean fetal weight is lower. The no effect dose is 0.25 g/kg body
weight/day. If 100% conversion of stevioside to steviol is assumed, this intake is equivalent
to 625 mg/kg/day of stevioside. This is about 80 times the acceptable daily intake calculated
by Xili *et al*. (1992).

Conclusions

Although much more information is needed, it does appear that steviol has the potential to be
mutagenic. However, two conditions have yet to be established for humans: (i) that *Stevia* glycosides are converted to steviol in meaningful amounts; and (ii) that steviol is further metabolized to the mutagenic substance. In the words of a recent paper, 'further work is necessary to
predict the genotoxic risk of steviol to human beings' (Matsui, Matsui *et al*. 1996).

OVERALL CONCLUSIONS

Much of the literature discussed above reports studies carried out on rats. Some of the findings, such as the conversion of stevioside to steviol, may reflect the idiosyncracies of that species. Studies on a variety of species would give better insight into potential risk for human
consumers. For example, such studies would include examination of human gut flora under
a variety of conditions, and the metabolic abilities of biopsy or autopsied samples of liver and
kidney.

Stevia extracts, although consisting largely of glycosides, are variable, complex mixtures,
partially due to differences in extraction technique and partially due to variability in plant
strain, cultivation conditions, and other factors. Extracts can have biological effects that are
not solely attributable to their content of stevioside or rebaudioside A [e.g. the inhibition of
2,4-dinitrophenol-stimulated ATPase in rat liver mitochondria (Kelmer Bracht, Kemmelmeier *et al*. 1985)]. The use of purified stevioside as a food additive thus appears preferable from
a public safety viewpoint.

Stevia glycosides interfere with energy metabolism at concentrations within the high μM
range. It is unclear how many of these effects are produced *in vivo*. However, these effects
may be of particular concern for children, a group with high metabolic rate, rapid
physiological growth, and a seemingly insatiable demand for sweet foods.

The potential risk:benefit ratio is also a function of the denominator. The potential benefits afforded by *Stevia* glycosides to certain groups may be considered to outweigh the potential risks. These benefits include better dietary control in diabetes and obesity. Claims have also been made that *Stevia*-containing products reduce risk of dental caries and periodontal disease.

There are perhaps two ways of investigating toxic potential. One is to administer the substance to experimental animals and then humans and look for toxic endpoints. The other is to examine the biological handling and mechanisms of action of the substance and look for potential problems. By the first approach, no problems have surfaced with stevioside and rebaudioside A (albeit that the literature on the latter compound is jejune). By the second approach, a number of concerns have surfaced, particularly in the areas of energy and carbohydrate metabolism, and in the metabolic potential of the glycosides themselves. Drawbacks with the first approach include the potential for missing susceptible groups or factors that predispose towards toxicity (e.g. nutrition, disease, age, ethnicity, drugs, etc.). Furthermore, even extensive, well-designed studies may miss low frequency but serious events, subtle effects or significant endpoints that are not examined for. Drawbacks with the second approach include the potential for fishing up red herrings: potential problems may be indicated that are never realized in exposed populations.

The aglycone, steviol, has the ability to penetrate cells, and has a number of potent effects on basic cell functions. The most important of these are inhibition of monosaccharide transport, and inhibition of oxidative phosphorylation, combined with disruption of a number of mitochondrial activities. In addition, steviol is bioactivated to a mutagenic metabolite, the nature of which is still unknown. The glycosides, stevioside and rebaudioside A, are assumed not to cross the plasma membrane into cells, and – except in rat intestine – not to be deglucosylated. Neither of these assumptions has been well tested experimentally. Although *Stevia* products have become widely used in certain countries within a short period without indications of toxicity surfacing, until these aspects of distribution and metabolism have been appropriately studied in humans, an assumption of universal safety appears inappropriate. The conclusion of Pezzuto (1986), although more than a decade old, is still relevant: 'It cannot be definitively stated on the basis of any results that are currently available that stevioside is harmful to human health. However, it does appear that additional safety assessments are required to resolve this question, as steviol, a likely hydrolysis product, is metabolized to active mutagenic species by human enzymes, and appears to covalently interact with DNA.' Phillip's comment 'that stevia has been in use for centuries does not necessarily make it safe' (Phillips 1987) is also relevant given that such use is often uncritically cited as evidence of safety. It is evidence of one sort, but there are examples of substances in wide use for centuries or even millennia – such as tobacco or comfrey (*Symphytum* spp.) – before their profound toxicity became apparent.

REFERENCES

Akashi, H. (1975) Safety of dried-leaves extracts of stevia (report of toxicological test). *Shokuhin Kogyo (Food Industry)* **18**, 1–4.

Akashi, H. and Yokoyama, Y. (1975) Dried-leaf extracts of Stevia. Toxicology tests. *Shokuhin Kogyo (Food Industry)* **18**, 34–43.

Alvarez, M., Bazzote, R. B., Godoy, G. L., Cury, R. and Espinoza, A. D. (1981) Hypoglycemic effect of *Stevia rebaudiana* Bertoni. *Arquivos Biologia e Tecnologia* **24**, 178.

Bakal, A. I. and O'Brien Nabors, L. (1986) Stevioside. In *Alternative Sweeteners*, L. O'Brien Nabors and R. C. Gelardi (Eds), Marcel Dekker, New York, pp. 295–307.

Boeckh, E. M. A. and Humboldt, G. (1981) Cardio-circulatory effects of total water extract in normal persons and of stevioside in rats. *Ciência e Cultura (São Paulo)* **32**, 208–210.

Cardoso, V. N., Barbosa, M. F., Muramoto, E., Mesquita, C. H. and Almeida, M. A. (1996) Pharmacokinetic studies of I^{131}-stevioside and its metabolites. *Nuclear Medicine and Biology* **23**, 97–100.

Compadre, C. M., Hussain, R. A., Nanayakkara, N. P. D., Pezzuto, J. M. and Kinghorn, A. D. (1988) Mass spectral analysis of some derivatives and *in vitro* metabolites of steviol, the aglycone of the natural sweeteners, stevioside, rebaudioside A, and rubusoside. *Biomedical and Environmental Mass Spectrometry* **15**, 211–222.

Crammer, B. and Ikan, R. (1987) Progress in the chemistry and properties of the rebaudiosides. In *Developments in Sweeteners-3*, T. H. Grenby (Ed.), Elsevier Applied Science, London, pp. 45–64.

Curi, R., Alvarez, M., Bazotte, R. B., Botion, L. M., Godoy, J. L. and Bracht, A. (1986) Effect of *Stevia rebaudiana* on glucose tolerance in normal adult humans. *Brazilian Journal of Medical and Biological Research* **19**, 771–774.

Fujita, H. and Edahiro, T. (1979) Safety and utilization of Stevia sweetener. *Shokuhin Kogyo (Food Industry)* **22**, 66–72.

Hanson, J. R. and De Oliveira, B. H. (1993) Stevioside and related sweet diterpenoid glycosides. *Natural Products Reports* **10**, 301–309.

Hubler, M. O., Bracht, A. and Kelmer-Bracht, A. M. (1994) Influence of stevioside on hepatic glycogen levels in fasted rats. *Research Communications in Chemical Pathology and Pharmacology* **84**, 111–118.

Humboldt, G. and Boeckh, E. M. (1977) Efeito do edulcorante natural (steviosideo) e sintético (sacarina) sobre o ritmo cardíaco em ratos. *Arquivos Brasileiros de Cardiologia* **30**, 275–277.

Ishidate, M., Jr, Sofuni, T., Yoshikawa, K., Hayashi, M. and Nohmi, T. (1984) Primary mutagenicity screening of food additives currently used in Japan. *Food and Chemical Toxicology* **22**, 623–636.

Ishii, E. L., Schwab, A. J. and Bracht, A. (1987) Inhibition of monosaccharide transport in the intact rat liver by stevioside. *Biochemical Pharmacology* **36**, 1417–1433.

Ishii-Iwamoto, E. L. and Bracht, A. (1995) Stevioside is not metabolized in the isolated perfused rat liver. *Research Communications in Molecular Pathology and Pharmacology* **87**, 167–175.

Kawamori, T., Tanaka, T., Hara, A., Yamahara, J. and Mori, H. (1995) Modifying effects of naturally occurring products on the development of colonic aberrant crypt foci induced by azoxymethane in F344 rats. *Cancer Research* **55**, 1277–1282.

Kelmer Bracht, A., Alvarez, M. and Bracht, A. (1985) Effects of *Stevia rebaudiana* natural products on rat liver mitochondria. *Biochemical Pharmacology* **34**, 873–882.

Kelmer Bracht, A. M., Kemmelmeier, F. S., Ishii, E. L., Alvarez, M. and Bracht, A. (1985) Effect of *Stevia rebaudiana* natural products on cellular and sub-cellular metabolism. *Arquivos Biologia e Tecnologia* **28**, 431–455.

Kim, K. K., Sawa, Y. and Shibata, H. (1996) Hydroxylation of ent-kaurenoic acid to steviol in *Stevia rebaudiana* Bertoni – purification and partial characterization of the enzyme. *Archives of Biochemistry and Biophysics* **332**, 223–230.

Kinghorn, A. D. and Soejarto, D. D. (1985) Current status of stevioside as a sweetening agent for human use. In *Economic and Medicinal Plant Research, Vol. 1*, H. Wagner, H. Hikino, and N. R. Farnsworth (Eds), Academic Press, London, pp. 1–52.

Kinghorn, A. D. and Soejarto, D. D. (1991) Stevioside. In *Alternative Sweeteners* (2nd edn, revised and expanded), L. O'Brien Nabors and R. C. Gelardi (Eds), Marcel Dekker, New York, pp. 157–171.

Klingenberg, M. and Heldt, H. W. (1982) The ADP/ATP translocation in mitochondria and its role in intracellular compartmentation. In *Metabolic Compartmentation*, H. Sies (Ed.), Academic Press, New York, pp. 101–122.

Kohda, A., Kasai, R., Yamasaki, K., Murakami, K. and Tanaka, O. (1976) New sweet diterpene glucosides from *Stevia rebaudiana*. *Phytochemistry* **15**, 981–983.

Lee, S. J., Lee, K. R., Park, J. R., Kim, K. S. and Tchai, B. S. (1979) A study on the safety of stevioside as a new sweetening source. *Korean Journal of Food Science and Technology* **11**, 224–231.

Levy, N. M., Bracht, A. and Kelmerbracht, A. M. (1994) Effects of *Stevia rebaudiana* natural products on the mitochondrial L-glutamate dehydrogenase. *Arquivos Biologia e Tecnologia* **37**, 673–680.

Matsui, M., Matsui, K., Nohmi, T., Mizusawa, H. and Ishidate, M. (1989) Mutagenicity of steviol: an analytical approach using the southern blotting system. *Bulletin of the National Institute of Hygenic Science, Tokyo, Japan* **107**, 83–87.

Matsui, M., Matsui, K., Kawasaki, Y. and Oda, Y. (1996) Evaluation of the genotoxicity of stevioside and steviol using six *in vitro* and one *in vivo* mutagenicity assays. *Mutagenesis* **11**, 573–579.

Matsui, M., Sofuni, T. and Nohmi, T. (1996) Regionally-targeted mutagenesis by metabolically-activated steviol: DNA sequence analysis of steviol-induced mutants of guanine phosphoribosyltransferase (gpt) gene of *Salmonella typhimurium* TM677. *Mutagenesis* **11**, 565–572.

Medon, P. J., Pezzuto, J. M., Hovanec-Brown, J. M., Nanayakkara, N. P. D., Soejarto, D. D., Kamath, S. K. and Kinghorn, A. D. (1982) Safety assessment of some *Stevia rebaudiana* sweet principles. *Federation Proceedings* **41**, 1568.

Medon, P. J. and Ziegler, M. M. (1986) Effect of acute and chronic administration of *Stevia rebaudiana* extracts on plasma glucose *in vivo*. *Pharmaceutical Research* **3**, 162S.

Melis, M. S. (1992a) Renal excretion of stevioside in rats. *Journal of Natural Products* **55**, 688–690.

Melis, M. S. (1992b) Stevioside effect on renal function of normal and hypertensive rats. *Journal of Ethnopharmacology* **36**, 213–217.

Melis, M. S. (1995) Chronic administration of aqueous extract of *Stevia rebaudiana* in rats: Renal effects. *Journal of Ethnopharmacology* **47**, 129–134.

Melis, M. S. (1996) A crude extract of *Stevia rebaudiana* increases the renal plasma flow of normal and hypertensive rats. *Brazilian Journal of Medical and Biological Research* **29**, 669–675.

Melis, M. S. and Sainati, A. R. (1991) Effect of calcium and verapamil on renal function of rats during treatment with stevioside. *Journal of Ethnopharmacology* **33**, 257–262.

Melis, M. S., Maciel, R. E. and Sainati, A. R. (1985) Effects of indomethacin on the action of stevioside on mean arterial pressure and on renal function in rats. *IRCS Medical Science* **13**, 1230–1231.

Melis, M. S., Sainati, A. R. and Maciel, R. E. (1986) Effects of two concentrations of stevioside on renal function and mean arterial pressure in rats. *IRCS Medical Science* **14**, 973–973.

Miquel, O. (1966) Un nuevo hipoglicemiante oral. *Revista Medica del Paraguay* **7**, 200–202.

Mori, K., Nakahara, Y. and Matsui, M. (1972) Diterpenoid total synthesis. 19. ($\pm$)Steviol and erythroxydiol A – rearrangements in bicyclooctane compounds. *Tetrahedron* **28**, 3217–3226.

Mori, N., Sakanoue, M., Takeuchi, M., Shimpo, K. and Tanabe, T. (1981) Effect of stevioside on fertility in rats. *Shokuhin Eiseigaku Zasshi* **22**, 409–414.

Nadamitsu, S., Segawa, M., Sato, Y. and Kondo, K. (1985) Effects of stevioside on the frequencies of chromosomal aberrations and sister chromatid exchanges in the D-6 cell of Chinese hamster. *Memoirs of the Faculty of Integrated Arts and Sciences, Hiroshima University* **10**, 57–62.

Nakayama, K., Kasahara, D. and Yamamoto, F. (1986) Absorption, distribution, metabolism, and excretion of stevioside in rats. *Shokuhin Eiseigaku Zasshi (Journal of the Food Hygienic Society of Japan)* **27**, 1–8.

Okamoto, H., Yoshida, D., Saito, Y. and Mizusaki, S. (1983) Inhibition of 12-*O*-tetradecanoylphorbol-13-acetate-induced ornithine decarboxylase activity in mouse epidermis by sweetening agents and related compounds. *Cancer Letters* **21**, 29–35.

Oliveira-Filho, R. M., Uehara, O. A., Minetti, C. A. S. A. and Valle, L. B. S. (1989) Chronic administration of aqueous extract of *Stevia rebaudiana* (Bert.) Bertoni in rats: endocrine effects. *General Pharmacology* **20**, 187–191.

Oviedo, C. A., Fronciana, G., Moreno, R. and Máas, L. C. (1970) Acción hipoglicemiante de la *Stevia rebaudiana* Bertoni (Kaá-hê-é). *Excerpta Medica* **209**, 92–92.

Panichkul, T., Glinsukon, T., Buddhasukh, D., Cheuvchit, P. and Pimolsri, U. (1988) The plasma levels of urea nitrogen, creatinine and uric acid and urine volume in rats and hamsters treated with stevioside. *Thai Journal of Toxicology* **4**, 47–52.

Pezzuto, J. M. (1986) Chemistry, metabolism and biological activity of steviol (*ent*-13-hydroxykaur-16-en-19-oic acid), the aglycone of stevioside. In *New Trends in Natural Products Chemistry*, Elsevier Science, Atta-ur-Rahman and P. W. Le Quesne (Eds), Amsterdam, pp. 371–386.

Pezzuto, J. M., Compadre, C. M., Swanson, S. M., Nanayakkara, N. P. D. and Kinghorn, A. D. (1985) Metabolically activated steviol, the aglycone of stevioside, is mutagenic. *Proceedings of the National Academy of Sciences of the United States of America* **82**, 2478–2482.

Pezzuto, J. M., Nanayakkara, N. P. D., Compadre, C. M., Swanson, S. M., Kinghorn, A. D., Guenthner, T. M. *et al.* (1986) Characterization of bacterial mutagenicity mediated by 13-hydroxy-*ent*-kaurenoic acid (steviol) and several structurally-related derivatives and evaluation of potential to induce glutathione *S*-transferase in mice. *Mutation Research* **169**, 93–103.

Phillips, K. C. (1987) Stevia: Steps in developing a new sweetener. In *Developments in Sweeteners – 3*, T. H. Grenby (Ed.), Elsevier Applied Science, London, pp. 1–43.

Planas, G. M. and Kuc, J. (1968) Contraceptive properties of *Stevia rebaudiana*. *Science* **162**, 1007–1007.

Procinska, E., Bridges, B. A. and Hanson, J. R. (1991) Interpretation of results with the 8-azaguanine resistance system in *Salmonella typhimurium*: no evidence for direct acting mutagenesis by 15-oxosteviol, a possible metabolite of steviol. *Mutagenesis* **6**, 165–167.

Sakaguchi, M. and Kan, T. (1982) The Japanese research on *Stevia rebaudiana* (Bert.) Bertoni and the stevioside. *Ciência e Cultura (São Paulo)* **34**, 235–248.

Shibata, H., Sawa, Y., Oka, T., Sonoke, S., Kim, K. K. and Yoshioka, M. (1995) Steviol and steviol-glycoside: glucosyltransferase activities in *Stevia rebaudiana* Bertoni – purification and partial characterization. *Archives of Biochemistry and Biophysics* **321**, 390–396.

Sincholle, D. and Marcorelles, P. (1989) Étude de l'activité anti-androgénique d'un extrait de *Stevia rebaudiana* Bertoni. *Plantes Médicinales et Phytothérapie* **23**, 282–287.

Suanarunsawat, T. and Chaiyabutr, N. (1997) The effect of stevioside on glucose metabolism in rat. *Canadian Journal of Physiology and Pharmacology* **75**, 976–982.

Suttajit, M., Vinitketkaumnuen, U., Meevatee, U. and Buddhasukh, D. (1993) Mutagenicity and human chromosomal effect of stevioside, a sweetener from *Stevia rebaudiana* Bertoni. *Environmental Health Perspectives* **101**(Suppl. 3), 53–56.

Suzuki, H., Kasai, T., Sumiiiara, M. and Sugisawa, H. (1977) Influence of oral administration of stevioside on levels of blood glucose and liver glycogen of intact rats. *Nogyo Kagaku Zasshi* **51**, 171–173.

Takaki, M., De Campos Takaki, G. M., De Santana Diu, M. B., De Andrade, M. S. S. and Da Silva, E. C. (1984) Antimicrobial activity in leaves extracts of *Stevia rebaudiana* Bert. *Revista Instituto de Antibióticos, Universidade do Recife* **22**, 33–39.

Toskulkao, C. and Sutheerawattananon, M. (1994) Effects of stevioside, a natural sweetener, on intestinal glucose absorption in hamsters. *Nutrition Research* **14**, 1711–1720.

Toskulkao, C., Deechakawan, W., Leardkamolkarn, V., Glinsukon, T. and Buddhasukh, D. (1994a) The low calorie natural sweetener stevioside – Nephrotoxicity and its relationship to urinary enzyme excretion in the rat. *Phytotherapy Research* **8**, 281–286.

Toskulkao, C., Deechakawan, W., Temcharoen, P., Buddhasukh, D. and Glinsukon, T. (1994b) Nephrotoxic effects of stevioside and steviol in rat renal cortical slices. *Journal of Clinical Biochemistry and Nutrition* **16**, 123–131.

Toskulkao, C., Sutheerawattananon, M. and Piyachaturawat, P. (1995a) Inhibitory effect of steviol, a metabolite of stevioside, on glucose absorption in everted hamster intestine *in vitro*. *Toxicology Letters* **80**, 153–159.

Toskulkao, C., Sutheerawattananon, M., Wanichanon, C., Saitongdee, P. and Suttajit, M. (1995b) Effects of stevioside and steviol on intestinal glucose absorption in hamsters. *Journal of Nutritional Science and Vitaminology* **41**, 105–113.

Toyoda, K., Matsui, H., Shoda, T., Uneyama, C., Takada, K. and Takahashi, M. (1997) Assessment of the carcinogenicity of stevioside in F344 rats. *Food and Chemical Toxicology* **35**, 597–603.

Usami, M., Sakemi, K., Kawashima, K., Tsuda, M. and Ohno, Y. (1995) Teratogenicity study of stevioside in rats. *Eisei Shikenjo Hokoku (Bulletin of National Institute of Hygienic Sciences, Tokyo, Japan)* **113**, 31–35.

Vignais, P. V., Duee, E. D., Vignais, P. M. and Huet, J. (1966) Effects of atractyligenin and its structural analogues on oxidative phosphorylation and on the translocation of adenine nucleotides in mitochondria. *Biochimica et Biophysica Acta* **118**, 465–483.

von Schmeling, G. A., Carvalho, F. N. and Espinoza, A. D. (1977) *Stevia rebaudiana* Bertoni. Evaluation of the hypoglucemic effect in alloxanized rabbits. *Ciência e Cultura (São Paulo)* **22**, 599–601.

Wasuntarawat, C., Temcharoen, P., Toskulkao, C., Mungkornkarn, P., Suttajit, M. and Glinsukon, T. (1998) Developmental toxicity of steviol, a metabolite of stevioside, in the hamster. *Drug and Chemical Toxicology* **21**, 207–222.

Wingard, R. E., Jr, Brown, J. P., Enderlin, F. E., Dale, J. A., Hale, R. L. and Seitz, C. T. (1980) Intestinal degradation and absorption of the glycosidic sweeteners stevioside and rebaudioside A. *Experientia* **36**, 519–520.

Xili, L., Chengjiny, B., Eryi, X., Reiming, S., Yuengming, W., Haodong, S. *et al.* (1992) Chronic oral toxicity and carcinogenicity of stevioside in rats. *Food and Chemical Toxicology* **30**, 957–965.

Yamada, A., Ohgaki, S., Noda, T. and Shimizu, M. (1985) Chronic toxicity study of dietary stevia extracts in F344 rats. *Shokuhin Eiseigaku Zasshi (Journal of the Food Hygienic Society of Japan)* **26**, 169–183.

Yamamoto, N. S., Kelmer Bracht, A. M., Ishii, E. L., Kemmelmeier, F. S., Alvarez, M. and Bracht, A. (1985) Effect of steviol and its structural analogues on glucose production and oxygen uptake in rat renal tubules. *Experientia* **41**, 55–57.

9 Use of *Stevia rebaudiana* sweeteners in Japan

Kenji Mizutani and Osamu Tanaka

INTRODUCTION

Worldwide, the most widespread use of *Stevia rebaudiana* sweeteners is in Japan. The leaves of *S. rebaudiana* contain several sweet *ent*-kaurene glycosides inclusive of stevioside, rebaudiosides A, C, D, and E, and dulcoside A, as shown in Figure 9.1 (Mosettig *et al.* 1963; Kohda *et al.* 1976; Sakamoto *et al.* 1977a; 1977b; Kobayashi *et al.* 1977). Since the latter half of the 1970s, extracts of the leaves of *S. rebaudiana* have been employed as sweetening agents, taste modifiers, and sugar substitutes in the food industry in Japan, and have been making slow but steady progress in their utilization and distribution. In Japan, sweet-tasting products from *S. rebaudiana* leaves are

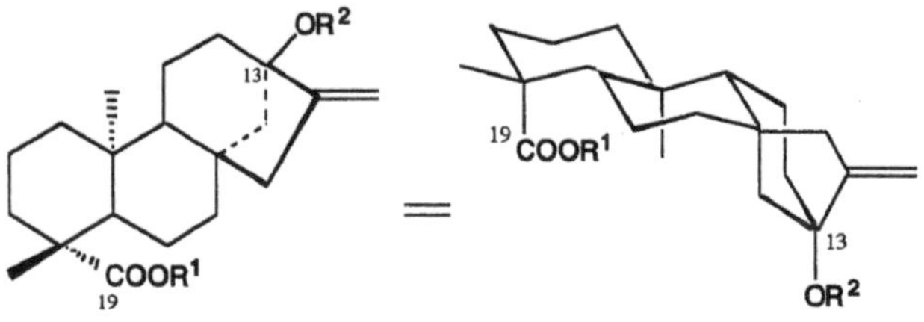

Steviol $R^1 = R^2 = $ -H

Glycoside	R^1	R^2
Stevioside	-Glc	-Glc-(2-1)-Glc
Rebaudioside A	-Glc	-Glc-(2-1)-Glc \(3-1)-Glc
Rebaudioside C (Dulcoside B)	-Glc	-Glc-(2-1)-Rha \(3-1)-Glc
Rebaudioside D	-Glc-(2-1)-Glc	-Glc-(2-1)-Glc \(3-1)-Glc
Rebaudioside E	-Glc-(2-1)-Glc	-Glc-(2-1)-Glc
Dulcoside A	-Glc	-Glc-(2-1)-Rha

Glc: β-D-glucose, Rha: α-L-rhamnose

Figure 9.1 Structures of sweet *ent*-kaurene glycosides from the leaves of *Stevia rebaudiana*.

called simply 'stevia sweeteners' and have been commercialized in three basic forms: (i) 'stevia extract'; (ii) 'sugar-transferred stevia extract'; and (iii) 'rebaudioside A-enriched stevia extract', which have been used in a variety of foods and beverages. This chapter deals with the practical use of stevia sweeteners in Japan, referring to their developmental background and sweetening properties relative to progress made in these uses. Information published previously in the English language concerning this subject may be obtained from reviews by Bakal and O'Brien Nabors (1985) and Kinghorn and Soejarto (1985).

BACKGROUND TO STEVIA SWEETENER DEVELOPMENT IN JAPAN

Stevia sweeteners in Japan were originally developed as alternative sweeteners for Japanese seasonings. Used in various Japanese foods for several hundred years, soy sauce is the most popular Japanese seasoning and is generally formulated with high levels of salt (sodium chloride) for preservation and flavoring. In the early part of the twentieth century, the suppressing effect of a decoction from roots of *Glycyrrhiza* species (licorice) on the saline taste of soy sauce was recognized. Since then, the use of licorice sweetener (glycyrrhizin) has expanded to various salty foods such as Japanese-style pickles, dried seafood, and miso (soybean paste). The use of licorice sweetener continued to grow until production was stopped during the Second World War. When production started again in 1950, saccharin and dulcin were already being used as inexpensive alternative sweeteners in response to the sugar shortage. Though cyclamate was also approved as a sweetener in 1956, the use of licorice sweetener was supported by the persistent popularity of soy source seasoned with it, and manufacturers of licorice sweetener continued operations by supplying a higher grade of glycyrrhizin manufactured by new refining techniques and facilities, which were later applied to the stevia sweeteners. The ban on dulcin and cyclamate in 1969 increased the demand for licorice sweetener significantly, but Japanese manufacturers of licorice sweetener were anxious about the import of licorice root from the Middle East, Russia, and the People's Republic of China due to the political situations in these countries, and they recognized the practical limits of using glycyrrhizin because of its latent sweetness. Manufacturers of licorice sweetener responded quickly to the successful cultivation of *S. rebaudiana* in Japan in the early 1970s. After the temporary ban on sodium saccharin in 1973, several companies competed by producing stevia extract as a new sweetener.

Thus, stevia extract was put onto the market in Japan with high hopes. But its limitations for practical usage were soon evident. When added to salted foods, it resulted in a sweet-salty taste. Soon manufacturers of sweeteners, aware of the suitability of glycyrrhizin for salted foods, developed combined preparations of stevia extract and glycyrrhizin, which became acceptable for their predominant taste-modifying effects, and demand increased significantly, especially in the soy sauce and Japanese-style pickle industries.

While stevia extract has a relatively quick onset of sweetness compared to glycyrrhizin, it is inferior to sucrose in its unpleasant sweetness aftertaste and bitterness. In order to reduce the aftertaste of stevia extract, a number of studies focused on the purification, formulation, and enzymatic modification of stevia sweeteners as well as on the breeding of *S. rebaudiana*. The result was enzymatically modified 'sugar-transferred stevia extract' and 'rebaudioside A-enriched stevia extract' which were commercialized as more sucrose-like sweeteners in terms of their taste. These improved stevia sweeteners could be used as sweeteners preferably for drinks and have come to be applied extensively in confectionery, ice cream, sherbet, dairy products, table-top sweeteners, etc. The use of 'rebaudioside A-enriched stevia extract' in the isotonic

Table 9.1 Japanese companies constituting the 'Stevia Kogyokai (Stevia Industrial Consortium)'

Dainippon Ink & Chemicals Co. Ltd. (Tokyo)
Fuji Chemical Co. Ltd. (Osaka)
Ikeda Tohka Industries Co. Ltd. (Fukuyama-city, Hiroshima)
Maruzen Pharmaceuticals Co. Ltd. (Onomichi-city, Hiroshima)
Morita Kagaku Kogyo Co. Ltd. (Osaka)
Nikken Chemicals Co. Ltd. (Tokyo)
Nippon Paper Industries Co. Ltd. (Tokyo)
Nichinoh Seiken Co. Ltd. (Kagoshima)
Tama Biochemical Co. Ltd. (Tokyo)
Tokiwa Phytochemical Co. Ltd. (Chiba)
Toyo Sugar Refining Co. Ltd. (Tokyo)

---

drink 'Pocari Sweat Stevia®' by Otsuka Pharmaceutical Co. Ltd. drew attention to the potential of stevia sweeteners in the beverage market.

Progress in the utilization of stevia sweeteners in Japan has continued due to a number of extensive studies on the chemistry, safety, stability, and application of stevia extracts and their sweet principles. Furthermore, the 'Stevia Konwakai', or the 'Stevia Consortium', organized by manufactures of stevia sweeteners, has promoted the distribution of stevia sweeteners, not only through cooperative work in obtaining permission and registration for use as a sucrose substitute but also through the exchange of information on production, marketing, and research. 'Stevia Konwakai' was renamed 'Stevia Kogyokai (Stevia Industrial Consortium)', which now constitutes the 11 companies shown in Table 9.1. Currently, the 11 companies of 'Stevia Kogyokai' and a few other companies are engaged in the production and distribution of stevia sweeteners, which are utilized by a number of food corporations in Japan.

STEVIA SWEETENER PRODUCTS

Of the stevia sweeteners used by the Japanese food industry, there are three basic forms plus their combined preparations with other food ingredients. In Table 9.2, some representative products which incorporate stevia sweeteners are shown.

'Stevia extract'

Soon after being introduced to the Japanese market, a simple, concentrated product of an aqueous decoction of *S. rebaudiana* leaves was employed, namely, 'stevia extract'. At present 'stevia extract' is made through several steps using absorbent resins to produce a white to pale yellowish powder or granule, containing stevioside, rebaudioside A, rebaudioside C, dulcoside A, and minor principles. According to the *Voluntary Specifications of Non-chemically Synthesized Food Additives*, published by the Japan Food Additive Association (1993), the content of total steviol glycosides including stevioside, rebaudioside A, rebaudioside C, and dulcoside A in a commercialized 'stevia extract' is standardized at a minimum level of 80%. Commercially available simple 'stevia extracts' commonly contain about 90% steviol glycosides which consist of around 50–55% stevioside, 20–25% rebaudioside A, 5–10% rebaudioside C, and 3–5% dulcoside A (Shibasato 1995). The available products are specified to have no more than 20 ppm of heavy metals (determined as Pb) and no more than 2 ppm of arsenic (determined as As_2O_3).

Table 9.2 Representative products in Japan containing stevia sweeteners

Trade name	Composition	Sweetness[a]	Characteristics and use
MST-90[b]	Stevia extract (Total steviol glycosides: not less than 90%)	250	Sweetness property is similar to that of sucrose. All kinds of foods.
Marumilon Pure[b]	Stevia extract (Total steviol glycosides: not less than 85.0 ± 3.0%)	200	Sweetness property is similar to that of sucrose. All kinds of foods.
Marumilon 50[b]	Stevia extract Dextrin	120	Applicability is facilitated by addition of dextrin. All kinds of foods.
Marumilon A[b]	Stevia extract Glycyrrhizin Sodium citrate Dextrin	30–50	Product in which the taste is modified by addition of glycyrrhizin to adapt for low salted foods. Low salted foods and sugar-reduced foods.
Marumilon S[b]	Stevia extract Glycyrrhizin Sodium citrate Dextrin	30–60	Product in which the taste is modified by addition of glycyrrhizin to apply to salty foods. Salty foods.
Marumilon Cool[b]	Stevia extract Glycyrrhizin Sodium citrate Glycerol fatty acid ester Dextrin	40	Product can be easily used in foods for which foams are not favorable. Ice creams and sherbets.
α-G Sweet PA[b]	Sugar-transferred stevia extract (not less than 95%)	170	Sweetness property is extremely similar to that of sucrose. All kinds of foods.
α-G Sweet H[b]	Sugar-transferred stevia extract Dextrin	60	Sweetness property is extremely similar to that of sucrose. Applicability is facilitated by addition of dextrin. All kinds of foods.
Chrysanta AR-P[c]	Stevia extract (as rebaudioside A: 50%) Dextrin	150	Sweetness property is extremely similar to that of sucrose. Applicability is facilitated by addition of dextrin. All kinds of foods.

Notes
a Relative sweetness to sucrose.
b Products of Maruzen Pharmaceuticals Co. Ltd.
c Product of Dainippon Ink and Chemicals Co. Ltd.

'Sugar-transferred stevia extract'

A number of studies on the enzymatic transglycosylation of steviol glycosides have been carried out to increase their sweetness and to improve their quality of taste (Tanaka 1997; see Chapter 7). Of these, the trans-α-1,4-glucosylated product of the 'stevia extract' is produced by the following procedure and is commercially utilized in Japan. On treatment of 'stevia extract' with cyclomaltodextrin-glucanotransferase (CGTase) and a soluble starch, α-glucosyl units of the soluble starch are transferred to the 4-position of the glucosyl moieties of sweet steviol glycosides in the 'stevia extract', thereby improving the quality of taste. Our research group has conducted a systematic study on the sweetness–structure relationship of steviol glycosides (Darise *et al*. 1984; Mizutani *et al*. 1989; Fukunaga *et al*. 1989; Ohtani *et al*. 1991). It was observed that the trans-α-1,4-glucosylation of too many glucosyl units to steviol glycosides led to a decrease in sweetness. Accordingly, a product prepared by treatment of the α-1,4-transglucosylated extract with β-amylase to shorten the poly-α-glucosyl chain was also developed. This product, an improved 'sugar-transferred stevia extract', is recommended as a better sweetener than the non-treated 'sugar-transferred stevia extract'.

In the volume *Voluntary Specifications of Non-chemically Synthesized Food Additives*, published by the Japan Food Additive Association (Second Edition, 1993), these 'sugar-transferred stevia extracts' are classified as 'enzymatically modified stevia' (or 'glucosyl stevia') and their total steviol glycosides and unglucosylated steviol glycosides are registered at levels of more than 85% and less than 15%, respectively. The heavy metals and arsenic in this product are restricted to maximum levels of 20 ppm and 2 ppm, respectively, as in the case of 'stevia extract'.

'Rebaudioside A-enriched stevia extract'

Among the sweet glycosides of *S. rebaudiana* leaves, rebaudioside A is the most potent sweet principle and has the most sucrose-like taste. 'Rebaudioside A-enriched stevia extract' is prepared from the leaves of an improved variety of *S. rebaudiana* containing greater proportions of rebaudioside A. Of the two commercialized types of 'rebaudioside A-enriched stevia extract', one is produced in the usual manner according to the method of production of 'stevia extract', and the other is further purified by subsequent recrystallization. In the *Voluntary Specifications of Non-chemically Synthesized Food Additives* (Second Edition 1993), these products are classified as 'stevia extracts'.

Combined preparations of stevia sweeteners

Several preparations of stevia sweeteners used with other food ingredients have been adapted for various foods. The most common preparations are made by mixing one of the basic forms of stevia sweeteners with dextrin and/or lactose, to facilitate use in foods. As preparations for modifying the taste of foods, especially salted foods, combinations of stevia extract and glycyrrhizin with dextrin, lactose, and an organic acid or amino acid are recommended. The content ratios of sweeteners to other ingredients in these preparations vary according to the sweetener manufacturer. More than 100 preparations containing stevia sweeteners are commercially available in Japan.

Analytical methods for steviol glycosides in stevia sweeteners

According to the *Voluntary Specifications of Non-chemically Synthesized Food Additives* (Second Edition 1993), the quantitative analysis of total steviol glycosides in stevia extracts is performed using high-performance liquid chromatography (HPLC) on an NH_2 column (normal-phase mode), in the following manner. About 60 to 120 mg of the sample is accurately weighed and dissolved in 80% (v/v) acetonitrile to make 100 ml of a test solution. About 50 mg each of stevioside and rebaudioside A, dried previously for 2 hours at 105 °C are accurately weighed, and the standard solutions of both compounds are prepared in the same fashion as the test solution. The test and standard solutions are injected into an HPLC apparatus under the following conditions:

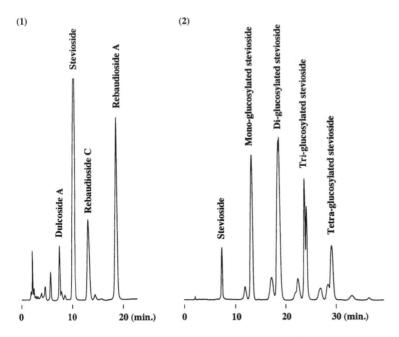

Figure 9.2 HPLC elution profiles of 'stevia extract' (1) and 'sugar-transferred stevia extract' after treatment with β-amylase (2).

detector: ultraviolet spectrometer (wavelength 210 nm);
packing material: NH_2 (aminopropyl)-modified silica or polymer gel;
column: A stainless column of 3.9–4.6 mm diameter and 15–30 cm length;
column temperature: 40 °C;
Mobile phase: acetonitrile-water (80:20);
flow rate: retention time of stevioside is adjusted to about 10 minutes;
injection volume: 10–20 μL.

Peak areas of stevioside, rebaudioside A, rebaudioside C and dulcoside A in the chromatogram of the test solution and those of standard stevioside and rebaudioside A in chromatograms of each standard solution are measured, and quantities of the four sweet principles of *S. rebaudiana* leaves are calculated as follows:

1 Stevioside (%) = (WS/WT) × (TS/SS) × 100
2 Dulcoside A (%) = (WS/WT) × (TD/SS) × 100
3 Rebaudioside A (%) = (WR/WT) × (TRA/SR) × 100
4 Rebaudioside C (%) = (WR/WT) × (TRC/SR) × 100

Where, WS: Weight (mg) of stevioside standard; WR: Weight (mg) of rebaudioside A standard; WT: Weight (mg) of the test sample; TS: Peak area of stevioside in the test sample chromatogram; TD: Peak area of dulcoside A in the test sample chromatogram; TRA: Peak area of rebaudioside A in the test sample chromatogram; TRC: Peak area of rebaudioside A in the test sample chromatogram; SS: Peak area of stevioside in the standard chromatogram; SR: Peak area of rebaudioside A in the standard chromatogram. The total amount of steviol glycosides is presented by the sum of the quantities of the above four principles.

In the *Voluntary Specifications of Non-chemically Synthesized Food Additives* (Second Edition 1993), 'sugar-transferred stevia extract' (enzymatically modified stevia) is registered by levels of both the total steviol glycosides and the unglucosylated steviol glycosides. The total steviol glycosides are calculated by the sum of the quantities of isosteviol and sugars in the acid hydrolysate of the steviol glycosides, while the quantity of unglucosylated steviol glycosides in this product is determined by using about 1.5 g of a test sample on an HPLC column according to the analytical method for the 'stevia extract'.

Figure 9.2 shows HPLC patterns of 'stevia extract' and 'sugar-transferred stevia extract' after treatment with β-amylase.

MARKETING IN JAPAN

More than 20 years have passed since the use of stevia sweeteners started in Japan. At first, stevia sweeteners were expensive products priced at 100 thousand yen per kilogram. Since then, from progress made in refinement, cultivation, and plant breeding techniques, along with the increase in supplies of raw materials and crude extracts from other countries, prices went down and some of these products are currently sold at less than 10 thousand yen per kilogram. Although in earlier days *S. rebaudiana* was cultivated throughout Japan, cultivation gradually spread to other countries of east and southeast Asia including Korea, Taiwan, the People's Republic of China, Vietnam, and Thailand. Now almost all *S. rebaudiana* for use in Japan is cultivated in the People's Republic of China.

According to the *Technical Journal on Food Chemistry and Chemicals* (Anonymous, 1996), it is estimated that about 200 metric tons of stevia sweeteners are used each year in Japan. This corresponds to about 2000 metric tons of dry leaves along with crude extracts for the production of stevia sweeteners. The total market value of stevia sweetener products in Japan is estimated to be around 2 to 3 billion yen a year.

Table 9.3 summarizes the current usage of stevia sweeteners, as surveyed by Maruzen Pharmaceuticals Co. Ltd. Formerly, the demand for stevia sweeteners in salted foods market increased continually. Currently, however, the demand for salted foods has leveled off and may be decreasing slightly. Nevertheless, the total demand for stevia sweeteners is increasing gradually due to the expansion of the utilization of 'sugar-transferred stevia extract' and

Table 9.3 Use of stevia sweeteners in the Japanese food industry as estimated by Maruzen Pharmaceuticals Co. Ltd. in 1995

Food item	Total (%)
Japanese-style pickles	28.1
Dried sea foods	12.6
Soy sauce and soypaste	6.3
Mashed and steamed fish and meat	4.9
Seasonings	4.8
Sea foods boiled down by soy sauce	2.7
Beverages and yogurt	17.1
Ice cream and sherbet	12.6
Confectioneries and bread	3.5
Table-top sweeteners, etc.	7.4

'rebaudioside A-enriched stevia extract' in non-salted foods such as ice cream, sherbet, dairy products, confectionery, and soft drinks.

PROPERTIES OF STEVIA SWEETENERS

Although stevia sweeteners possess a slightly latent sweetness compared to sucrose, their use has been developing steadily. The wide use of stevia sweeteners is due to various characteristics as follows:

1 Approximately 200 times sweeter than sucrose and costing comparatively less than sugar
2 Sweetness-quality superior to sucrose in mildness and refreshment
3 Sweetness is intensified by combining with salts and organic acids
4 The slightly latent sweetness is improved with the addition of sugars and organic acids
5 A 'salt-softening' property is evident and this effect is improved with glycyrrhizin
6 Stable at high temperatures and across a wide pH range
7 Substantially non-nutritive and noncariogenic.

In addition to these characteristics, stevia sweeteners are more manageable than sugar in the following cases:

1 Fermentation and spoiling in soy sauce, soybean paste, pickles, etc.
2 Burning in bread, cookies, fried foods, etc.
3 Coloring in foods containing amino acids through the Maillard's reaction
4 Hardening in baking of bread, cookies, etc.
5 Absorption of moisture in dried foods and baked foods such as cookies
6 Depression of freezing point in ice creams, sherbets and frozen foods.

While stevia sweeteners can be used in a wide variety of products, it is first necessary to identify the purpose for using these materials, i.e. as a sweetener, taste modifier, sugar defect eliminator or calorie reducer, before deciding on the type and amount of sweetener to be used.

Sweetness intensity

The sweetness intensity of a sweetener varies according to purity, temperature, pH, the content ratio, the presence of other food ingredients, and the concentration of a comparable sugar solution. In general, the sweetness multiples of high-intensity sweeteners may be presented in terms of relative sweetness to around 2 to 6% sucrose in an aqueous solution. As shown in Table 9.4, in a sensory evaluation of relative sweetness to a 4% sucrose solution by trained taste-panelists at Maruzen Pharmaceuticals Co. Ltd., pure stevioside and rebaudioside A were determined as 160 times and 308 times as sweet as sucrose, respectively (Saizuka 1982; Anonymous 1994). The sweetness multiple of commercialized simple 'stevia extracts' was evaluated to be almost equal to that of isolated stevioside because they usually contain rebaudioside A and other less sweet principles together with about 50–60% stevioside. The sweetness potency of enzymatic α-1,4-transglucosylated stevia extract is considered to vary with products based on the conditions of the enzymatic reaction. As mentioned previously, the sweetness of α-1, 4-transglucosylated products having a long α-1,4-transglucosyl chain becomes weak, so products prepared by treatment of α-1,4-transglucosylated stevia extract with β-amylase to

Table 9.4 Sweetness intensities of stevioside and
rebaudioside A in aqueous solution

Comparable sucrose concentration (%)	Relative sweetness (to sucrose = 1)	
	Stevioside	Rebaudioside A
2	267	385
4	160	308
6	133	261
8	118	250
10	111	208

shorten the poly α-glucosyl chain are sold commercially in addition to simple α-1,4-trans-glucosylated stevia extracts. According to Aikawa and Miyata (1990), one of these products, SK Sweet Z3® (Nippon Paper Industries Co. Ltd.), is about 170 times as sweet as a 4% sucrose solution.

By mixing with salts or organic acids such as citric acid, acetic acid, lactic acid, malic acid, and tartaric acid, the sweetness of stevia sweeteners is intensified, while their unpleasant aftertaste is diminished. For example, the sweetness of Marumilon 50® (Maruzen Pharmaceuticals Co. Ltd.), which consists of 'stevia extract' and dextrin, is 308 times that of sucrose in a 5% saline solution, and 111 times that of sucrose in aqueous solution, comparable to a 4% sucrose concentration, as shown in Table 9.5 (Anonymous 1994). This effect is due to the fact that the sweetness intensity of stevia sweeteners increases to some degree as the sweetness of sucrose is diminished by mixing with salts or acids (Yokoyama 1981).

Sweetness quality

Figure 9.3 shows the sweetness-taste profiles of stevioside, rebaudioside A, and other sweeteners including saccharin, sucrose, and glycyrrhizin. When placed in the mouth, stevioside gives a slightly latent sweet sensation with an aftertaste. Nevertheless, the taste profile of stevioside more closely resembles that of sucrose compared with the triterpene glycoside sweetener, glycyrrhizin, and the sweetness quality of stevioside is superior to that of sucrose in mildness and freshness. The commercially available simple sweetener, 'stevia extract' has a slightly latent sweetness as a result of the stevioside content and offers a few problems in its

Table 9.5 Sweetness intensity of Marumilon 50® in aqueous
solution and 5% saline solution[a]

Comparable sucrose concentration (%)	Relative sweetness (to sucrose = 1) of Marumilon 50	
	In aqueous solution	In 5% saline solution
2	143	333
4	111	308
6	90	273
8	60	200
10	50	125

Note
a Marumilon 50® is a product of Maruzen Pharmaceuticals Co. Ltd.,
   which consists of 60% stevia extract and 40% dextrin.

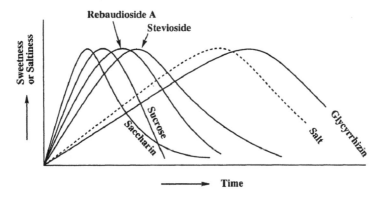

Figure 9.3 Sweetness-taste profiles of sweeteners at equal sweetness intensity and a saltiness-taste profile of salt.

practical use. When the stevia sweetener is used for foods, especially non-salted foods such as confectionery and drinks, it is generally formulated with other sweeteners. It was observed that the sweetness property of stevioside is improved and becomes similar to that of sucrose when combined with sucrose, glucose, and fructose, and its sweet aftertaste is also reduced (Ishima and Katayama 1976). There seems to be no optimum ratio for combing stevia sweeteners with sugars. When combined with sugar alcohols, these effects are not so significant (Ishima and Katayama 1976). The sweetness-taste profiles of 'sugar-transferred stevia extract' and 'rebaudioside A-enriched stevia extract' are more similar to those of sucrose and are preferable compared to the simple 'stevia extract' with respect to quality of sweetness.

Solubility

The solubility of stevia sweeteners has not been reported in detail. When comparing stevioside and rebaudioside A, the former is difficult to dissolve in water, so, only 2 g can be dissolved in 100 ml of water at 30 °C while the latter can easily be dissolved in water (Yokoyama 1981). However, there is no problem with the practical use of stevia sweeteners because of their highly sweet intensities.

Stability

Thermostability

Thermostability is an important factor when considering the applicability of sweeteners in heat-treatment such as cooking and disinfection. The sweetness property of an aqueous solution of stevioside does not change when heated to 95 °C for 2 hours. When heated at 95 °C for 8 hours, the sweetness falls slightly (Saizuka 1982). It is commonly known that stevia sweeteners do not decompose during their practical use.

Stability in acidic and alkaline solution

An acidic solution containing 0.02% stevioside at over pH 3 maintained at 95 °C for 1 hour shows no significant degradation, but a 12% solution of stevioside at pH 2 under these conditions decomposes (Saizuka 1982). According to Fujita and Edahiro (1979), stevioside is stable across the pH range 3–9 at 100 °C for 1 hour. It is considered that the major sweet principles, stevioside and rebaudioside A, of 'stevia extract' decompose into steviolbioside and rebaudioside B, respectively, in a strong alkaline solution and thus the sweetness intensity decreases. However, stevia sweeteners are not used under such extreme conditions.

PRACTICAL APPLICATIONS OF STEVIA SWEETENERS

Formulations of orange juice and coffee-flavored jelly containing a simple 'stevia extract' known as Marumilon Pure® (Maruzen Pharmaceuticals Co. Ltd.), which is 200 times sweeter than sucrose, are shown in Table 9.6. In both examples, the amounts of Marumilon Pure® to be added are 0.008% (w/v) and 0.0184% (w/w), respectively, of the formulations, and the sweetener

Table 9.6 Formulation of orange juice, coffee-flavored jelly, and a cookie mix containing stevia extract

Food type and ingredients	Weight or volume used
Orange juice	
Sucrose	10 kg
Marumilon Pure[a]	8 g
1/5 Concentrated orange juice	6 kg
Orange essence	100 ml
Citric acid	250 g
Ascorbic acid	80 g
Salt	50 g
Water	to 100 kg
Coffee-flavored jelly	
Sucrose	14.7 kg
Marumilon Pure[a]	18.4 g
Coffee extract powder	560 g
Carrageenan	600 g
Citric acid	48 g
Caramel	28 g
Flavor	Proper quantity
Water	to 100 l
Cookie	
Wheat flour	100 kg
Skim milk	5 kg
Shortening	16 kg
Sucrose	32 kg
Marumilon 10[b]	320 g
Baking powder	500 g
Baking soda	500 g
Salt	1.5 kg
Egg	8 kg
Flavor	Proper quantity
Water	25 l

Notes
a Product of Maruzen Pharmaceuticals Co. Ltd. (see Table 9.2).
b Product of Maruzen Pharmaceuticals Co. Ltd. which consists of 'stevia extract' and dextrin.

substitutes for about 20% of the commonly used sugar in these foods. Stevia sweeteners also have flavor-enhancing and aromatizing properties and give palatable tastes, although those properties of stevia sweeteners are not as effective as glycyrrhizin. In the orange juice example, Marumilon Pure® not only intensifies the flavor but also suppresses the unripe smell of the juice. In the cookie example (seeTable 9.6), Marumilon 10®, which can easily be used in foods by mixing with dextrin, gives soft and palatable tastes and prevents burning in the baking process.

For salted foods, 'stevia extract', as a simple sweetener, is not always acceptable. In fact, the addition of 'stevia extract' to highly salted foods leads to the subsequent sweet-salty sensation in products. When consideration is given to the salt-softening effect of a sweetener, it may be seen that the taste curve of the salt solution is similar to that of the sweetener, although one is salty and the other is sweet. As shown in Figure 9.3, the taste curve of salt is similar to that of glycyrrhizin rather than that of stevioside, i.e. stevioside does not match glycyrrhizin's 'salt-softening' capability. It has been found that combined sweeteners of stevioside and glycyrrhizin, of which the former has the better sweetness taste and the latter has the more effective 'salt-softening' property, are extremely useful in salted foods. For highly salted foods, addition of the preparation containing greater proportions of glycyrrhizin gives more harmonized tastes. Marumilon S® (Maruzen Pharmaceuticals Co. Ltd.) is one preparation used in highly salted foods like soy sauce, soy paste, and tangles boiled down in soy sauce (seeTable 9.7). When used in such salty foods, this product suppresses their unpleasant saline taste and harmonizes well with amino acids in foods.

Table 9.7 Formulation of tangles boiled down in soy sauce and orange-sherbet containing a combined preparation of stevia extract and glycyrrhizin

Food type and ingredients	Weight or volume used
Tangles boiled down in soy sauce	
Tangles	100 kg
Amino acid solution	30 l
Soy sauce	11 l
Sucrose	7.5 kg
Marumilon S[a]	83 g
Sorbit	4.2 g
Sodium glutamate	900 g
Nucleic acids	90 g
Mirin (sweet type of sake)	1.7 l
Caramel	500 g
Preservative	100 g
Water	14 l
Kamaboko (boiled fish paste)	
Minced fish	100 kg
Potato starch	1.0 kg
Albumen	5.0 kg
Sucrose	6.0 kg
Marumilon A[a]	150 g
Mirin	3.0 kg
Sodium glutamate	1.0 kg
Salt	2.5 kg
Ice water	10 kg

Note
a Products of Maruzen Pharmaceuticals Co. Ltd. (seeTable 9.2).

On the other hand, combined preparations containing greater proportions of 'stevia extract' have been used extensively in low-salted foods or sugar-reduced foods. A typical preparation, Marumilon A®, is useful for low-salted foods such as pickled vegetables, 'kamaboko' (fish meat paste) (see Table 9.7), and sausage. Its use in pickled vegetables helps to prevent fermentation and discoloration, in addition to improving taste. In fish meat products such as 'kamaboko', discoloration by the Maillard reaction is reduced and a popular plain taste is obtained. In these preparations of 'stevia extract' and glycyrrhizin, an organic acid is added to improve the sweetness-aftertaste of glycyrrhizin (see Table 9.2).

Marumilon Cool®, which is a blended 'stevia extract' and glycyrrhizin with citric acid and glycerol fatty acid ester, is recommended for use in ice cream and sherbet. Although the sweetness of the products with sucrose only lacks a refreshing feeling and punch, addition of this preparation increases the amplitude of the sweetness and flavors, and prevents freezing point depression. In the formulation example of diet ice cream shown in Table 9.8, Marumilon Cool® and maltitol complement the defects in each other's sweetness properties: i.e. maltitol

Table 9.8 Formulation of ice cream and orange-sherbet containing a combined preparation of stevia extract and glycyrrhizin

Food type and ingredients	Weight or volume used
Diet ice cream	
Shortening	3.2 kg
Margarine	2 kg
Non-fat sweetened condensed skim milk	10 kg
Non-fat dry milk	5 kg
Maltitol	5.4 kg
Marumilon 50[a]	10 g
Marumilon Cool[a]	130 g
Glucose syrup	15 kg
Salt	150 g
Stabilizer	400 g
Emulsifier	250 g
Vanilla essence	100 ml
Cream essence	30 ml
Water	to 100 kg
Orange-sherbet	
Non-fat sweetened condensed skim milk	5.0 kg
Sucrose	9.0 kg
Powdered glucose syrup	2.0 kg
Glucose	6.0 kg
Marumilon Cool[a]	200 g
Citric acid	50 g
Sodium citrate	100 g
1/5 Concentrated orange juice	2.5 kg
Stabilizer	200 g
Emulsifier	100 g
Flavor	100 ml
Water	to 100 kg

Note
a Products of Maruzen Pharmaceuticals Co. Ltd. (see Table 9.2).

Table 9.9 Formulation of bread and a carbonated drink containing 'enzymatically modified stevia' and 'rebaudioside A-enriched stevia extract'

Food type and ingredients	Weight or volume used
Bread	
Wheat flour	5 kg
Milk	3.3 kg
Sucrose	250 g
Salt	50 g
Butter	500 g
Dry yeast	80 g
α-G Sweet PA[a]	20 g
Carbonated drink	
Granulated sugar (sucrose)	1.5 kg
High fructose corn sweetener	7.6 kg
Chrysanta AR-P[b]	13.6 g
Citric acid	140 g
Sodium citrate	10 g
Cider essence	100 ml
Carbonated water	to 100 l

Notes
a Product with 'enzymatically modified stevia' (Maruzen Pharmaceuticals Co. Ltd.) (see Table 9.2).
b Product with 'rebaudioside A-enriched stevia extract' (Dainippon Ink & Chemicals Co. Ltd.) (see Table 9.2).

gives the sweetness-body and Marumilon Cool® gives the durable sweet sensation which is preferred in ice creams or sherbets.

The sweetness properties of 'sugar-transferred stevia extract' and 'rebaudioside A-enriched stevia extract' are similar to that of sucrose. The present developments in the use of stevia sweeteners are due to the use of these two preparations in non-salted foods and beverages. Table 9.9 shows a formulation example each for bread and a carbonated drink. Although these sweeteners replace up to about 50% of the sweetness of sugar, their demand is currently increasing for use in modifying the taste of foods. The plain and simple sweetness is suitable for the sour taste of low sugar yogurt. In sugarless foods and drinks, combinations with sugar alcohols are increasing. For example, the sweetness of sugarless chewing gum is created by combining maltitol (having a good sweetness-body), erythritol (having a refreshing sweetness), and stevia sweeteners (having a durable sweetness). In candies or chewables containing vitamin C, this combination of sweeteners ensures the desired taste.

Stevia sweeteners are also used as taste modifiers for dietary-supplement health foods. Table 9.10 shows formulations of Gingko extract tablets (Fuji Pharmaceuticals Co. Ltd.) and Collagen P granules (Zenyaku Kogyo Co. Ltd.). In these cases, stevia sweetener suppresses the bitter taste of the gingko extract and the peculiar taste of the liquid collagen peptide to palatable levels. Stevia sweeteners are especially useful as taste modifiers when used in the tablet form. In 1998, the Ministry of Health and Welfare of Japan authorized the use of 'stevia extract' as a ingredient in pharmaceuticals, based on the results of the recent chronic toxicity and carcinogenicity study in animals performed at the National Institute of Health Sciences, Tokyo, Japan (Toyoda *et al.* 1997).

Table 9.10 Formulation of dietary-supplement health foods containing stevia sweetener

Health supplement food Ingredients
Gingko extract[a] *(300 mg/tablet)*
Gingko extract powder
Powdered cellulose
Ascorbic acid
Powdered vitamin E
Cili extract powder
Apple flavor
Sweetener (stevia)
β-Carotene
Glycerin fatty acid ester
Shellac
Collagen P b (2 g, granule)
Liquid collagen peptide
Erythritol
Dietary fiber
Vitamin C
Vitamin B_6
Sour agent
Sweetener (stevia, licorice)
Flavor

Notes
a Product of Fujimi Pharmaceuticals Co. Ltd.
b Product of Zenyaku Kogyo Co. Ltd.

SAFETY AND TOXICITY ASSESSMENT OF STEVIA SWEETENERS IN JAPAN

Wingard *et al*. (1980) reported that stevioside and rebaudioside A are both decomposed to the aglycone, steviol, by rat intestinal microflora *in vitro*, and steviol is absorbed from the lower bowel of the rat. Steviol was demonstrated to be mutagenic after metabolic activation in the forward mutation assay, the umu test, the chromosomal aberration test, and a gene mutation assay (Pezzuto *et al*. 1985; Matsui *et al*. 1989), while no mutagenic activity has been observed for stevioside and 'stevia extract'. In Japan, the safety and toxicity of stevia sweeteners have been investigated since the 1970s (see Table 9.11). A chronic toxicity study of 'stevia extract' containing 74.54% stevioside and 16.27% rebaudioside A (total steviol glycosides: 95.2%) has been carried out using F344 rats for 22 months in the case of males and 24 months in the case of females (Yamada *et al*. 1985). After about 2 years of feeding of up to 1% stevioside (up to 550 mg/kg/day) in the diet, there were no significant dose-related changes in hematological and blood biochemical findings, organ weights, or macroscopic and microscopic findings. In a similar study on chronic toxicity, it was observed that up to 5% stevioside (95.6% purity, up to ±617 mg/kg/day) in the diet, when administered to male and female F334 rats for 104 weeks (about two years), exerts no carcinogenic activity (Toyoda *et al*. 1997).

In the development of stevia sweeteners in Japan, the largest problem has been a report on the antifertility effect of *S. rebaudiana* decoction by Planas and Kuć (1968). These authors claimed that Paraguayan Matto Grosso Indian tribes use *S. rebaudiana* as an oral contraceptive

Table 9.11 Principal toxicity assessment studies of stevia sweeteners carried out in Japan

Acute oral toxicity

Stevia extract (stevioside: 20.4%)[a]	$LD_{50} > 17.073$ g/kg (DDY-N mouse)
Stevia extract (stevioside: 41.4%)[a]	$LD_{50} > 42$ g/kg (DDY-N mouse)
Stevioside (93.5%)[a]	$LD_{50} > 15$ g/kg (ICR-SCLC mouse)
Stevioside[b]	$LD_{50} > 8.2$ g/kg (ICR mouse, Wister rat)
Enzymatically modified stevia[c]	$LD_{50} > 60$ g/kg (mouse)

Subacute and subchronic oral toxicity

Stevia extract (stevioside: 53.1%)[a]	No toxicity (SLC-Wister rat, up to 7%/diet [up to 5.6 g/kg/day] for 13 weeks)
Stevioside[b]	No toxicity (Wister rat, up to 2500 mg/kg/day, for 1 month)
Enzymatically modified stevia[c]	No toxicity (rat, 5%/diet for 13 weeks)
Stevioside (95.6%)[d]	No toxicity (F334 rat, up to 5%/diet, for 13 months)

Chronic toxicity

Stevia extract (stevioside: 74.54% rebaudioside A: 16.27%)[e]	No toxicity (F334 rat, up to 1%/diet [up to 550mg/kg/day], for 22 to 24 months)
Stevioside (95.6%)[f]	No toxicity (F334 rat, up to 5%/diet [up to 1997 ± 617 mg/kg/day], for 104 weeks)

Mutagenicity

Stevia extract (stevioside: about 18%)[g]	Negative[h]
Stevia extract (stevioside: about 55%)[g]	Negative[h]
Stevioside (95–98%)[g]	Negative[h]
Enzymatically modified stevia[c]	Negative

Notes and Sources
a From Akashi and Yokoyama (1975).
b From Katayama *et al.* (1976).
c From Shibasato (1995).
d From Aze *et al.* (1991).
e From Yamada *et al.* (1985).
f From Toyoda *et al.* (1997).
g From Okumura *et al.* (1978).
h Evaluation by using *Escherichia coli* strain WP2 and *Salmonella typhimurium* strain TA1535, TA100, TA1537, TA1538 and TA98 in the presence or absence of S-9 mix.

and demonstrated that fertility is reduced 57 to 79% in female rats by imbibing a decoction of 5% (w/v) dry plant in water. Since later investigations by other groups have been unable to confirm the antifertility effect of *S. rebaudiana* extracts and stevioside (e.g. Akashi and Yokoyama 1975; Mori *et al.* 1981), few consumers have felt uneasy about the side effects of products containing stevia sweeteners. In a more recent study it was reported that *S. rebaudiana* decoction prepared in the same manner as discribed by Planas and Kuć (1968) had no effect on fertility in rats under the same conditions (Shiotsu 1996). Furthermore, no adverse effects were observed in either pregnant rats or rat fetuses at the maximum dose of 1000 mg/kg/day of stevioside (Usami *et al.* 1995). According to Kinghorn and Soejarto (1985), inquiries made in several locations in north-eastern Paraguay did not confirm the use of *S. rebaudiana* extracts for contraceptive purposes. This theme is expanded on the chapter in this volume on the ethno-botany of *S. rebaudiana* by Soejarto. Also, a more thorough discussion on the toxicology of stevioside, rebaudioside A, and steviol is provided in Chapter 8.

The leaves of *S. rebaudiana* have been used to sweeten beverages such as 'maté' by natives in Paraguay since at least before the begining of the twentieth century (Kinghorn and Soejarto 1985). The safety of stevia sweeteners indicated by long-term human exposure in Paraguay has been confirmed scientifically in Japan, and the use of these products in foods and beverages has been making steady progress in the Orient, especially in Japan.

REFERENCES

Aikawa, M. and Miyata, T. (1990) Current status on improvement of taste quality of stevia. *Technical Journal on Food Chemistry and Chemicals* No. 4, 217–222.

Akashi, H. and Yokoyama, Y. (1975) Safety of dried-leaf extracts of *Stevia rebaudiana*. *Shokuhin Kogyo* **18**(20), 34–43.

Anonymous (1994) *Application of Glycyrrhizin and Stevioside in Foods*, Maruzen Pharmaceuticals Co. Ltd., Onomichi-city, Hiroshima, Japan, pp. 1–12.

Anonymous (1996) Mini mini marketing – stevioside. *Technical Journal on Food Chemistry and Chemicals* No. 9, 140.

Aze, Y., Toyoda, K., Imaida, K., Hayashi, S., Imazawa, T., Hayashi, Y. *et al.* (1991) Subchronic oral toxicity study of stevioside in F344 rats. *Bulletin of the National Institute of Hygienic Sciences, Tokyo, Japan* **109**, 48–54.

Bakal, A. I. and O'Brien Nabors L. (1985) Stevioside. In *Alternative Sweeteners*, L. O'Brien Nabors and R. C. Gelardi (Eds), Marcel Dekker, New York, pp. 295–307.

Darise, M., Mizutani, K., Kasai, R., Tanaka, O., Okada, S., Ogawa, S. *et al.* (1984) Enzymic transglucosylation of rubusoside and structure-sweetness relationship of steviol-bisglucosides. *Agricultural and Biological Chemistry* **48**, 2483–2488.

Fujita, H. and Edahiro, T. (1979) Safety and utilization of *Stevia rebaudiana* sweetener. *Shokuhin Kogyo* **22**(20), 66–72.

Fukunaga, Y., Miyata, T., Nakayasu, N., Mizutani, K., Kasai, R. and Tanaka, O. (1989) Enzymic transglucosylation products of stevioside: Separation and sweetness-evaluation. *Agricultural and Biological Chemistry* **53**, 1603–1607.

Ishima, N. and Katayama, O. (1976) Sensory evaluation of stevioside as a sweetener. *Report of the Natural Food Research Institute* No. 31, 80–85.

Japan Food Additive Association (1993) Stevia extract. In *Voluntary Specifications of Non-Chemically Synthesized Food Additives* (2nd edn), Japan Food Additive Association (Ed.), Tokyo, pp. 119–124.

Katayama, O., Sumida, T., Hayashi, K. and Mitsuhashi, H. (1976) Safety of stevioside. In *Applicability of Stevia and Data on Research and Development*, Isu Co. Ltd., Tokyo, pp. 225–281.

Kinghorn, A. D. and Soejarto, D. D. (1985) Current status of stevioside as a sweetening agent for human use. In *Economic and Medicinal Plant Research, Volume 1*, H. Wagner, H. Hikino, and N. R. Farnsworth (Eds), Academic Press, London, pp. 1–52.

Kobayashi, M., Horikawa, S., Degrandi, I. H., Ueno, J. and Mitsuhashi, H. (1977) Dulcosides A and B, new diterpene glycosides from *Stevia rebaudiana*. *Phytochemistry* **16**, 1405–1408.

Kohda, H., Kasai, R., Yamasaki, K., Murakami, K. and Tanaka, O. (1976) New sweet diterpene glycosides from *Stevia rebaudiana*. *Phytochemistry* **15**, 981–983.

Matsui, M., Matsui, K., Nohmi, T. and Ishidate, M. (1989) Mutagenicity of steviol: An analytical approach using the Southern blotting system. *Bulletin of the National Institute of Hygienic Sciences, Tokyo, Japan* **107**, 83–87.

Mizutani, K., Miyata, T., Kasai, R., Tanaka, O., Ogawa, S. and Doi, S. (1989) Study on improvement of sweetness of steviol bisglycosides: Selective enzymatic transglucosylation of the 13-*O*-glucosyl moiety. *Agricultural and Biological Chemistry* **53**, 395–398.

Mosettig, E., Beglinger, U., Dolder, F., Lichti, H., Quitt, P. and Waters, J. A. (1963) The absolute configuration of steviol and isosteviol. *Journal of American Chemical Society* **85**, 2305–2309.

Mori, N., Sakanoue, M., Takeuchi, M., Shimpo, K. and Tanabe, T. (1981) Effect of stevioside on fertility in rats. *Shokuhin Eiseigaku Zasshi* **22**, 409–414.

Ohtani, K., Aikawa, Y., Ishikawa, H., Kasai, R., Kitahata, S., Mizutani K., *et al*. (1991) Further study on the 1,4-α-transglucosylation of rubusoside, a sweet steviol-bisglucoside from *Rubus suavissimus*. *Agricultural and Biological Chemistry* **55**, 449–453.

Okumura, M., Fujita, Y., Imamura, M. and Aikawa, K. (1978) Studies on the safety of stevioside with rec-assay and reversion test. *Shokuhin Eiseigaku Zasshi* **19**, 486–490.

Pezzuto, J. M., Compadre, C. M., Swanson S. M., Nanayakkara N. P. D. and Kinghorn, A. D. (1985) Meta-bolically activated steviol, the aglycone of stevioside, is mutagenic. *Proceedings of the National Academy of Sciences of the United States of America* **82**, 2478–2482.

Planas, G. M. and Kuć, J. (1968) Contraceptive properties of *Stevia rebaudiana*. *Science* **162**, 1007.

Saizuka, H. (1982) Current status of stevioside and its application to food processing. *Japan Food Science* **21**, 24–30.

Sakamoto, I., Yamasaki, K. and Tanaka, O. (1977a) Application of ^{13}C-NMR spectroscopy to chemistry of natural glycosides. Rebaudioside C, new sweet diterpene glycoside from *Stevia rebaudiana*. *Chemical and Pharmaceutical Bulletin* **25**, 844–846.

Sakamoto, I., Yamasaki, K. and Tanaka, O. (1977b) Application of ^{13}C-NMR spectroscopy to chemistry of plant glycosides. Rebaudioside-D and -E, new sweet diterpene glycosides from *Stevia rebaudiana* Bertoni. *Chemical and Pharmaceutical Bulletin* **23**, 3437–3439.

Shibasato, M. (1995) Current status of stevia sweeteners and its applications. *Japan Food Science* No. 12, 51–58.

Shiotsu, S. (1996) Fertility study of stevia decoction in rats. *Technical Journal on Food Chemistry and Chemicals* No. 4, 108–113.

Tanaka, O. (1997) Improvement of taste of natural sweeteners. *Pure and Applied Chemistry* **69**, 675–683.

Toyoda, K., Matsui, H., Shoda, T., Uneyama, C., Takada, K. and Takahashi, M. (1997) Assessment of the carcinogenicity of stevioside in F344 rats. *Food and Chemical Toxicology* **35**, 597–603.

Usami, M., Sakemi, K., Kawashima, K., Tsuda, M. and Ohno, Y. (1995) Teratogenicity study of stevioside in rats. *Bulletin of the National Institute of Hygienic Sciences, Tokyo, Japan* **113**, 31–35.

Wingard, R. E., Jr, Brown, J. P., Enderlin, F. E., Dale, J. A., Hale, R. L. and Seitz, C. T. (1980) Intestinal degradation and absorption of glycosidic sweeteners stevioside and rebaudioside A. *Experientia* **36**, 519–520.

Yamada, A., Ohgaki, S., Noda, T. and Shimizu, M. (1985) Chronic toxicity study of dietary stevia extracts in F344 rats. *Journal of Food and Hygenic Science of Japan* **26**, 169–183.

Yokoyama, Y. (1981) New intention on sweeteners (10) – Stevia sweeteners. *Food Processing* **16**, 44–55.

10 Use of stevioside and cultivation of *Stevia rebaudiana* in Korea

Jinwoong Kim, Young Hae Choi and Young-Hee Choi

INTRODUCTION

Sugar cane and sugar beet, which are plant sources of sugar, are not cultivated in Korea, so most crude materials for the production of sucrose have had to be imported from abroad. Recently, the demand for low-calorie sweeteners has prompted researchers to investigate alternative sweet compounds which could be substituted for the use of sugar to some degree (Suh 1979). The consumption of the major sweeteners used currently in Korea is listed in Table 10.1 (Seon 1995).

To date, several synthetic sweeteners have been considered as alternatives to sucrose in Korea. However, due to the possibility of health hazards such as carcinogenicity, the intake of such compounds is either limited or even banned in several countries. Therefore, it is natural that there has been a strong and ever-increasing demand for a harmless sweetener in Korea. For this purpose, stevioside has been proposed as a promising sweetener due to its low calorie content and relatively low toxicity (Fujita and Edahiro 1979). Stevioside was introduced to Korea in 1973 and its use as an alternative sweetener was first considered officially in 1976, but it was not approved as a food additive until September 1984. Its ultimate approval as a safe alternative sweetener was based on three examinations by the Korea Food and Hygienic Consideration Committee, and its use in liquor has been permitted since January 1991. However, potential problems with the safety of stevioside were raised in October 1994 when the Australian government denied the clearance of *soju*, a traditional Korean distilled liquor, due to the presence of stevioside. This incident prompted the Korean Congress to demand that the Korea Consumer Protection Board provide a clear answer on the safety of stevioside at the 177th parliamentary inspection of the administration of government offices. As a result of this debate on the safety of stevioside, the Food Additives Division under the Ministry of Health and

Table 10.1 Consumption of major sweeteners in Korea during the period 1988–1991 (in metric tons)

Sweetener	1988	1989	1990	1991
Sucrose	596,000	654,000	655,000	698,000
Glucose	161,000	201,929	210,740	248,450
Fructose	219,290	244,077	270,501	276,429
Aspartame	18	12	36	40
Stevioside	8	10	40	60

Source: Seon (1995).

Welfare drew the conclusion, based on much scrutiny, that stevioside is safe for human intake. Thus, at present, despite the various criticisms on the safety of stevioside that have arisen, it is used restrictively as a food additive in Korea.

In this chapter, we present details on how stevioside is defined in Korea, its market demand, and its types of use, and also information in terms of the cultivation of *S. rebaudiana* leaf needed to obtain the crude materials to produce stevioside.

DEFINITION AND STANDARDIZATION OF STEVIOSIDE IN KOREA

Stevioside is defined as those compounds having the skeleton of steviol originating from *S. rebaudiana*, by the Korean Standard of Food Additives published by the Ministry of Health and Welfare (Ministry of Health and Welfare 1996). In addition to this definition, the product must contain 98.0% of stevioside $(C_{38}H_{60}O_{18})$ as the major component by quantitative analysis, after being dried for two hours at 100 °C. Stevioside is permitted as a food additive in distilled liquors, unrefined rice wines, confectioneries, soybean sauces, and pickles. However, it cannot be used in breads, baby foods, candies, and dairy products, or as a table-top sweetener.

USE OF STEVIOSIDE IN KOREA

Stevioside occupies 40% of the sweetener market in Korea (Seon 1995). Table 10.1 shows the increasing consumption of stevioside in Korea over the period 1988–1991. Presently, the majority of the *S. rebaudiana* leaves for the isolation of stevioside are imported from the People's Republic of China. Five companies in Korea manufacture stevioside from *S. rebaudiana* leaves and products containing this sweet compound. The total output of stevioside per year in Korea is estimated to be 200–250 metric tons (Korea Consumer Protection Board 1996a).

Most of the stevioside produced in Korea is used in the food industry. In particular, a large percentage of stevioside has been used in the production of *soju*, the traditional Korean distilled liquor made from sweet potatoes. As shown in Tables 10.2 and 10.3, stevioside was used at a level of 20 metric tons when it was first permitted for use in *soju*, but this figure has increased to 116 metric tons in 1995 with about a 50% share of the total consumption of this sweetener in Korea. In general, stevioside can be added to liquor since most countries do not differentiate

Table 10.2 The use of stevioside in the food industry during the years 1992 and 1993 in Korea (in metric tons)

Use	1992	1993
Soju (distilled liquor, 25%)	20	20
Unrefined rice wine	8	10
Confectionery	7	10
Medicine additive	15	18
Soybean sauce	30	33
Pickle	5	10
Others (ice cake, etc.)	15	29
Total	100	130

Source: Seon (1995).

Table 10.3 Amounts of stevioside used by Korean companies in 1995 in the manufacturing of *soju* (in metric tons)

Company	Amount
Jinro Brewery Co. Ltd. (Seoul)	50.0
Doosan Kyung Woul Co. Ltd. (Kangnung)	26.0
Dae Sun Distilling Co. Ltd. (Pusan)	11.0
Muhak Brewery Co. Ltd. (Masan)	7.0
Keum Bok Ju Co. Ltd. (Seoul)	6.0
Sun Yang Brewery Co. Ltd. (Taejon)	6.0
Bohae Brewery Co. Ltd. (Mokpo)	5.2
Bobae Co. Ltd. (Iksan)	4.5
Chungbuk Co. Ltd. (Chungju)	0.7
Total	116.4

Source: Korea Consumer Protection Board (1996b).

food from liquor. However, there are actually just two nations, Korea and Japan, where stevioside is now used in the manufacturing of liquor. The amounts of stevioside in *soju* used by various Korean companies is listed in Table 10.3. Two major companies producing *soju* (Jinro and Doosan Kyung Woul) use over 75% of the total amount of stevioside between them for this purpose, while the remaining companies utilize 0.7–11 metric tons of stevioside each. Although stevioside is used as an additive in *soju*, it is not used alone but is employed together with other sweeteners such as various oligosaccharides, honey, and sucrose (Korea Consumer Protection Board 1996a). However, stevioside is not added to *soju* exported to other countries (e.g. Australia and the United States), where the use of stevioside is not permitted. Table 10.4 shows the composition of sweeteners used in the manufacture of *soju* by Korean companies for both domestic and export purposes.

In October 1996, the Korea Consumer Protection Board (1996b) reported the content of stevioside in 45 kinds of domestic *soju* manufactured by ten companies. The concentrations of

Table 10.4 Sweeteners used in *soju* for domestic use in Korea and for export purposes

Sweeteners	Jinro	Bohae	Keum Bok Ju	Muhak	Dae Sun	Bobae	Doosan Kyung Woul	Sun Yang	Chung-buk
Domestic use									
Stevioside	+	+	+	+	+	+	+	+	+
Oligosaccharide	+	−	+	+	−	+	+	+	+
Fructose	+	+	−	−	−	−	+	−	+
Honey	−	−	−	+	−	−	−	−	−
Sucrose	−	−	−	−	−	+	−	+	−
Aspartame	−	−	−	−	−	−	−	−	+
Sorbitol	−	+	−	−	−	−	+	−	−
Export use									
Stevioside	+	−	−	−	−	−	−	−	−
Oligosaccharide	+	−	−	−	−	+	−	−	−
Fructose	+	−	−	−	+	−	+	−	−
Sucrose	−	+	+	+	−	+	−	−	−
Aspartame	−	−	−	−	−	−	+	−	−
Sorbitol	−	+	−	−	−	−	+	−	−

Source: Korea Consumer Protection Board (1996b).

stevioside were determined as from 5 to 33 mg/ml for an average of 12.0 mg/ml (Table 10.5). Stevioside is used mainly to reduce the production costs of *soju*. The cost of stevioside in one bottle (360 ml) of *soju* is about 0.15 cents (1$ = 1,200 won), which is about 0.59% of the total cost. However, if oligosaccharides are substituted for stevioside, 0.43 cents must be consumed per bottle (1.72% of the total cost). A comparison of the costs of using stevioside with those of oligosaccharides in the production of *soju* products of various manufactures is shown in Table 10.6.

EVALUATION OF THE SAFETY OF STEVIOSIDE IN KOREA

While there has been considerable doubt expressed on the safety of stevioside in Korea, as in several other countries, a recent report of the Korean Food and Hygienic Consideration Committee of the Ministry of Health and Welfare concluded provisionally that consumption of stevioside is regarded as safe for human health (Korea National Institute of Health 1996). In addition to stevioside, steviol, the aglycone of stevioside, has also been reported to have toxic effects (Pezzuto *et al.* 1985; Pezzuto *et al.* 1986). Thus, the possibility of stevioside being transformed to steviol was investigated. First, this was investigated in alcoholic solution. The Korea Consumer Protection Board (1996b) examined the percentage of changes of steviol from stevioside every five days, after 100 ppm of stevioside was added to a 25% alcoholic solution, which is the same percentage found in a commercial *soju*. As a result, it was found that stevioside was not transformed to steviol at all under these conditions (Table 10.7). In addition, the Korea Consumer Protection Board (1996b) reported that steviol was not detected in the commercial *soju* of 45 different types produced by ten companies. Since Koreans often drink *soju* after soaking with fruits such as grapes and botanical products such as ginseng, another examination was carried out as to whether steviol was produced from stevioside under these conditions. This was felt to be desirable in particular because fruit enzymes may potentiate the hydrolysis of stevioside to steviol. However, steviol was not detected in the fruit extracts of *soju*. From various systematic examinations of the toxicity of stevioside and the study of its possible chemical change to steviol, the Korea Consumer Protection Board has concluded provisionally as follows:

> On the basis of the results by previous researchers, there are two divided opinions as to whether stevioside is safe or not. Thus, there are a few governments which permit the use of stevioside in food. In particular, these governments are not sure if stevioside may change to steviol and are not sure of the safety of the compound. However, our report on safety has shown through several quantitative examinations that there is no evidence that stevioside is harmful to human health or that stevioside may change to steviol in alcoholic solution.
>
> (Korea Consumer Protection Board 1996a)

CULTIVATION OF *STEVIA REBAUDIANA* IN KOREA

Although in recent years *S. rebaudiana* has been imported to Korea from the People's Republic of China, several trials were conducted on the cultivation of this species after it was introduced to Korea in 1973. The yield of *Stevia* leaves per 10 metric acre was assumed to be 200–250 kg in Korea, which is slightly higher than that in its country of origin (Chung and Lee 1978). This plant is indigenous to subtropical zones in Paraguay located in latitudes of 21–27° south. The

Table 10.5 Content of stevioside in *soju* products used domestically in Korea

Company	Product number	Alcohol (%)	Bottle size (ml)	Stevioside content (mg/L)
Jinro	1	25	360	17
	2	25	360	16
	3	25	300	16
Bohae	1	25	300	13
	2	23	360	Not detected
	3	23	360	33
	4	25	360	Not detected
	5	15	300	16
	6	25	360	30
	7	30	1,800	33
	8	35	1,800	11
Keum Bok Ju	1	25	360	Not detected
	2	25	360	15
	3	25	360	14
	4	25	640	14
	5	25	1,800	14
Muhak	1	23.5	360	19
	2	23.5	360	17
	3	25	360	32
	4	23	360	30
	5	25	360	31
	6	25	1,800	29
	7	30	1,800	33
Dae Sun	1	25	360	18
	2	23	360	15
	3	25	360	15
	4	23	360	10
	5	30	1,800	11
	6	25	1,800	13
Bobae	1	30	1,800	18
	2	35	1,800	Not detected
	3	25	360	16
	4	25	360	15
	5	20	300	10
Doosan Kyung Woul	1	25	355	Not detected
	2	25	360	Not detected
	3	25	360	20
Sun Yang	1	25	350	20
	2	25	360	Not detected
	3	25	1,800	18
Baekhak Brewery Co. Ltd. (Cheju)	1	25	360	6
	2	25	375	5
	3	25	640	6
	4	25	360	Not detected
Haitai Co. Ltd. (Seoul)	1	25	640	Not detected

Source: Korea Consumer Protection Board (1996b).

Table 10.6 Comparison of cost expenditures for stevioside versus oligosaccharides in *soju* (in Korean won) (1$ = 1,200 won)

Company	Cost of sweeteners	Cost for stevioside (A)	Cost for Oligosaccharides (B)	Added cost for using oligosaccharides (B—A)
Jinro	2.88	1.89	6.26	4.37
Doosan	2.11	1.57	5.56	3.99
Muhak	3.12	2.25	6.68	4.43
Dae Sun	6.85	3.02	9.50	6.48
Bobae	4.99	2.32	7.22	4.90
Sun Yang	2.74	2.33	5.96	3.63
Bohae	2.96	2.26	4.25	1.99
Keum Bok Ju	2.48	1.84	6.07	4.23
Average (%)	3.51	2.19	6.43	4.24

Source: Korea Consumer Protection Board (1996b).

Table 10.7 Content of stevioside in 25% alcoholic solution and in commercial *soju* with time after addition at the level of 100 ppm

Period (days)	Alcoholic solution (25%)	Commercial soju
0	99.5	98.8
5	98.5	99.0
10	98.8	98.4
15	98.9	98.9

Source: Korea Consumer Protection Board (1996b).

differences in temperature of the various seasons and in day and night is higher in these sub-tropical areas but the difference of day-length as well as the amount of precipitation is less than in Korea. Thus, optimization of the cultivation conditions of *S. rebaudiana* in Korea had to be conducted. The established conditions for *S. rebaudiana* in Korea are as follows (Chung and Lee 1978; Lee *et al.* 1980; Kang and Lee 1981): The critical day-length for the plant seems to be approximately 12 hours. The growth of the plants is severely retarded at day-lengths of less than 12 hours. The plant height, the number of branches, and the dry weight of *S. rebaudiana* are reduced if the cutting date is delayed from March 20 to May 20. The optimum population density of the plant for the highest yield of dry leaves is 20,000 plants/1,000 square meters, since when planted in a less dense fashion, the cultivation process is not economical. The highest content of stevioside has been noted to be in the upper leaves of the plant while the lowest occurs in plant parts 20 cm above ground level. The leaf dry weight and the stevioside yield are contributed mainly by plant parts 60 to 120 cm above the ground but the varietal differences are also significant. Delayed harvests until the time of flower bud formation increase the leaf dry weight remarkably. However, there are insignificant changes of yield when harvests are made at any time after flower bud formation. Stevioside levels are the highest at the time of flower bud formation and lower at times preceding and following flower bud formation. The optimum harvesting time, as determined by the leaf dry weight and stevioside content, are at periods from flower bud formation to just before flowering, which is from September 10 to September 15 in Korea.

SUMMARY AND CONCLUSIONS

Korea does not produce sucrose from either sugar cane or sugar beet grown domestically. Thus, it has become necessary to investigate alternative sweeteners that can be substituted for sugar. Among these alternative sweeteners, stevioside is considered as a promising sucrose substitute. *Stevia rebaudiana*, the source of stevioside, was introduced to Korea in 1973 and the use of stevioside as a sweetener has been permitted since 1984. Thus, there has been a significant effort to cultivate *S. rebaudiana* in Korea. Presently, stevioside occupies 40% of the sweetener market in Korea and it is being used more broadly in the food industry as a sugar substitute than any other alternative sweetener in products such as ice cream, ice cakes, confectioneries, gum, pickles, sauces, non-caloric diet foods, and beverages. In particular, 50% of the total consumption of stevioside in Korea is made through in the alcoholic liquor, *soju*. Although stevioside has been highlighted as an alternative sweetener and used broadly in the food industry, there has also been much criticism in terms of its safety ever since the use of compound was first permitted. Due to this criticism, the Korean government has concluded tentatively that stevioside is not toxic to human health and, as a result, its rate of use is greatly increasing in the domestic market at present.

REFERENCES

Chung, M. H. and Lee, M. Y. (1978) Studies on the development of *Hydrangea* and *Stevia* as a natural sweetening product. *Korean Journal of Pharmacognosy* **9**, 149–156.

Fujita, H. and Edahiro, T. (1979) Safety and utilization of Stevia sweetener. *Shokuhin Kogyo (Food Industry)* **22**, 66–72.

Kang, K. H. and Lee, K. W. (1981) Physio-ecological studies on *Stevia* (*Stevia rebaudiana* Bertoni). *Hanjakji* **26**, 69–87.

Korea Consumer Protection Board (1996a) Reports on the safety of stevioside as an natural sweetener. *Reports of the Korea Consumer Protection Board*, Seoul, Korea, pp. 1–35.

Korea Consumer Protection Board (1996b) Detection of stevioside and steviol in commercial *soju*. *Reports of the Korea Consumer Protection Board*, Seoul, Korea, pp. 41–52.

Korea National Institute of Health (1996) *Report on the Safety of Stevioside*, Seoul, Korea, pp. 1–5.

Lee, J. I., Kang, K. H., Park, H. W. and Ham, Y. S. (1980) Effect on the new sweetening resources plant *Stevia* (*Stevia rebaudiana* Bertoni M.) in Korea. *Nongupshinbo* **22**, 138–144.

Ministry of Health and Welfare (1996) *Standard of Food Additives*, Seoul, Korea, pp. 268–270.

Pezzuto, J. M., Compadre, C. M., Swanson, S. M., Nanayakkara, N. P. D. and Kinghorn, A. D. (1985) Metabolically activated steviol, the aglycone of stevioside is mutagenic. *Proceedings of the National Academy of Sciences of the United States of America* **82**, 2478–2482.

Pezzuto, J. M., Nanayakkara, N. P. D., Compadre, C. M., Swanson, S. M., Kinghorn, A. D., Guenthner, T. M. *et al.* (1986) Characterization of bacterial mutagenicity mediated by 13-hydroxy-*ent*-kaurenoic acid: evaluation of potential to induce glutathione *S*-transferase in mice. *Mutation Research* **169**, 93–103.

Seon, J. H. (1995). Stevioside as natural sweetener. *Report of Pacific R & D Center* October, Seoul, Korea, pp. 1–8.

Suh, K. B. (1979) Advancement of stevioside use. *Food Science* **1**, 38–42.

Index

Milton Keynes UK
Ingram Content Group UK Ltd.
UKHW051952071024
449327UK00026B/2276